# Introduction to
# Quantum Electronics

# Introduction to Quantum Electronics

PAUL HLAWICZKA

*Department of Electronics and Electrical Engineering,
University of Glasgow, Scotland*

1971

ACADEMIC PRESS: LONDON AND NEW YORK

ACADEMIC PRESS INC. (LONDON) LTD
24/28 Oval Road
London NW1 7DX

*U.S. Edition published by*
ACADEMIC PRESS INC.
111 Fifth Avenue,
New York, New York 10003

PRINTED IN GREAT BRITAIN BY
WILLIAM CLOWES & SONS LIMITED
LONDON, COLCHESTER AND BECCLES

# Preface

The title and contents of this book are based on the conviction that a new engineering discipline is emerging. An accumulation of knowledge and techniques can only be described as a discipline if it has a well defined conceptual unity. In the present case this unity is founded on quantum physics, from which the various subdivisions and facets of the new discipline are more or less directly derived.

Several names have been used to denote the new subject, but most of them are inaccurate or unsuitable in some respect. The term "Electrical Materials" has a connotation which is far too passive to convey the dynamism of the many devices and processes which continue to emanate from the new subject. "Solid State Electronics" is clumsy and too narrow, as it excludes the properties and applications of gases, which are now central to the development of lasers. Also, by implication, it omits the dielectric and viscoelastic properties of liquids. "Physical Electronics" is altogether too vague.

By contrast the term "Quantum Electronics" conveys tersely both the quantum mechanical parentage of the new discipline, and suggests its dynamism through the word "Electronics". It is unfortunate that in recent years this excellent, accurate name has come to denote the more specialist aspects of maser and laser devices. It is one of the aims of the present book to correct this aberration and to claim "Quantum Electronics" for the general use for which it is intended.

This book is the result of an attempt to assemble an elementary but comprehensive and thoroughly integrated text on quantum electronics which can be used as an undergraduate textbook, and as a reference on matters of principle by postgraduate students and practising engineers. The book is based on my experience of teaching the subject at undergraduate level and is also influenced by my research activities, past and present, in the field of communications, masers, ferrite devices and superconductivity. The organization of the contents has been largely dictated by the requirements of a course given to honours students, and the subject matter divides naturally into three parts.

The first part covers semiconductors, electron emission phenomena, gas discharges and dielectrics at a strictly elementary level. The treatment is carefully adapted to the application of these topics which the student is likely to encounter in his working career. Although the subject matter is ultimately based on quantum physics, the presentation avoids direct reference to the latter and relies on simplified statements of a few indispensable principles, e.g. the wave nature of electrons, the Fermi-Dirac statistics and the density of states function.

The second part of the book introduces the fundamentals of quantum mechanics, as applicable to quantum electronics. No effort is spared at this stage to deal with basic concepts and methods while avoiding excessive preoccupation with mathematical procedures, which are unlikely to be necessary to the future electronics engineer. The treatment is taken far enough to supply quantum mechanical derivations of the principles used in the first part of the book and thus affords an opportunity to revise the material from a more mature standpoint. In addition, the coverage provides the basis for a quantum mechanical treatment of magnetism.

The third part of the book deals with paramagnetism within the framework of quantum theory, including microwave spectroscopy and the three level maser. Ferromagnetism is introduced on the basis of the exchange interaction and the basic properties relevant to applications are dealt with, including magnetic hysteresis. Chapters on ferrimagnetism and ferromagnetic resonance cover the fundamentals of ferrite microwave devices. The book closes with chapters on lasers and superconductivity. The treatment of these topics is carefully matched to the requirements of engineers and concentrates on aspects relevant to applications, rather than on the type of discussion which would be more suitable for physicists aiming at fundamental research in these fields.

Since the material contained in this book will be covered by undergraduate students over a period of, say, two years it must be expected that the level of treatment will vary as the subject unfolds. Readers will therefore understand that an increasing degree of intellectual maturity is required to appreciate the text as the subject is developed.

In a book of this type it is most difficult to strike a balance between the depth of treatment of basic concepts and the elaboration of applications. In the last analysis the contents are the result of the author's interests and judgement, and of teaching expediency. However, I have endeavoured at every stage to seek out and emphasize the general principles which seem likely to have permanent validity, while leaving aside technological details which, however exciting at the moment, may give way to other fashionable developments.

The question of listing additional reading has caused me some considerable worry. It is generally agreed that the quoting of an excessive number of references does not assist students, or professionals seeking information of a basic kind on topics outside their speciality. For these reasons I have given no specific references in the text and I have limited the list of supplementary reading to a minimum. In an attempt to adopt a rational principle of selection I have been led to view references on the basis of a hierarchy. A general elementary text like the present is followed by treatments in depth of the major subdivisions of the subject, e.g. semiconductors, magnetism, quantum mechanics, etc. Then come specialist monographs dealing with individual aspects such as microcircuits, photoconductivity, ferrite devices and others. Finally there are research papers. The list of reading at the end of this book is almost entirely restricted to books occupying the same or the nearest layer of the bibliographic hierarchy.

There is a short collection of examples at the end of the book. The primary aim is to provide typical numerical illustrations of the textual material, but in a few cases an element of extension is included.

In a book like the present one it is impossible to have a consistent system of symbols without going to extreme lengths to find esoteric characters for some of the countless quantities which must be discussed. However, every effort has been made to ensure that the meaning of symbols is clear from the context. In one or two cases different symbols are used to denote the same entity in different contexts in order to conform to accepted usage. Thus the imaginary unit $\sqrt{-1}$ is denoted by $i$ in chapters dealing with mathematical theory while in the context of applications the symbol $j$ is used. The electronic charge is treated as a positive physical constant and is explicitly preceded by a negative sign in electrostatic or current flow equations.

It is a pleasure to acknowledge my indebtedness to several colleagues who have read through substantial parts of the manuscript, offered constructive criticism, eliminated many errors and helped with proof correction. Their names are: Dr. K. A. M. Arton, Dr. A. J. Barlow, Dr. W. Duncan, Mr. R. Hutchins and Dr. C. D. W. Wilkinson.

*September 1971*                                    P. HLAWICZKA

# Contents

Preface    .   .   .   .   .   .   .   .   .   .   .     v

## PART I

### 1. Energy Bands in Solids

| | | |
|---|---|---|
| 1.1 | Introduction    .   .   .   .   .   .   . | 3 |
| 1.2 | Energy Bands by the Tight Binding Approach  .   .   . | 4 |
| 1.3 | Energy Bands on the Free Electron Model    .   .   . | 8 |
| 1.4 | Dynamics of Electrons .   .   .   .   .   .   . | 14 |
| 1.5 | Outline of Intrinsic Conduction in Pure Semiconductors   . | 15 |

### 2. Impurities in Semiconductors

| | | |
|---|---|---|
| 2.1 | Classification of Crystal Defects    .   .   .   .   . | 19 |
| 2.2 | Donor Impurities in Semiconductors; $n$-Type Material.   . | 20 |
| 2.3 | Acceptor Impurities in Semiconductors; $p$-Type Material   . | 21 |
| 2.4 | Outline of Extrinsic and Intrinsic Conduction in Semiconductors with Impurities    .   .   .   .   .   . | 23 |

### 3. Populations of Conduction Electrons and Holes in Semiconductors

| | | |
|---|---|---|
| 3.1 | The Fermi-Dirac Distribution .    .   .   .   .   . | 24 |
| 3.2 | Approximate Forms of the Fermi-Dirac Distribution .   . | 28 |
| 3.3 | Density-of-States Function    .   .   .   .   .   . | 30 |
| 3.4 | Charge Carrier Concentrations in Semiconductors    .   . | 34 |
| 3.5 | The $n$-$p$ Product .    .   .   .   .   .   . | 37 |
| 3.6 | Combined Donor and Acceptor Impurities    .   .   . | 38 |
| 3.7 | Variation of the Fermi Level with Temperature, $E_F(T)$    . | 41 |
| 3.8 | Variation of the Fermi Level with Impurity Density   .   . | 42 |
| 3.9 | Effect of Temperature on Conduction Processes in Semiconductors .    .   .   .   .   .   .   . | 43 |
| 3.10 | Optical Phenomena in Semiconductors    .   .   . | 45 |
| 3.11 | The Hall Effect    .   .   .   .   .   .   . | 47 |

### 4. Charge Continuity in Semiconductors    .   .   .   .   .

| | | |
|---|---|---|
| 4.1 | Diffusion Currents in Semiconductors    .   .   .   . | 52 |
| 4.2 | Recombination of Charge Carriers .    .   .   .   . | 53 |
| 4.3 | Sources and Sinks    .   .   .   .   .   . | 54 |

4.4     The Equation of Charge Continuity .     .     .     .     .     54
4.5     Applications of the Equation of Charge Continuity   .     .     56
4.6     The Einstein Relation   .     .     .     .     .     .     .     62

## 5. The *p-n* Junction

5.1     Qualitative Charge and Energy Relations at a *p-n* Junction   .     66
5.2     Calculation of the Contact Potential     .     .     .     .     68
5.3     The *p-n* Junction in Equilibrium     .     .     .     .     .     68
5.4     The Reverse Biased *p-n* Junction     .     .     .     .     .     72
5.5     The Forward Biased *p-n* Junction and the Rectifier
        Equation     .     .     .     .     .     .     .     .     .     74
5.6     Barrier Capacitance     .     .     .     .     .     .     .     75
5.7     The Photovoltaic Effect     .     .     .     .     .     .     77
5.8     Breakdown of a *p-n* Junction     .     .     .     .     .     78

## 6. The Work Function and Electron Emission Processes

6.1     The Work Function     .     .     .     .     .     .     .     81
6.2     Thermionic Emission and its Applications .     .     .     .     82
6.3     Contact Potential of Two Solids     .     .     .     .     .     85
6.4     Photoelectric Emission .     .     .     .     .     .     .     86
6.5     Metal to Semiconductor Contacts     .     .     .     .     .     90
6.6     Other Emission Phenomena     .     .     .     .     .     .     95

## 7. The Transistor

7.1     The *p-n* Junction as an Emitter of Charge Carriers     .     .     97
7.2     The Junction Triode     .     .     .     .     .     .     .     100
7.3     Properties of the Base Layer .     .     .     .     .     .     103
7.4     Current Gain of a Transistor .     .     .     .     .     .     107
7.5     Alloyed Junctions     .     .     .     .     .     .     .     110
7.6     The Field Effect Transistor     .     .     .     .     .     .     111

## 8. Semiconductor Devices

8.1     Thermistors     .     .     .     .     .     .     .     .     114
8.2     Hall Effect Devices     .     .     .     .     .     .     .     116
8.3     Tunnel Diodes     .     .     .     .     .     .     .     .     117
8.4     Opto-Electronic Devices     .     .     .     .     .     .     120
8.5     Semiconductor Controlled Rectifiers     .     .     .     .     125

## 9. Electrical Properties of Gases

9.1     Collisions in a Gas     .     .     .     .     .     .     .     130
9.2     Collision Probability and Cross-section     .     .     .     .     132
9.3     Energy Transfer in Elastic Collisions     .     .     .     .     134
9.4     Inelastic Collisions     .     .     .     .     .     .     .     136
9.5     Interaction between Gas Particles and Electro-Magnetic
        Radiation     .     .     .     .     .     .     .     .     .     137

9.6    The Function of Electrodes in Gas Discharges    .        .        .    138
9.7    Basic Concepts of Charge Transport in Gases     .        .        .    139
9.8    Ohmic Conduction and Saturation in a Weakly Ionized Gas .    140
9.9    The Townsend Discharge          .        .        .        .        .        .    141
9.10   The Townsend Criterion of Breakdown and the Self-Sus-
       taining Discharge .        .        .        .        .        .        .        .    146
9.11   The Ionization Coefficients and Paschen's Law .        .        .    146
9.12   The Glow Discharge       .        .        .        .        .        .        .    149
9.13   The Arc     .        .        .        .        .        .        .        .        .    152
9.14   The Effect of High Pressures .        .        .        .        .        .    153
9.15   A.C. Discharges  .        .        .        .        .        .        .        .    154
9.16   Some Applications of Gas Discharges       .        .        .        .    155

10. Properties of Dielectrics
10.1   Introduction      .        .        .        .        .        .        .        .    158
10.2   Polar and Non-polar Atoms and Molecules      .        .        .    160
10.3   Static Dielectric Constant of Gases .        .        .        .        .    162
10.4   A.C. Dielectric Properties of Gases .        .        .        .        .    165
10.5   Static Dielectric Properties of Solids and Liquids        .        .    168
10.6   A.C. Permittivity of Liquid and Solid Dielectrics—
       Relaxation Processes     .        .        .        .        .        .        .    169
10.7   Breakdown in Dielectrics        .        .        .        .        .        .    171

PART II

11. Schroedinger's Equation
11.1   Introduction of Schroedinger's Equation .        .        .        .    177
11.2   Elimination of the Time Variable     .        .        .        .        .    179

12. Basic Concepts of Quantum Mechanics and the Linear Harmonic Oscillator
12.1   The Equation of Motion of the Linear Oscillator and its
       Solutions: Position Probability Distributions       .        .        .    181
12.2   Dynamical Variables as Operators and their Average Values .    187
12.3   Operators, Matrices and their Physical Interpretation .        .    191
12.4   Dirac's Notation and Summary of Chapter 12    .        .        .    195

13. The Hydrogen Atom
13.1   The Equation of Motion of an Electron in a Central Field of
       Force and its Solution  .        .        .        .        .        .        .    197
13.2   Discussion and Interpretation of the Hydrogen Atom Wave
       Functions .        .        .        .        .        .        .        .        .    205
13.3   Energy Levels of the Hydrogen Atom—Degeneracy    .        .    207
13.4   Dirac's Notation for States Characterized by more than one
       Quantum Number   .        .        .        .        .        .        .        .    209

14. **Angular Momentum and Magnetic Moment**
14.1    Orbital Angular Momentum Operators—their Eigenstates
        and  Eigenvalues .       .       .       .       .       .       .       .       211
14.2    Orbital Magnetic Moment       .       .       .       .       .       .       216
14.3    Commuting Operators and Simultaneous Eigenstates .       .       218
14.4    Electron Spin       .       .       .       .       .       .       .       .       221

15. **Magnetic Energy**
15.1    Normal Zeeman Effect .       .       .       .       .       .       .       224
15.2    Spin-Orbit Coupling       .       .       .       .       .       .       227
15.3    The Principle of Superposition       .       .       .       .       .       228

16. **Many-Electron Atoms**
16.1    The Equation of Motion of a System of Many Particles       .       233
16.2    The Hamiltonian of the Many-Electron Atom       .       .       .       234
16.3    The Central Field Approximation .       .       .       .       .       235
16.4    The Eigenstates and Eigenvalues of Many-Electron Atoms—
        the Exclusion Principle .       .       .       .       .       .       237
16.5    The Periodic Table of Elements       .       .       .       .       .       240
16.6    Addition of Angular Momenta       .       .       .       .       .       246
16.7    Magnetic Moments of Many-Electron Atoms       .       .       .       251
16.8    The Stern-Gerlach Experiment       .       .       .       .       .       253

17. **Electrons in Solids**
17.1    The Hamiltonian of Valence Electrons in a Solid       .       .       255
17.2    Electron Waves in One Dimension .       .       .       .       .       256
17.3    Electrons in a Three-Dimensional Potential Well       .       .       261
17.4    Density-of-States Function       .       .       .       .       .       .       265
17.5    Thermionic Emission Equation       .       .       .       .       .       268

18. **Motion of Electron Beams and Individual Electrons—Uncertainty Principle**
18.1    Free Electron Beams and Matter Waves       .       .       .       .       271
18.2    Reflection of Electrons at a Potential Step       .       .       .       273
18.3    Tunnelling of Electrons Through a Potential Barrier       .       .       277
18.4    Expression for Current Flow .       .       .       .       .       .       279
18.5    Motion of Charged Particles in Magnetic Fields .       .       .       280
18.6    The Principle of Uncertainty .       .       .       .       .       .       281

19. **Statistics of Energy Level Occupation**
19.1    Fermions and Bosons       .       .       .       .       .       .       .       283
19.2    Possibilities of Quantum State Occupation by Electrons       .       286
19.3    The Fermi-Dirac Distribution .       .       .       .       .       .       289
19.4    The Maxwell-Boltzmann Distribution       .       .       .       .       292
19.5    The Bose-Einstein Statistics .       .       .       .       .       .       295

## PART III

20. **Paramagnetism and the Maser**
    20.1   Macroscopic Description of Magnetic Materials .      .       .      299
    20.2   Paramagnetic Energy Levels  .       .       .       .       .      301
    20.3   Static Paramagnetic Susceptibility  .       .       .       .      302
    20.4   The Effect of Time Varying Fields on Energy Level
           Populations   .       .       .       .       .       .       .      309
    20.5   Spectroscopy of Paramagnetic Solids   .       .       .       .      312
    20.6   Stimulated Emission and the Three Level Maser .       .       .      315

21. **Ferromagnetism**
    21.1   Exchange Forces as the Basis of Ferromagnetism    .       .      321
    21.2   Crystalline Anisotropy .       .       .       .       .       .      323
    21.3   Magnetic Domains and Domain Walls    .       .       .       .      323
    21.4   Magnetisation of Ferromagnetic Materials      .       .       .      326
    21.5   Magnetic Hysteresis   .       .       .       .       .       .      329
    21.6   The Molecular or Exchange Field   .       .       .       .      331
    21.7   Temperature Variation of Magnetization  .       .       .      333
           Appendix to Chapter 21—The Exchange Interaction  .       .      338

22. **Ferrimagnetism**
    22.1   Antiferromagnetism    .       .       .       .       .       .      341
    22.2   The Indirect Exchange Interaction .       .       .       .      343
    22.3   Ferrites and Garnets   .       .       .       .       .       .      345

23. **Ferromagnetic Resonance**
    23.1   Larmor Precession     .       .       .       .       .       .      348
    23.2   Ferromagnetic Resonance   .       .       .       .       .      350
    23.3   A.C. Susceptibility    .       .       .       .       .       .      352
    23.4   Some Ferrite Microwave Devices   .       .       .       .      355

24. **Diamagnetism**
    24.1   Electromagnetic Induction as Diamagnetism     .       .       .      359
    24.2   Diamagnetism   .       .       .       .       .       .       .      360

25. **Superconductivity**
    25.1   Some Basic Concepts   .       .       .       .       .       .      363
    25.2   The London Theory   .       .       .       .       .       .      365
    25.3   Perfect Diamagnetism and Superconductivity    .       .       .      366
    25.4   Flux Quantization and Electron Pairs   .       .       .      369
    25.5   The Superconducting Phase Transition and the Critical Field     371
    25.6   Intermediate State of Type I Superconductors .       .       .      376

25.7    Interphase Boundaries and their Surface Energy    .    .    377
25.8.   Type II Superconductors    .    .    .    .    .    .    379
25.9    Some Applications of Superconductors    .    .    .    .    384
25.10   Outline of the Microscopic Theory of Superconductivity    .    386

26. Lasers
26.1    Population Inversion at Light Frequencies    .    .    .    392
26.2    Optical Resonators    .    .    .    .    .    .    .    398
26.3    Laser Oscillators and Amplifiers    .    .    .    .    .    402
26.4    Pumping Methods    .    .    .    .    .    .    .    404
26.5    Some Typical Lasers    .    .    .    .    .    .    .    407

Short List of Collateral and Further Reading    .    .    .    .    .    413

Some Physical Constants    .    .    .    .    .    .    .    .    415

Examples    .    .    .    .    .    .    .    .    .    .    .    417

Index    .    .    .    .    .    .    .    .    .    .    .    429

# Part I

*Chapter 1*

# Energy Bands in Solids

Many of the electrical properties of solids are construed in terms of energy concepts, just like the properties of individual atoms. In both cases the electron plays a dominant role, but whereas in atoms it is confined to discrete energy levels, in solids its energies spread over *energy bands*. Since the latter provide the conceptual framework for the discussion of conduction, we start our study with a description of electron motion in solids to see how this gives rise to energy bands.

The treatment in Part I of the book is based on the Bohr model of the atom with which the student is assumed to be familiar. Readers who may not be content with this elementary approach are invited to proceed to Part II, in which the atomic background is formulated on the basis of quantum mechanics. This may be read first, or else it may be used at a later stage as a means of revision and consolidation of the basic concepts from a more mature stand-point.

## 1.1 Introduction

As our starting point we recall the dual nature of the electron. On this view the electron is both a particle and a wave. The relation between the wavelength of moving electrons and their momentum is given by de Broglie's equation.

$$p = \frac{h}{\lambda} = \hbar k \qquad (1.1.1)$$

where:      $p$ = momentum

$\lambda$ = wavelength

$k = \dfrac{2\pi}{\lambda}$ = wavenumber (phase constant)

$\hbar = \dfrac{h}{2\pi} = \dfrac{1}{2\pi} \times$ Planck's constant

3

Our interest will be mainly in crystalline solids but many of the results we shall obtain will also apply to polycrystalline samples.

The study of solids will be approached in two ways. In the first place a crystal will be thought of as the result of many atoms being brought close together by stages, and the consequent modifications of the electronic energy levels will be considered. In the second place the regular array of atoms in a lattice will be visualized as providing the electrical forces subject to which some of the electrons are free to move throughout the crystal.

We begin in the following section with the former model.

## 1.2 Energy Bands by the Tight Binding Approach

Since samples of solids are aggregates of many atoms (some $10^{23}$ atoms per $cm^3$) it should be instructive to consider what happens to the discrete energy levels as individual atoms are brought close together to form molecules and then macroscopic crystals.

It is easier to understand the problem once it is realized that the orbital motions of electrons are in the nature of resonant phenomena. Bearing this in mind the analogy of two identical resonant circuits provides a useful guide. Figure 1.1(a) shows two tuned circuits, sufficiently far apart not to interact

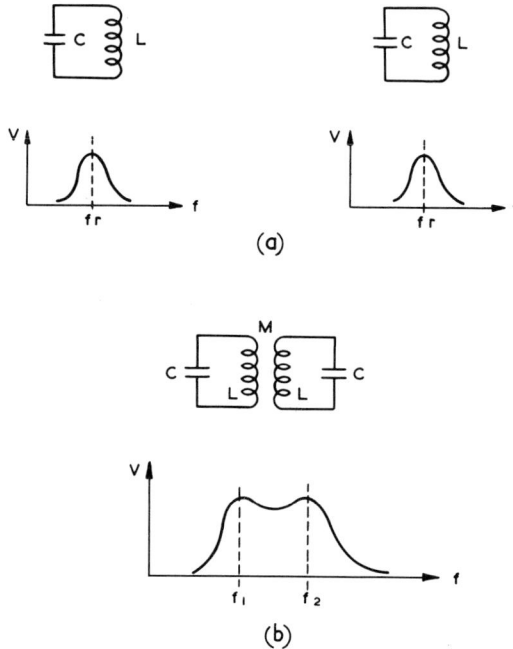

Fig. 1.1. Resonant properties of coupled circuits; $V$ is the voltage across the capacitor.

inductively, with their resonance curves shown diagrammatically on separate graphs. If the two circuits are brought sufficiently close together to become a coupled pair they cease to be separate dynamical entities. Instead they become a single dynamical system with properties different from their constituents. In fact, the resonant behaviour of the *coupled circuit* is complicated by the appearance of two resonant frequencies $f_1$ and $f_2$ which differ from the original common resonance frequency $f_r$. This state of affairs is depicted schematically in Fig. 1.1(b).

If a greater number of tuned circuits are coupled together, inductively or otherwise, their resonant properties follow from an extension of the above observation. In general there will be as many resonant frequencies as there are individual circuits, with the proviso that on occasion some of the resonant frequencies may coincide.

The discrete energy levels of atoms have properties similar to the foregoing example. As, say, two atoms are brought sufficiently close together to interact dynamically, that is for the forces of one atom to influence the other, their *energy levels split* as indicated schematically in Fig. 1.2(a). The nature of the splitting and its dependence on the interatomic distance will vary from element

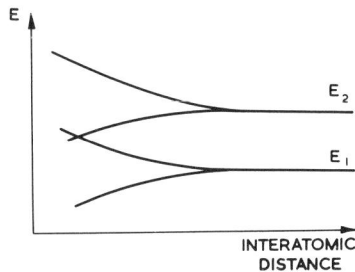

(a)

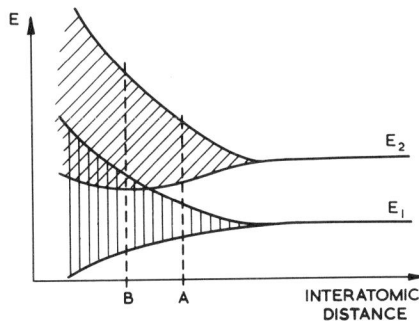

(b)

Fig. 1.2.

to element. In some cases the split levels may cross over for certain values of interatomic distance, and separate again for others.

As more than two identical atoms are brought together the process of energy level splitting continues until a multitude of levels, very closely spaced, form in bands of energy with which the atomic electrons are allowed to move. The resultant situation is shown diagrammatically in Fig. 1.2(b).

The relation of the dotted lines $A$ and $B$ to the energy bands should be noted with reference to the interatomic spacing. If the atoms of the element in question settle in a crystal at a spacing represented by $A$ the solid will exhibit two distinct *allowed bands,* corresponding to the original discrete levels $E_1$ and $E_2$, separated by a *forbidden gap.* If, on the other hand, the lattice spacing corresponds to $B$ the solid will have a single allowed band, partly including overlapping levels derived from both $E_1$ and $E_2$.

The foregoing hypothetical example contains the most important features of the energy bands in solids as derived from the *tight-binding approach.* As atoms are brought together to form a crystal the energy levels associated with the outermost electron orbits split to form bands. The bands derived from energy levels atually occupied by electrons in individual atoms, will themselves be occupied. However, in many cases the bands cannot be simply identified with energy level and consequently their occupation by electrons cannot be predicted on this basis.

As regards the levels of the innermost electrons in many-electron atoms, it should be noted that they do not split into bands. This is because they are prevented from interaction with neighbouring atoms by the screening effect of the electrons in the outermost shells.

Before going on to describe specific examples it is important to bear in mind that there are still discrete levels within each allowed band. The spacing between them is minute compared with the extent of the bands. This consideration will prove important in the discussion of the occupation of the energy bands by electrons, and of electrical conductivity.

Diamond provides an example of a crystal for which the energy bands have been evaluated by the tight-binding approach. The result is shown in Fig. 1.3. Although the 2s and 2p levels of the carbon atom form the starting point, the resulting energy bands cannot be identified with the levels. In the diamond crystal, having a lattice spacing of some 1·7Å, the bands are separated by a large gap, and the lower band is fully occupied by electrons, while the upper band is empty. Under normal conditions of temperature and pressure the probability that electrons could be excited from the lower to the upper band is quite negligible, for reasons which will become apparent as our treatment unfolds. The lower band which is fully occupied by electrons is usually referred to as the *valence band,* while the upper empty band is called the *conduction band.* They are separated by a forbidden energy gap, or briefly *energy gap,* which in diamond has a value of 6 eV. The width of this gap is closely related to the insulating properties of diamond in a manner to be discussed in a later section.

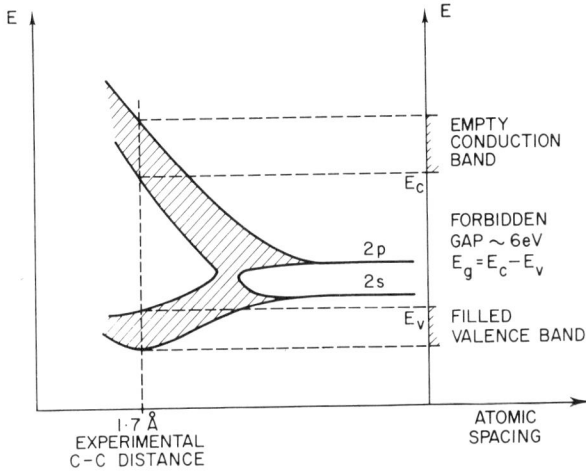

Fig. 1.3. Energy bands in diamond.

Since the energy values corresponding to the top of the valence band and the bottom of the conduction band will recur frequently in our study we agree to denote them by the symbols $E_v$ and $E_c$ respectively. Their difference, or the extent of the energy gap, will be denoted by $E_g$. Thus:

$$E_g = E_c - E_v \qquad (1.2.1)$$

An altogether different example of band structure is provided by copper which is a typical metal or conductor. The main features of its conduction band are shown in Fig. 1.4. Here the allowed energy band, which corresponds to the actual lattice spacing of a copper crystal, results from the overlap of three distinct bands. These in turn originate in the discrete levels $3d$ $4s$, and $4p$ of the Cu atom. In the ground state the $3d$ level is fully occupied by electrons, the $4s$ level is half occupied while the $4p$ level is empty. Hence the resulting conduction band is approximately half occupied by electrons.

The examples of copper and diamond are, from our point of view, extreme. In between there are materials with energy gaps much smaller than that of diamond, say about 1 eV. These are the *semiconductors* with which we shall be mainly concerned in the chapters to follow. In all cases our interest will centre on that range of the energy axis within which there is a change from a full occupation of available states by electrons to completely vacant states. It is found that only electrons in this energy range enter into conduction processes in solids.

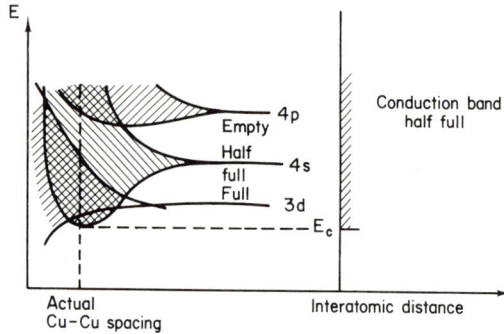

Fig. 1.4. Conduction band in copper.

## 1.3  Energy Bands on the Free Electron Model

In this approach the starting point is the crystalline solid rather than individual atoms. The assumption is that charged atomic nuclei, situated at lattice sites, are the sources of an electrostatic potential in the field of which the electrons perform their motions. The electrons in the innermost orbits are largely within the sphere of influence of the nucleus to which they belong. Their movements remain independent of the field of force of neighbouring atoms and their energy levels remain discrete. The outermost electrons, however, are affected by the neighbouring atoms and their movements are significantly modified. Some of them may become completely detached from their parent atoms, to become *free electrons*. Such free electrons move throughout the macroscopic extent of a crystal, subject to lattice forces.

To obtain some insight into the movements of free electrons it is necessary to note some of the features of the fields of force inside the lattice. The starting point is the potential of an electron in the field of a nucleus of atomic number $Z$, shown in Fig. 1.5(a). When a hypothetical situation is created whereby two such atoms are brought to within a lattice spacing, the potential of an electron in their combined field of force will be as shown in Fig. 1.5(b). The dotted lines indicate the potentials of the individual atoms when separated, while the solid lines indicate the potential of the pair of atoms obtained by superposition. It should be noted that the potential between the atoms is considerably lower than in the outward direction. Hence a free electron will tend to occupy a low energy position between the atoms, while being repelled by the high-energy barrier at the extremities of this configuration.

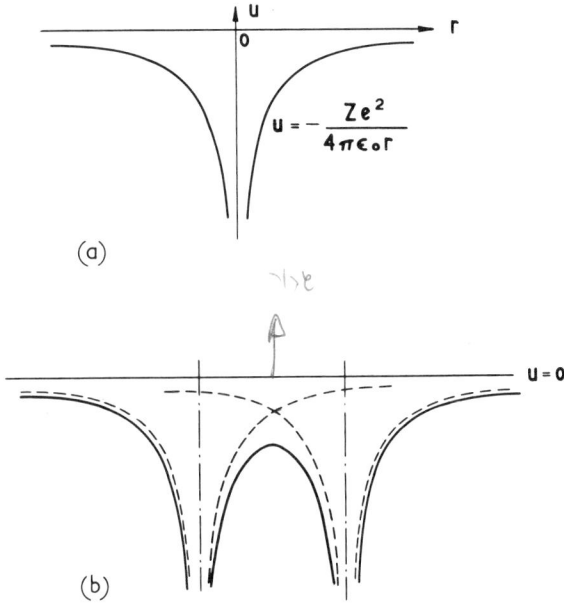

**Fig. 1.5.** (a) Electron potential in the field of a nucleus. (b) Electron potential in the field of two closely spaced nuclei.

The situation inside a crystal consisting of many atoms can now be visualized as an extension of the foregoing example. The potential *along a line of atoms* in a lattice will vary as shown in Fig. 1.6(a), while the potential *off a line of atoms* will follow the line of Fig. 1.6(b).

The electrons can be subdivided into three categories. First there are the *core electrons* whose movements are *screened* from the effect of neighbouring atoms. Secondly there are the *valence electrons* which are to some extent affected by neighbouring atoms but remain in bound states in the vicinity of their parent atoms; these electrons are responsible for the cohesion of the solid. Thirdly there are the *free or conduction electrons* which become detached from their parent atoms, and which move about the crystal subject to thermal excitations imparted to them by the lattice, and subject to external forces such as applied electric or magnetic fields. The free electrons are responsible for the conductivity of solids. Since we shall be greatly concerned with this property, we now consider the motions of the free electrons quantitatively, on the basis of a highly idealized model.

In this model the *periodic lattice potential* of Fig. 1.6 is simplified by ignoring the regular variations inside and retaining only the repelling potential barrier at the boundary of the crystal. Indeed, the lattice is further simplified by assuming that it is vertical and very high. The concept thus arrived at is that of a *potential well* shown in Fig. 1.7 for one dimension.

Free electrons move within the potential well in a manner analogous to billiard balls on a frictionless table, bouncing off the opposite barriers and on occasion colliding with each other. Although this picture is of some help it cannot be made the basis of a mathematical description, however simple. The reason is that this classical model would yield results which would not agree with experimental observations.

To formulate a more viable description it is necessary to take into account the wave nature of electrons. On this view the electrons perform motions between the potential barriers analogous to standing waves on a string, fixed at both ends, as suggested in Fig. 1.7. The standing waves are subject to the restriction that only multiples of half a wavelength can fit into the one-dimensional box. If the length of the potential well is $L$ this condition is stated in the equation

$$\lambda_n = \frac{2L}{n}; \; n = 1, 2, 3, \ldots. \tag{1.3.1}$$

De Broglie's relation, eqn (1.1.1), is now used to obtain the wave number and momentum of the electron waves and hence their energy. The wave number is

$$k_n = \frac{2\pi}{\lambda_n} = \frac{\pi}{L} n. \tag{1.3.2}$$

Hence the momentum

$$p_n = \hbar k_n = \frac{\pi\hbar}{L} n = \frac{h}{2L} n. \tag{1.3.3}$$

Finally the kinetic energy is

$$T_n = \frac{p_n^2}{2m_e} = \frac{\hbar^2}{2m_e} k_n^2 = \frac{\hbar^2}{8m_e L^2} n^2 \tag{1.3.4}$$

where $m_e$ is the electron mass.

As it is assumed that the potential energy of electrons in the well is constant, and since every potential contains an arbitrary constant, we are allowed to set its

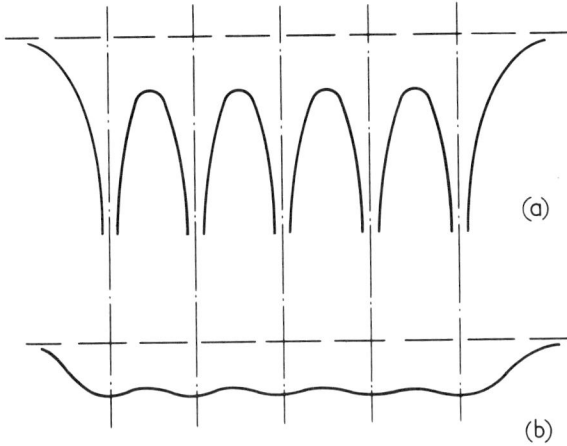

**Fig. 1.6.** (a) Lattice potential along a line of atoms. (b) Lattice potential off a line of atoms.

value equal to zero. Hence the kinetic energy obtained above is the total energy of the electrons in the box.

$$E_n = \frac{h^2}{8m_e L^2} n^2. \qquad (1.3.5)$$

The foregoing results show that the wave numbers and momenta, and hence the energy levels of electrons in the potential well, are quantized. A graph of great practical importance is obtained by plotting the energy as a function of wave number or momentum as shown in Fig. 1.8. The first point to note is that the discrete values of *momentum or wave number are evenly spaced* on the horizontal axis. This, of course is a direct consequence of the standing wave nature of electron motion in the box. The corresponding values of the *energy are not evenly* spaced on the energy axis because of the quadratic dependence of energy on wave number.

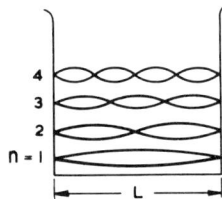

**Fig. 1.7.** Electron waves in a one-dimensional potential well.

If the potential well is to represent a solid, its length must be assumed to be of macroscopic dimensions, say 1 cm. Since Planck's constant $h$ is an extremely small quantity, it follows that the spacing of the energy levels on the energy axis is very small indeed. For this reason the discreet nature of the energy spectrum is disregarded in calculations of the conductivity of solids. Instead the concept of the *density of levels or states* is used. *This is defined as the number of states per unit energy range,* and will be explained at length and applied to semiconductors in Chapter 3.

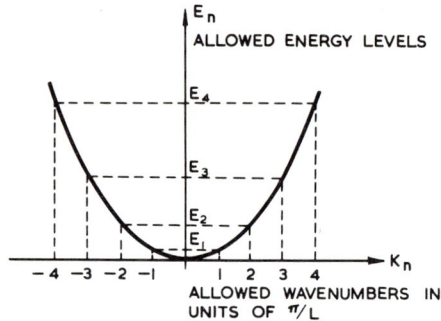

**Fig. 1.8.** Energy as a function of wavenumber for electrons in a one-dimensional potential well.

The occupation of states by electrons is severely restricted by Pauli's *exclusion principle* which is dealt with on the basis of quantum mechanics in Chapter 18. For our present purpose it is sufficient to note that according to the exclusion principle each electron, which is part of a single dynamical system, must have a separate state all to itself. Thus there must be enough distinct states in a macroscopic potential well or chunk of metal to accommodate within a reasonable energy range ($\cong 1$ eV) all free electrons, of which there are some $10^{22}$ per $cm^3$.

Let us now consider how the motions of electrons, that is their energy levels, are modified if a periodic variation is superimposed on the constant potential of the idealized well, as suggested in Fig. 1.6(b). It is beyond our scope at the present stage to attempt a mathematical analysis of any simplified models. Instead a graphical description of some general features of wave motion in periodic structures will be given, and their relation to the results obtained from the tight binding approach will be explained.

The superposition of a periodically varying term on the flat potential well modifies the energy versus wave-number graph in a manner indicated in Fig. 1.9. The parabolic relationship is distorted by the appearance of *gaps* in the energy level distribution. Allowed electron energy levels are available only over energy ranges marked accordingly in the diagram. The first discontinuity in the energy spectrum appears at a value of the wave number at which half the electron wavelength equals the period of the lattice potential, that is the lattice spacing. The second discontinuity occurs at twice this wave number or one wavelength equal to the lattice spacing etc. For these particular values of wavelength the electrons are inhibited from propagation along the lattice, that is there are no available electron states.

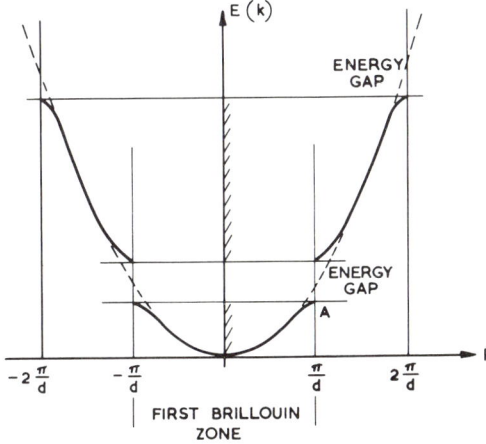

Fig. 1.9. Energy band scheme for electron states in a periodic potential. Individual electron energy levels are densely packed within the allowed bands.

The process may be visualized on the basis of Bragg reflection in the limiting case when the waves are incident along a lattice line. Under this condition $\theta = 90°$ in the Bragg formula $2d \sin \theta = n$ and the condition for reflection is

$$\lambda = \frac{2d}{n} \quad \text{or} \quad k = \frac{\pi}{d} n \tag{1.3.6}$$

where $d$ = lattice spacing and $n = 1, 2, 3, \ldots$.

The above values of the wave number divide the $k$-axis into equal regions called *Brillouin zones*. The lowest energy band corresponds to the *first Brillouin zone*.

Thus we have arrived again at the energy band concept of solids by considering electron waves in the periodic potential of the lattice. Although the foregoing results have been obtained on the basis of highly idealized one-dimensional models, the concepts derived are found to form a valid foundation for the discussion of many electrical properties of solids.

Table 1.1 lists some typical semiconductors and the energy gaps between the valence and conduction bands.

**Table 1.1**

Some semiconductors and their energy gaps

| Material | Chemical Symbol | Energy Gap [eV] |
|----------|-----------------|-----------------|
| *Elemental Semiconductors* | | |
| Grey Tin | Sn | 0·1 |
| Tellerium | Te | 0·38 |
| Germanium | Ge | 0·803 |
| Silicon | Si | 1·12 |
| Boron | B | 1·1 |
| Grey Selenium | Se | 1·8 |
| Diamond | C | 5·5 |
| *Compound Semiconductors* | | |
| Indium Antimonide | In Sb | 0·16 |
| Gallium Arsenide | Ga As | 1·43 |
| Cadmium Sulphide | Cd S | 2·4 |
| Zinc Oxide | Zn 0 | 3·2 |

### 1.4 Dynamics of Electrons

In this section we introduce some additional concepts relating to the motion of electrons in solids. They will be required subsequently in the discussion of conduction phenoma in semiconductors.

Let us consider more closely the form of the function $E(k)$ near the edges of the energy gaps, in the vicinity of point $A$, say, in Fig. 1.9. Comparing the downward deviation of the energy from the parabolic curve with eqn (1.3.4) we note that this must be visualized in terms of a change in the electron mass $m$, if the classical expression for the energy is to be retained. This is indeed the procedure usually adopted when dealing with semiconductors. The electron energy is written in the form

$$E = \frac{p^2}{2m^*} \tag{1.4.1}$$

where $m^*$ is a modified or *effective mass* of the electrons.

Now, it is interesting to note that the effective mass, and also the velocity, can be related to the shape of the $E(k)$ curve. Differentiation of the above equation with respect to $p$ yields the velocity

$$\frac{dE}{dp} = v.$$

Using de Broglie's relation, eqn (1.1.1), to make a change of variable we find

$$v = \frac{dE}{dk}\frac{dk}{dp} = \frac{1}{\hbar}\frac{dE(k)}{dk}. \tag{1.4.2}$$

Thus the electron velocity is proportional to the slope of the $E(k)$ curve. Differentiating eqn (1.4.1) with respect to $p$ a second time and once again changing variables we obtain

$$\frac{1}{m^*} = \frac{1}{\hbar^2}\frac{d^2 E(k)}{dk^2}. \tag{1.4.3}$$

Hence the effective mass of the electrons is related to the curvature of the $E(k)$ curve. Equations (1.4.2) and (1.4.3) can be used to derive the electron velocity and mass from $E(k)$ curves obtained experimentally.

The foregoing results by no means imply that the mass of a fundamental particle like the electron undergoes actual changes in the course of its motion through the lattice. The results mean that in certain states (close to a band edge) the lattice exerts forces on electrons which have the same effect as hypothetical changes of mass. Sometimes it is nevertheless convenient to speak about "heavy" or "light" electrons in solids.

## 1.5  Outline of Intrinsic Conduction in Pure Semiconductors

Before going on to discuss the modifications of the band structure caused by lattice defects, we include a preliminary description of conduction in pure semiconductors. The subject of electrical conduction will come up for an extensive discussion in a later chapter, when phenomena in semiconductors with impurities will also be included.

The band structure of semiconductors is characterized by the fact that there are just enough states in the valence band to accommodate all electrons. At the absolute zero of temperature all electrons are indeed in the valence band. Under such conditions it is impossible to cause a flow of electric current by the application of an electric field. Motion in the direction of the applied field would mean that the electrons have acquired some additional energy and have, therefore, occupied slightly higher energy levels. Since none are available such a process is impossible. There are, of course, vacant levels in the conduction band, but at the lowest temperatures electrons cannot be excited into them.

Occupation of the excited levels in the conduction band can only take place at higher temperatures at which thermal vibrations of the lattice are energetic enough to impart sufficient energy to valence electrons. The conduction electrons are free to move about the lattice with energies corresponding to the levels they occupy. Their motions are random, subject to changes of direction (momentum) and energy as a result of collisions between themselves and with the lattice atoms. The application of an external field imposes a *drift velocity* on the electrons which is small compared with their thermal velocities, but which does constitute an electric current. The intensity of the current will depend on the number of conduction electrons and on their ability to respond to the applied force, that is on their *mobility*.

There will be a further contribution to the conduction current from *holes* in the valence band. Since some levels in the latter have been vacated, the valence electrons acquire some freedom of movement and can contribute towards the conduction current. However, this process is best discussed in terms of the vacant states or holes.

To visualize what happens let us refer to Fig. 1.10(a). This represents diagrammatically a semiconductor lattice, say germanium. The valence electrons, of which four belong to every atom, are symbolized by black dots. At normal temperatures practically all the valence electrons will remain in bound states in the vicinity of their parent atom but now and again one of them will receive a sufficient amount of energy to free itself and move about the lattice. It will leave behind a *vacant state or hole* which can be filled by any one of the neighbouring electrons. This propensity of electrons to fill a nearby hole constitutes a motion of the hole which under equilibrium conditions is as random as that of free electrons. On the energy-level diagram the position is represented by Fig. 1.10(b).

The application of an electric field or potential difference to the semi-conductor is usually symbolized by Fig. 1.11. The energy-band scheme is tilted by the differing potential at the opposite ends of the semiconductor and the charge carriers are induced to move down the potential gradient in directions consistent with their polarity. The motion is in the nature of a slow drift, superimposed on the random thermal movements of both electrons and holes, amounting to a conduction current. The resulting current density is the sum of electron- and hole-current densities and can be written in terms of the conductivity $\sigma$ as follows:

$$J = J_n + J_p = e(\mu_n n + \mu_p p)\mathscr{E} = \sigma\mathscr{E} \qquad (1.5.1)$$

where   $J_n$ = electron-current density
　　　   $J_p$ = hole-current density
　　　   $n$　= free- or conduction-electron density

$p$ = hole density

$\mu_n, \mu_p$ = mobilities of electrons and holes defined as average drift velocities per unit applied field

$e$ = electronic charge.

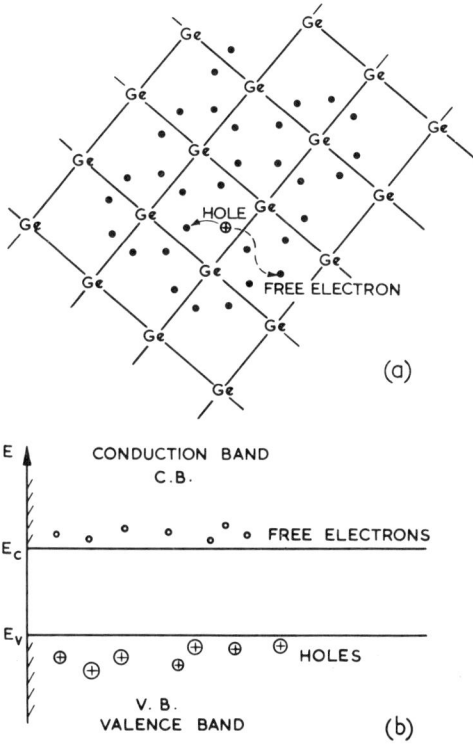

Fig. 1.10. (a) Diagrammatic representation of free electrons and holes in a germanium lattice. (b) Intrinsic electrons and holes on the energy level diagram.

In the above equation, and throughout this book, the electronic charge $e$ is treated as a positive physical constant. The current densities are treated in the present context as scalar magnitudes, to be added or substracted as the situation demands.

In the case of a pure semiconductor considered here, the number of holes in the valence band is the same as the number of electrons in the conduction band and the transport of electric current is said to be *intrinsic*.

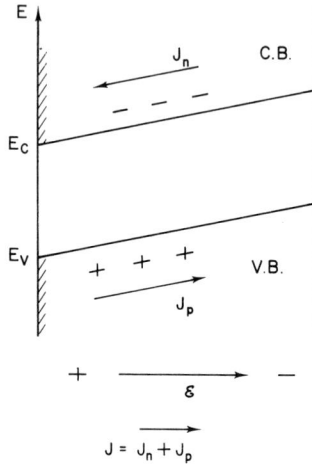

Fig. 1.11. Intrinsic conduction in a pure semiconductor.

Using eqn (1.5.1) the conductivity of a sample can be evaluated if the number of charge carriers are given and if the mobilities can be ascertained theoretically or experimentally. These questions will be the subject matter of subsequent chapters.

*Chapter 2*

# Impurities in Semiconductors

## 2.1 Classification of Crystal Defects

The presence of defects in crystal lattices modifies the energy band scheme of solids and has a profound influence on the conducting properties of semiconductors. As a preliminary to a discussion of the latter we list and illustrate lattice defects.

(a) Substitutional impurities. Occasional lattice sites are occupied by foreign atoms as indicated in Fig. 2.1(a). This type of crystal defect is deliberately introduced in impurity-type semiconductors under conditions carefully controlled for transistor manufacture. It will form the main subject of the following chapters.

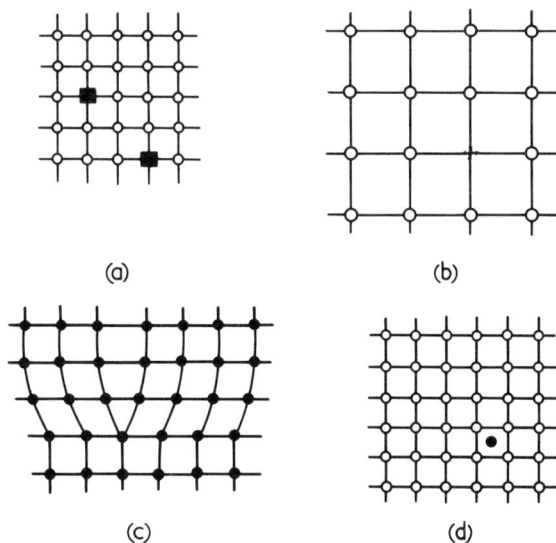

Fig. 2.1. Two-dimensional representation of lattice defects: (a) substitutional impurities; (b) vacancies or Schottky defects; (c) dislocations; (d) interstitial impurities or Frenkel defects.

19

(b) Vacancies or Schottky defects. An occasional atom is missing from a lattice site as shown in Fig. 2.1(b). These defects enter into the operation of copper-oxide rectifiers and oxide-coated thermionic cathodes.

(c) Dislocations. These are irregularities in the structure of the crystal lattice unconnected with either vacancies or foreign atoms. An example is shown diagrammatically in Fig. 2.1(c). It should be borne in mind that a real three-dimensional lattice can exhibit a variety of structural dislocations.

(d) Interstitial impurities or Frenkel defects. These differ from substitutional impurities in that additional atoms are squeezed between the atoms of the regular lattice. Fig. 2.1(d) suggests an interstitial impurity.

## 2.2  Donor Impurities in Semiconductors; $n$-Type Material

Substitutional impurities in semiconductors fall into two categories: (a) elements of valency higher than the host crystal are called *donor impurities* because they are able to release the extra electron from their outermost shell to occupy an empty level in the conduction band of the host crystal; (b) elements of valency lower than the host crystal are called *acceptor impurities* because they are able to accomodate an extra electron in their outer shell, which has vacated a level in the valence band and thus created a hole. The details of the changes in the energy-band scheme are best discussed on the specific example of a ypical semiconductor, say silicon. We deal first with donor impurities.

Silicon is an element of valency 4 which forms crystals of the diamond type. The replacement of an occasional silicon atom by an element of valency 5 upsets the covalent-bonding scheme by adding a surplus electron to the orbital configuration, as suggested in Fig. 2.2(a). The impurities usually used for this purpose are phosphorus (P), arsenic (As) or antimony (Sb). The concentration of these impurities is very low; of the order of 1 in $10^6$ of the host atoms being replaced. As the extra electron associated with an impurity does not properly fit into the bonding scheme it occupies an energy level well above the top of the valence band. Indeed its state is just below the bottom of the conduction band, because it requires very little energy to release it into the conduction band. Since the impurity atoms are spaced far apart on he scale of atomic distances, they do not interact with each other and the surplus electrons all occupy the same discreet level marked $E_d$, the *donor level,* in Fig. 2.2(b).

The high energy of the extra electron makes it susceptible to be detached from its parent atom and to be excited into the conduction band where it becomes a free conduction electron, called a *donor electron.* The thermal energy required to accomplish this equals $E_c - E_d$ which is much less than $E_c - E_v$. The impurity atom, deprived of one electron, becomes ionized and constitutes an *immobile positive charge* in the lattice, which cannot contribute towards any conduction current. It thus differs fundamentally from a positive hole created in the valence band by excitation of valence electrons. Depending on temperature few, some or practically all of the donor atoms become ionized, thus contributing in increasing measure to the conductivity of the sample as the temperature is increased.

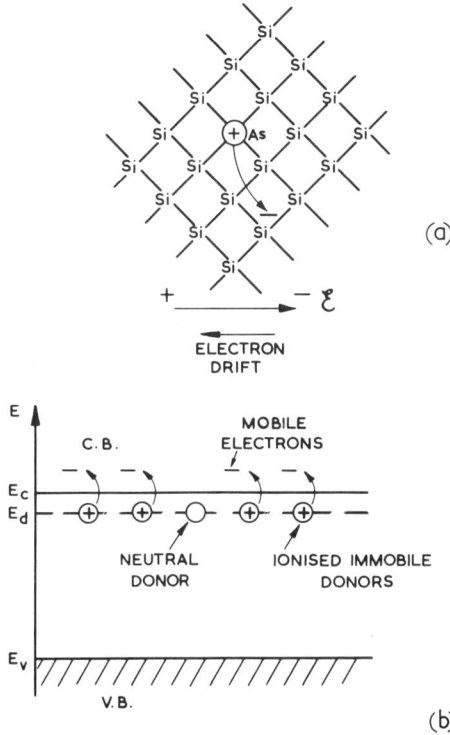

Fig. 2.2. (a) Diagrammatic representation of the silicon lattice showing an ionized arsenic atom. (b) Donor impurities shown on the energy level diagram.

Semiconductors containing trace impurities of the above type are said to be *doped* to form *n-type material,* because conduction of electricity takes place through the medium of negative charge carriers—the electrons.

## 2.3 Acceptor Impurities in Semiconductors; $p$-type Material

A different situation arises when impurities of lower valency are substituted for an occasional atom of the host crystal lattice. Taking germanium as an example which has valency 4 and forms crystals similar to silicon, we consider a trace impurity of valency 3, e.g. boron (B). The resulting situation is indicated in Fig.

2.3. The impurity atom creates a deficiency of one electron in the covalent-bonding scheme with the result that the valence electrons of neighbouring germanium atoms have a tendency to fill it. To do this they only need a small amount of energy, $E_a - E_v$, which is much less than the energy gap $E_c - E_v$, as

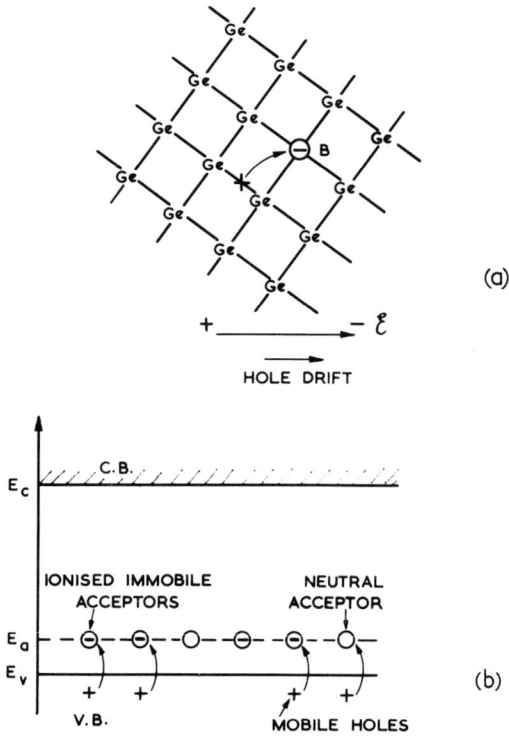

Fig. 2.3. (a) Diagrammatic representation of the germanium lattice showing an ionized acceptor boron atom. (b) Acceptor impurities on the energy-level diagram.

indicated in Fig. 2.3(b). The additional electron attached to the impurity atom occupies the discrete energy level $E_a$, situated just above the top of the valence band. The impurity atom becomes ionized by accepting the extra electron into its outer orbit and is therefore referred to as an *acceptor* atom. It carries a *negative charge which is immobile*. The electrons accepted by the impurity atoms leave vacant states or holes in the valence band which are mobile, and hence capable of carrying an electric current.

Semiconductors doped in the above manner form *p-type material* because they conduct electricity through the medium of positive-charge carriers—the holes. The number of holes in the valence band depends on the fraction of acceptors which are ionized, and this in turn depends on temperature in the manner to be described in the next chapter.

## 2.4  Outline of Extrinsic and Intrinsic Conduction in Semiconductors with Impurities

Doped semiconductors differ from pure semiconductors in that the number of electrons in the conduction band is not necessarily equal to the number of empty states or holes in the valence band. This is true at temperatures high enough for a good fraction of the impurities to be ionized, but too low for a significant number of valence electrons to be excited into the conduction band by the thermal activation across the energy gap.

In *n-type material* under such conditions the number of *negative charge carriers* is much greater than the number of positive charge carriers. The position is described by saying that electrons are *majority carriers* and holes are *minority carriers*. The converse observation applies to *p*-type material, where holes are the majority carriers. A further distinction in terminology is frequently made between charge carriers originating in impurities and those due to thermal excitation of valence electrons. The former are called *extrinsic carriers* while the latter are termed *intrinsic carriers*.

The conduction processes in semiconductors can now be classified in terms of the concepts explained above.

(i) Extrinsic conduction by majority electrons provided by donor impurities in *n*-type material.

(ii) Extrinsic conduction by majority holes created in the valence band by the absence of electrons which have been excited into the acceptor impurities in *p*-type material.

(iii) Intrinsic conduction by intrinsic (or thermal) electrons and holes.

In general all three processes are present, but depending on the type of material and temperature one of them may be dominant. In practical semiconductors at room temperature the extrinsic processes invariably predominate.

The conductivity equation, eqn (1.5.1), applies in the general case with the important proviso that the numbers of electrons and holes are not necessarily equal. Hence we can write

$$J = \sigma \mathscr{E} = e(\mu_n n + \mu_p p)\mathscr{E} \qquad (2.4.1)$$

where $p \neq n$ in most cases.

It is emphasized that all three conduction processes listed above require the thermal excitation of electrons to higher levels, although the word "thermal" is sometimes applied to intrinsic electrons only.

*Chapter 3*

# Populations of Conduction Electrons and Holes in Semiconductors

### 3.1 The Fermi-Dirac Distribution

In order to evaluate the conductivity of semiconductors according to eqn (2.4.1) it is necessary to know the populations of free electrons and holes and their mobilities. In the present chapter we shall be concerned with the populations of charge carriers, leaving the question of mobilities until a later stage.

The problem of evaluating the number of free charge carriers splits into two parts. On the one hand it is necessary to find the total number of available electron states within a volume of solid; this is done in terms of the density-of-states function, to be dealt with in the next section. On the other hand it is necessary to know how many of the available states are actually occupied by charge carriers. As it is usually more convenient to consider this question first, we proceed to deal with it now.

The Fermi-Dirac function which governs the distribution of electrons among the available states of a dynamical system is derived on the basis of quantum statistics. Such a derivation will be given in a later chapter, but since we wish to apply the Fermi-Dirac distribution to semiconductors now, we state it without proof and explain how it is used.

The distribution function has the following form

$$f(E) = \frac{1}{1 + e^{(E-E_F)/kT}}. \tag{3.1.1}$$

*It gives the fraction f(E) of available states at the energy E actually occupied by electrons, under conditions of thermal equilibrium.* The significance of the symbols used is as follows:

e = base of natural logarithms, to distinguish it from the electronic charge *e*

$k$ = Boltzman's constant (= $1 \cdot 380 \times 10^{-23}$ joule/deg)

24

$T$ = absolute temperature ($^\circ$ Kelvin)

$E_F$ = Fermi level.

The Fermi level is a constant parameter of the Fermi-Dirac distribution having the property that for $E = E_F$, $f(E_F) = 0.5$. A graphical representation of the Fermi-Dirac function is given in Fig. 3.1. The following features of the graphs should be noted.

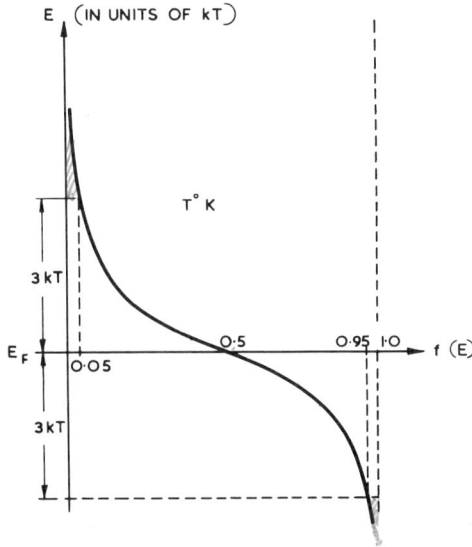

**Fig. 3.1(a).** Fermi-Dirac distribution in normalized form.

(i) Although $E$ is the argument of the function, it is plotted on the vertical axis to allow easy correlation with energy level diagrams.

(ii) The Fermi level $E_F$ provides a natural reference line to which to refer other properties of the distribution. Above $E_F$ the distribution tails off rapidly, indicating that few of the available states in this upper energy range are occupied by electrons. Below $E_F$ the function tends asymptotically to 1, indicating that in this energy range practically all the available states are occupied by electrons. At the Fermi level exactly half the available states are occupied and half are empty.

(iii) If the energy axis is calibrated in units of $kT$ above and below the Fermi level, the reader will easily verify the following quantitative observations. At a height $E - E_F = 3kT$ above $E_F$ only some 5% of the available states are occupied; at a depth of $E - E_F = -3kT$ below $E_F$ some 95% of the states are full.

Figure 3.1(a) shows the Fermi-Dirac distribution in a normalized form applicable at any temperature. In Fig. 3.1(b) several curves, corresponding to different temperatures, are plotted against an energy axis calibrated in electron-volts relative to the Fermi level. The trend of these curves shows the form the distribution assumes at 0 °K. It becomes a step, located on the Fermi level, above which all states are empty and below which all states are occupied by electrons.

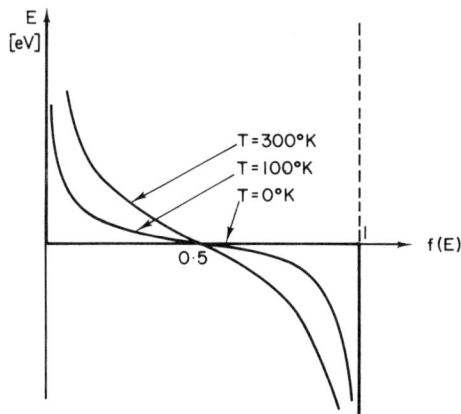

**Fig. 3.1(b).** Family of Fermi-Dirac functions for various temperatures.

The use of the Fermi-Dirac distribution is best illustrated on the example of a solid with a partly filled allowed band, say metallic sodium or copper. Under conditions of thermal equilibrium, at a reasonable temperature such as 300 °K, practically all states up to a certain level in the conduction band are occupied by electrons. Above that level practically all states are empty. Over a range of energies around the level in question there is a gradual transition from all levels occupied to all levels empty. The precise form of this transition follows the Fermi-Dirac function. The particular value of energy at which half the states are occupied *is the Fermi level*. The situation is shown diagrammatically in Fig. 3.2.

It is important to compare the range of energies around the Fermi level, where significant fractions of electron states are partly occupied, with the full extent of the conduction band in a typical metal. For sodium this is 3·16 eV while $6kT$, the range of energy around $E_F$ which covers occupation percentages from 5% to 95%, is only 0·075 eV at 300 °K. This comparison demonstrates that it is quite justifiable to speak about a sea of electrons in the conduction band of a metal which is very deep compared with the extent of surface irregularities at the Fermi level. Indeed at the absolute zero of temperature there is a sudden change from full occupation of states below $E_F$ to completely empty states

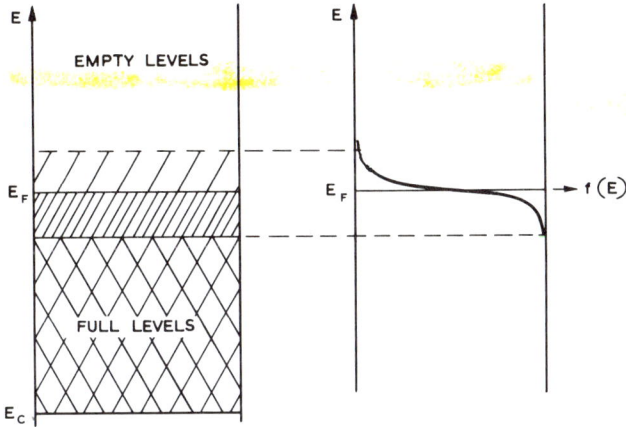

**Fig. 3.2.** Partly filled conduction band of a metal and the Fermi-Dirac distribution of electrons.

above $E_F$. The last observation provides perhaps the best interpretation of the Fermi level.

Let us now consider how the Fermi-Dirac function is applied to semiconductors. Figure 3.3 shows the position in an intrinsic semiconductor, say germanium, at room temperature. Under these conditions very few of the available states in the conduction band are occupied by electrons, excited from the valence band. Since the number of electrons equals the number of holes per unit volume of material in this case, the Fermi level must be positioned near or at the centre of the energy gap as shown. The shaded areas adjoining the distribution curve in the conduction and valence bands are intended to suggest the equal numbers of electrons and vacant states. A comparison of the energy

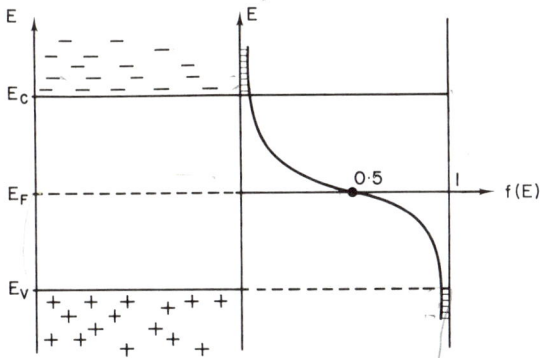

**Fig. 3.3.** Distribution of electrons and holes in an intrinsic semiconductor.

gap of germanium, 0·8 eV, with the energy range $3kT$ at room temperature demonstrates the minute percentage of levels occupied by electrons in the conduction band. The Fermi level in semiconductors is defined as the central point of the Fermi-Dirac curve; it has no such straightforward significance as in metals.

The reader will now be able to visualize how the Fermi-Dirac function must be positioned to represent the distribution of charge carriers in an impurity semiconductor. In $n$-type material, for example, the number of electrons in the conduction band is much greater than the number of holes in the valence band. If the distribution curve is to represent this state of affairs, the Fermi level must be much higher than the centre of the energy gap. Its precise location will depend on the concentration of impurities and temperature.

Mathematical equations expressing quantitatively the relations outlined above will be derived in subsequent sections.

### 3.2 Approximate Forms of the Fermi-Dirac Distribution

In calculations on semiconductors it is rarely necessary to use the full form of the Fermi-Dirac function. The approximate forms, which are sufficiently accurate in most cases, will be described now.

Let us first consider the upper part of the Fermi-Dirac curve, usually used to describe the occupation of conduction band states by electrons. Whenever $E - E_F \gg kT$ the complete function of eqn (3.1.1) simplifies to the form

$$f(E) \cong e^{-(E-E_F)/kT}. \qquad (3.2.1)$$

The relation of this expression to the full Fermi-Dirac function is shown diagrammatically in Fig. 3.4(a).

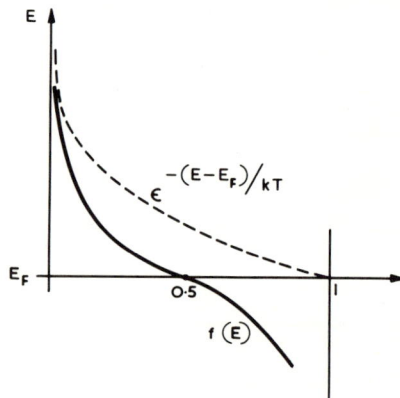

Fig. 3.4(a)

The above exponential form is the Maxwell-Boltzman distribution, well known in classical thermodynamics and applicable to the distribution of energy among molecules of a classical gas. It should be noted that the condition $E - E_F \gg kT$ applies to energy ranges in which very few of the available states are actually occupied by electrons. Hence the classical distribution applies only under such conditions.

The population of the valence band is discussed in terms of probability of occupation of states by holes. Mathematically this is given by the expression

$$1 - f(E) = 1 - \frac{1}{1 + e^{(E-E_F)/kT}}$$

$$= \frac{1}{1 + e^{(E_F-E)/kT}} \qquad (3.2.2)$$

For energy levels well below the Fermi level the inequality $E_F - E \gg kT$ holds and the above relation simplifies to the form

$$1 - f(E) \cong e^{-(E_F-E)/kT}. \qquad (3.2.3)$$

Figure 3.4(b) shows diagrammatically the relationship between the approximate form eqn (3.2.3) and the exact distribution of eqn (3.2.2). However, it is emphasized again that the above expressions give the fraction of *empty states* (occupied by holes). Hence the exponential form gives the distance between the right-hand vertical line and the dotted line of Fig. 3.4(b).

The remarks made about the classical distribution of electrons among sparsely populated states apply by anology to holes.

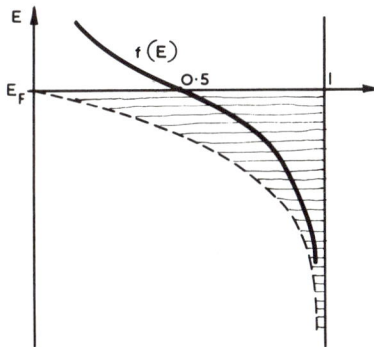

Fig. 3.4(b). The factor $e^{-(E_F - E)kT}$ gives the distance between the dotted line and the vertical line at 1.

### 3.3 Density-of-States Function

Our aim is to obtain quantitative relations giving the density of free electrons and holes in a semiconductor crystal. The Fermi-Dirac distribution takes us one step in this direction by stating the fraction of available states actually occupied by electrons at a given energy in the conduction band and by holes in the valence band. To make progress it is necessary to know the number of available states. This is evaluated with the help of a function giving the *density of states* per unit energy range in the allowed bands.

The concept of the density of states was touched upon in Section 1.3, in connection with electrons in a one-dimensional potential well. To obtain expressions which can be expected to agree with experiment it is necessary to deal with a three-dimensional model. This problem will be dealt with in Chapter 17 on the basis of quantum theory. For the moment we shall merely quote the necessary results, explain their significance in the present context and apply them to the problem of evaluating the populations of conduction electrons and holes in semiconductors.

The *density-of-states function* has the form (in the conduction band)

$$g(E) = 4\pi/h^3 (2m)^{3/2} (E - E_c)^{1/2}. \tag{3.3.1}$$

The function is shown diagrammatically in Fig. 3.5, which also suggests its physical significance. According to this the expression $g(E)dE$ gives the available electron states per unit volume of crystal, in the small energy range $dE$ located at the height $E$ on the energy axis. The expression allows for the fact laid down by the exclusion principle, that two electrons can occupy the same state, provided they have opposite spins.

In Fig. 3.5 the function $g(E)$ is plotted in the conventional way against a horizontal energy axis $E$. For our purpose it is more convenient to use a vertical energy axis to facilitate comparison with energy-level diagrams of semiconductors. This is done in Fig. 3.6 which also shows the density-of-states function in the valence band. The latter is parabolic, like the function for the conduction band, but its sign is reversed.

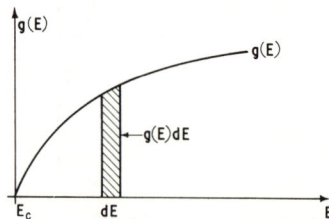

**Fig. 3.5.** The density-of-states function.

Apart from the sign the two density-of-levels functions differ in a more fundamental way. As explained in Section (1.4) the dynamics of electrons in states near the energy gap can only be understood if the electrons are assumed to have a modified *effective mass*. It is this effective mass which must be substituted in the function $g(E)$ if agreement with experimental data is to be preserved. We shall use the symbol $m_n$ to denote the effective mass of electrons in the conduction band. The same considerations apply in the valence band, but the

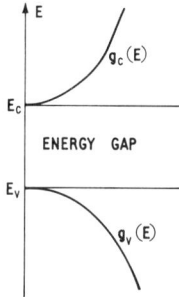

Fig. 3.6. Density-of-states functions in a semiconductor.

effective mass of electrons there is in general different from $m_n$. Moreover, the transport of charge carriers in he valence band is usually described in terms of positive holes. Hence we shall denote the effective mass of charge carriers in the valence band by $m_p$. The density-of-states functions can now be written in the form

$$g_c(E) = 4\pi/h^3 (2m_n)^{3/2} (E - E_c)^{1/2}$$

$$g_v(E) = 4\pi/h^3 (2m_p)^{3/2} (E_v - E)^{1/2} \tag{3.3.2}$$

where $m_n$ = effective mass of electrons,

$m_p$ = effective mass of holes,

The effective masses are usually stated in the form of ratios. Values for the typical semiconductors are given in Table 3.1.

The actual charge-carrier densities are now obtained as a product of the density-of-states function and the Fermi-Dirac distribution. The expression

$$g(E) f(E) dE = \rho(E) dE \tag{3.3.3}$$

gives the concentration of charge carriers occupying states in the small range $dE$, per unit volume of the material.

## Table 3.1

### Some properties of typical semiconductors at room temperature

| Properties | Ge | Si | Ga As |
|---|---|---|---|
| Energy gap (eV) | 0·803 | 1·12 | 1·43 |
| Effective density of states $N_c(\text{cm}^{-3})$ | $1·04 \times 10^{19}$ | $2·8 \times 10^{19}$ | $4·7 \times 10^{17}$ |
| Effective density of states $N_v$ (cm$^{-3}$) | $6·1 \times 10^{18}$ | $1·02 \times 10^{19}$ | $7·0 \times 10^{18}$ |
| Intrinsic carrier concentration (cm$^{-3}$) | $2·5 \times 10^{12}$ | $1·06 \times 10^{10}$ | $1·1 \times 10^{7}$ |
| Atoms/cm$^3$ | $4·42 \times 10^{22}$ | $5·0 \times 10^{22}$ | $2·2 \times 10^{22}$ |
| Electron effective mass $(m_n/m_e)$ | 0·22 | 0·33 | 0·068 |
| Hole effective maas $(m_p/m_e)$ | 0·39 | 0·55 | 0·5 |
| Mobilities (cm$^2/V$ sec) | | | |
| $\mu_n$(electrons) | 3900 | 1500 | 8500 |
| $\mu_p$(holes) | 1900 | 600 | 400 |
| Minority carrier lifetime (sec) | $10^{-3}$ | $2·5 \times 10^{-3}$ | $10^{-8}$ |
| Work function, $\phi$, ($V$) | 4·4 | 4·8 | 4·7 |
| Electron affinity, $\chi$, ($V$) | 4·0 | 4·05 | 4·07 |
| Melting point (°C) | 937 | 1420 | 1238 |
| Density (g/cm$^3$) | 5·33 | 2·33 | 5·32 |
| Dielectric constant | 16 | 11·8 | 10·9 |
| Lattice constant (Å) | 5·66 | 5·43 | 5·65 |
| Crystal structure | Diamond | Diamond | Zinc blende |
| Constant $M_c$ | 4 | 6 | |
| Constant $M_v$ | 1 | 1 | |

The total number of free electrons per unit volume of a semiconductor or metal is obtained by integrating the above expression over the energy range of the conduction band.

$$n = \int_{E_c}^{E_0} g_c(E) f(E) \, dE \qquad (3.3.4)$$

where $E_0$ is the energy of an electron just outside the sample.

An analogous expression applies to positive charge carriers or holes in the valence band of a semiconductor. A detailed evaluation of these expressions for semiconductors will be carried out in following sections. However, as an immediate example we apply eqn (3.3.4) to the conduction electrons in a metal.

The position is represented diagrammatically in Fig. 3.7. The dotted line represents the position at a finite temperature, while the solid line applies at

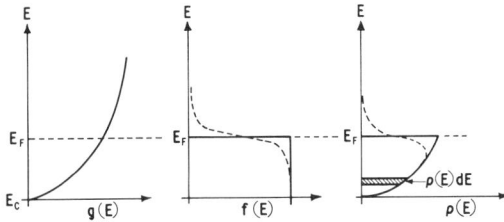

Fig. 3.7. Conduction electrons in a metal.

$0\,°K$. It is found in practice that the latter provides a sufficiently close approximation at normal ($300\,°K$) temperatures. Hence the Fermi-Dirac distribution takes on the form of a step function. Consequently the integral (3.3.4) assumes the simple form

$$n = \int_{E_c}^{E_F} g(E)\,dE$$

$$= \frac{8}{3h^3}(2m_e)^{3/2}(E_F - E_c)^{3/2}. \qquad (3.3.5)$$

Given the position of the Fermi level $E_F$ in relation to $E_c$ the above expression can be evaluated. It should be noted that the true electronic mass $m_e$ has been used. It is verified by experiment that in the typical metallic conductors this is accurate enough.

Before the density-of-states function can be applied to semiconductors a small but important modification of eqn (3.3.2) must be introduced. This is caused by the fact that the periodicity of the lattice varies from one crystalline direction to another and hence the band structure of the energy levels must be expected to vary as well. In particular the relative positions of levels $E_c$ and $E_v$ on the energy level axis must be expected to vary for the different lattice directions. Thus, referring to Fig. 3.8, the two hypothetical energy level diagrams shown may refer to two different crystal directions of *the same*

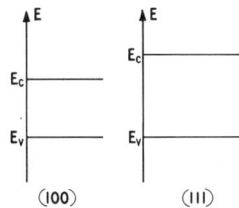

Fig. 3.8. Energy band structure for two directions of a hypothetical crystal.

*material.* In general it is found that coincidence of $E_c$ (or $E_v$) for all possible directions is the exception rather than the rule.

Hence it is clear that electrons moving along the [100] direction are much more likely to be excited across the narrower gap than those moving along the [111] direction. Moreover, if there are several directions for which the energy level scheme is like that for [100] the opportunities for electrons to be excited will be correspondingly greater. This fact is reflected in a modification of the density-of-states functions of eqn (3.3.2) by the inclusion of an additional factor $M_c$ or $M_v$.

$$g_c(E) = M_c \, 4\pi/h^3 (2m_n)^{3/2} (E - E_c)^{1/2}$$

$$g_v(E) = M_v \, 4\pi/h^3 (2m_p)^{3/2} (E_v - E)^{1/2}. \qquad (3.3.6)$$

For the typical semiconductors the factors $M_c$ and $M_v$ assume simple integral values. A few of them are listed in Table 3.1. They can be calculated by a theoretical analysis of the crystal structure.

## 3.4  Charge Carrier Concentrations in Semiconductors

Applying eqn (3.3.4) to the conduction band of an intrinsic semiconductor we can write

$$n_i = \frac{4\pi}{h^3} M_c (2m_n)^{3/2} \int_{E_c}^{E_0} \frac{(E - E_c)^{1/2}}{1 + e^{(E - E_F)/kT}} \, dE. \qquad (3.4.1)$$

To evaluate the integral without difficulty two assumptions are made: (i) the approximate form of the Fermi-Dirac function, eqn (3.2.1) is used; (ii) the upper limit of integration is extended to infinity. The latter modification is made possible by the extremely sparse population of states above $E_0$, which will make a negligible contribution to the integral. Hence

$$n_i = \frac{4\pi}{h^3} M_c (2m_n)^{3/2} e^{-(E_c - E_F)/kT} \int_{E_c}^{\infty} (E - E_c)^{1/2} e^{-(E - E_c)/kT} \, dE. \qquad (3.4.2)$$

The change of variable $E - E_c = kTx$ reduces the integral to the form

$$\int_0^{\infty} x^{1/2} e^{-x} \, dx = \sqrt{\pi}/2,$$

where the final result can be extracted from a table. The number of intrinsic electrons per unit volume is thus given by the expression

$$n_i = N_c e^{-(E_c - E_F)/kT} \tag{3.4.3}$$

where

$$N_c = 2M_c \left(\frac{2\pi m_e k}{h^2}\right)^{3/2} \left(\frac{m_n}{m_e}\right)^{3/2} T^{3/2}.$$

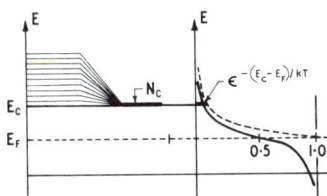

Fig. 3.9. Reduction of the conduction band to $N_c$ states at the level $E_c$.

The physical interpretation of eqn (3.4.3) is illustrated in Fig. 3.9. The available electron states in the conduction band can be visualized as reduced to the number $N_c$ situated at the bottom of the conduction band $E_c$. The exponential factor then gives the fraction of $N_c$ actually occupied by electrons, provided the position of the Fermi level $E_F$ is known. It has been argued in Section 3.1 that $E_F$ is situated roughly in the centre of the energy gap in intrinsic semiconductors. A more precise calculation of $E_F$ will be carried out shortly. In the meantime we note that the use of exponential approximation to $f(E)$ is justified for typical semiconductors at normal temperatures (300 °K) since $kT \ll \frac{1}{2}(E_c - E_v)$.

The concentration of holes $p_i$ in an intrinsic semiconductor is evaluated by a similar procedure applied to the valence band.

$$p_i = \int_{-\infty}^{E_v} (1 - f(E)) g_v(E)\, dE. \tag{3.4.4}$$

where $1 - f(E)$ is the probability of occupation of valence-band states by holes (eqn (3.2.3)), and the lower limit of integration is extended to $-\infty$ for the same reasons as above. After simplification and a change of variable the same basic integral is arrived at, yielding the same numerical value. Hence the final result is

$$p_i = N_v e^{-(E_F - E_v)/kT}$$

where                                                                                         (3.4.5)

$$N_v = 2M_v \left(\frac{2\pi m e k}{h^2}\right)^{3/2} \left(\frac{m_p}{m_e}\right)^{3/2} T^{3/2}.$$

The interpretation of this result is similar to that in he conduction band. It is illustrated in Fig. 3.10.

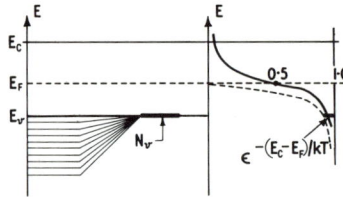

Fig. 3.10. Reduction of the valence band to $N_V$ states at the level $E_V$.

The eqns (3.4.3) and (3.4.5) can be used to evaluate the concentrations of electrons and holes in an intrinsic semiconductor if the position of the Fermi level is known. To answer this question we recall that $n_i = p_i$. Hence the above formulae lead to the result

$$E_F = \frac{E_c + E_v}{2} + \tfrac{1}{2}kT \log \frac{N_v}{N_c}.$$                          (3.4.6)

The Fermi level thus lies in the centre of the energy gap except for the logarithmic term. The latter is sufficiently small to be negligible in many applications of the most common semiconductors. However, there are exceptions, notably indium antimonide. In Sb, which has a rather small energy gap, $E_c - E_v \cong 0.2$ eV, combined with a large ratio $N_v/N_c \cong 20$. The Fermi level is thus shifted significantly towards the conduction band.

The logarithmic correction is due to unequal densities of states in the conduction and valence bands, reflected in the effective masses of electrons and holes and in the factors $M_c$ and $M_v$. Substitution for $N_v$ and $N_c$ yields the alternative expression

$$E_F = \frac{E_c + E_v}{2} + \tfrac{1}{2}kT \log \frac{M_v}{M_c} \left(\frac{m_p}{m_n}\right)^{3/2}.$$    (3.4.7)

The foregoing analysis is applicable to impurity semiconductors under two assumptions:

(a) The Fermi level remains sufficiently far below $E_c$ and above $E_v$ for the exponential approximations to the Fermi-Dirac distribution to be applicable.

(b) All impurities are ionized.

The first assumption means that the necessary integrations can be carried out by the same method as was applied to intrinsic semiconductors. The second assumption means that the numbers of free electrons in an $n$-type sample and holes in a $p$-type sample can be equated to $N_d$ and $N_a$, the concentrations of donor and acceptor impurities respectively. This is only possible if $N_d$ and $N_a$ are much greater than the populations of intrinsic electrons $n_i$ and holes $p_i$. The foregoing assumptions usually hold at normal temperatures for the typical semiconductors, provided impurity concentrations are in the range $10^{14}$ to $10^{19}$ impurity atoms per cm$^3$ of material. *Extrinsic saturation* is the term commonly used to describe the implications of assumptions (a) and (b).

Taking the case of an $n$-type sample first, we find $(N_d \gg n_i)$

$$n = N_d = N_c \, e^{-(E_c - E_F)/kT}. \tag{3.4.8}$$

Hence the position of the Fermi level relative to the bottom of the conduction band is

$$E_c - E_F = kT \, \log \frac{N_c}{N_d}. \tag{3.4.9}$$

For a $p$-type semiconductor we find $(N_a \gg p_i)$

$$p = N_a = N_v \, e^{-(E_F - E_v)/kT} \tag{3.4.10}$$

while the position of the Fermi level relative to the top of the valence band is

$$E_F - E_v = kT \, \log \frac{N_v}{N_a}. \tag{3.4.11}$$

It will be clear from the foregoing relationships that the impurity concentration is the crucial item of information when dealing with doped semiconductors, particularly in saturated extrinsic conditions.

### 3.5 The $n$-$p$ Product

Although eqns (3.4.8) and (3.4.10) were explicitly written down for $n$-type and $p$-type samples respectively they are also applicable to a single sample simultaneously. Thus one can write *for any semiconductor sample in equilibrium*

$$n = N_c \, e^{-(E_c - E_F)/kt}$$
$$p = N_v \, e^{-(E_F - E_v)/kT}. \tag{3.5.1}$$

In the case of an intrinsic sample $n$ and $p$ will be equal to $n_i$ and $p_i$, in the case of an $n$-type sample $n$ can be equated to $N_d$ while in the case of a $p$-type sample $p$ can be equated to $N_a$, provided that the extrinsic saturated condition applies.

A very useful relationship is obtained on forming the product of eqns (3.5.1).

$$np = N_c N_v e^{-(E_c - E_v)/kT}$$

$$= N_c N_v e^{-E_g/kT} \tag{3.5.2}$$

where $E_g = E_c - E_v$ is the extent of the energy gap. Comparison with eqns (3.4.3) and (3.4.5) shows that the above relation applies to an intrinsic sample as well. Hence we write

$$n_i p_i = n_i^2 = p_i^2 = np. \tag{3.5.3}$$

This relation shows that the product $np$ is the same for a given semiconductor at a given temperature, regardless of whether it is pure or doped, as long as it remains with the limitations stated on p. 37.

It is instructive to reflect on the quantities $n$ and $p$ in an extrinsic semiconductor by comparison with an intrinsic sample. Since in an $n$-type sample $n$ will be much greater than $n_i$, $p$ must be correspondingly less than $p_i$. This is accounted for by the fact that many states in the conduction band are occupied by donor electrons, thus leaving fewer available states for intrinsic electrons from the valence band. Hence the probability of thermal excitation of electrons across the energy gap is reduced and the number $p$ is correspondingly less than $p_i$. An analogous argument applies to $p$-type samples.

## 3.6  Combined Donor and Acceptor Impurities

As the next step we consider the general case when a sample of semiconductor is doped by donor and acceptor impurities simultaneously. To begin with let us summarize symbol definitions to be used; all symbols represent quantities per unit volume, or densities.

$n$ = electrons in C.B.
$p$ = holes in V.B.
$N_d$ = donor density
$N_a$ = acceptor density
$n_d$ = electrons in bound states round donor atoms (unionized donors)
$p_a$ = holes in bound states round acceptor atoms (unionized acceptors)
$2N = N_d - N_a$ = difference of donor and acceptor densities

The free electrons in the conduction band together with ionized acceptors form a distribution of negative charge whose density is

$$n + (N_a - p_a).$$

Similarly the free holes in the valence band together with ionized donors form a distribution of positive charge of density

$$p + (N_d - n_d).$$

Since we are considering an isolated sample in equilibrium and assume it to be uncharged, the condition of electrical neutrality is stated by equating the above expressions

$$n + (N_a - p_a) = p + (N_d - n_d)$$

or
$$2N = N_d - N_a = (n + n_d) - (p + p_a). \tag{3.6.1}$$

In the present treatment we limit our considerations to the *saturated extrinsic case* in which all impurities are ionized. Under these conditions $n_d = 0$ and $p_a = 0$ and eqn (3.6.1) reduces to the simple form

$$2N = N_d - N_a = n - p. \tag{3.6.2}$$

Again the *np*-product relation, eqn (3.5.3), applies in the present case. Substituting for $p$ in eqn (3.6.2) in terms of $n_i^2$, we obtain the quadratic equation

$$n^2 - 2Nn - n_i^2 = 0$$

whose solution is

$$n = N + \sqrt{N^2 + n_i^2} \tag{3.6.3}$$

where the negative root has been omitted as it does not represent a physical situation. Equation (3.6.3) gives the concentration of conduction electrons in terms of quantities which are either known experimentally ($N$) or can be calculated by the methods of Section (3.4) ($n_i$).

The concentration of holes is obtained on substitution of eqn (3.5.3) into eqn (3.6.2)

$$p = -N + \sqrt{N^2 + n_i^2}. \tag{3.6.4}$$

The foregoing relations are now analysed with a view to ascertaining the characteristics of a material doped by both donor and acceptor impurities. We consider a few clear cut cases.

*(i) Complete compensation*

This is the case when $N_d = N_a$ or $N = 0$ whence

$$n = n_i = p. \tag{3.6.5}$$

From the point of view of charge carrier densities the material has *intrinsic characteristics*. However, it differs from an ideally pure sample by the presence of possibly large numbers of ionized impurities of both kinds (up to $10^{18}$ per $cm^3$). The two types of impurities simply exchange charge: the donor electrons occupy all acceptor states since these are much lower in energy than conduction band levels. The ionized impurities, which are immobile and may be quite dense, *scatter* free current carriers causing *low mobility*. Hence the resistivity of compensated samples must be expected to be higher than that of a very pure sample. An ideally pure intrinsic sample is a hypothetical entity in any case.

### (ii) n-type case

When a sample contains donor impurities which exceed in concentration the acceptor impurities by one order of magnitude or more, ($N_d = 10^{16}$, $N_a = 10^{15}$ per $cm^3$, say) it assumes the characteristics of *n*-type material. This follows from eqns (3.6.3) and (3.6.4) when the substitution is made $n = \frac{1}{2}(N_d - N_a) \cong \frac{1}{2}N_d$. We find that $n \cong N_d$ while $p \cong 0$ provided that $N_d \gg n_i$.

### (iii) p-type case

This occurs when acceptors are more numerous than donors. The condition $N_a \gg N_d$ enables us to make the substitution $N = \frac{1}{2}(N_d - N_a) \cong -\frac{1}{2}N_a$ which yields the results $p \cong N_a$ and $n \cong 0$.

The above results provide the theoretical basis of practical procedures for the doping of semiconductors and formation of *p-n* junctions by diffusion of impurities into the crystal lattice. When semiconductor crystals are grown it is impossible to ensure that very pure intrinsic specimens are obtained. This being so it is preferable to introduce desired impurities in carefully controlled quantities deliberately. Crystals grown in this way have *p*-type or *n*-type characteristics in the first place. To change a *p*-type specimen into an *n*-type specimen it is only necessary to introduce impurities of the latter type in sufficient numbers to make them dominant, according to the principle explained above. This can be done by diffusion. The specimen is placed in a high-temperature furnace and surrounded by a vapour atmosphere containing the desired impurities, which diffuse gradually into the crystal lattice. Such processes can be accurately controlled and samples having predictable characteristics can be prepared as required.

To secure materials having saturated extrinsic characteristics at normal temperatures the concentration of impurities must be within the limits imposed on the one hand by the populations of intrinsic charge carriers ($N_d, N_a \gg n_i$) and on the other hand by the densities of available states ($N_c, N_v$). In the latter case the exponential approximations to the Fermi-Dirac distribution will not apply and the relationships derived so far will be invalid unless $N_c \gg N_d$ and $N_v \gg N_a$. Numerical values of these quantities are listed in Table 3.1 for some typical semiconductors.

## 3.7 Variation of the Fermi Level with Temperature, $E_F(T)$

It will be apparent from the results of Section 3.4 that the position of the Fermi level will vary with temperature and also with the concentration of impurities. Since it is both instructive and useful to have some insight into these variations, the subject will be discussed in the present and next sections.

For the sake of concreteness we consider an $n$-type semiconductor having $N_d$ donor impurities at the donor energy level $E_d$, shown in Fig. 3.11. The absolute

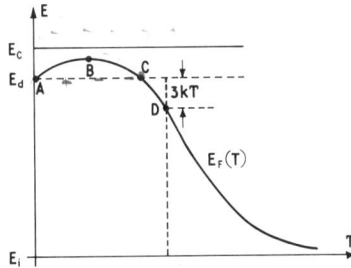

Fig. 3.11. Variation of Fermi level with temperature in an $n$-type semiconductor.

temperature $T$ is plotted along the horizontal axis. Typical situations are marked by points $A$, $B$, $C$, $D$ in Fig. 3.11, and the discussion is formulated in terms of these points.

$A$. At 0 °K the electrons settle in the lowest available states which means that all states in the valence band are occupied and *all donor atoms are unionized*. Even if there were any hypothetical states an *infinitesimal* height above $E_d$ they would all be empty. Hence the Fermi level of the "step" Fermi-Dirac function is situated at $E_d$ and we can write $E_F(0) = E_d$.

$B$. At a temperature a little above 0 °K a very small percentage of donors are ionized and the corresponding electrons are in the conduction band, occupying a minute proportion of the states available there. This situation is consistent with the Fermi level being located somewhere between $E_c$ and $E_d$; $E_d < E_F < E_c$.

$C$. As the temperature rises, more of the donor atoms become ionized, until exactly 50% of them have been vacated by electrons. At this temperature the Fermi level coincides with $E_d$.

$D$. The ionization of donors continues and at the temperature at which only 5% of them are still occupied by electrons, the Fermi level is located a distance of $3kT$ below the donor level. From this temperature upwards the semiconductor is definitely in the saturated extrinsic condition and the relationships derived in Section (3.4) can be applied to calculate $E_F(T)$. For the typical semiconductors this happens at sufficiently low temperatures for $3kT$ to be a very small fraction of the energy gap $E_g$ (about 50 °K for Si and Ge).

At high temperatures (well above 300 °K in the case of Si and Ge) the intrinsic electrons excited from the valence band become comparable in numbers with the impurity electrons and the saturated extrinsic approximation breaks down. Ultimately the intrinsic electrons swamp the impurity electrons and the material becomes effectively intrinsic. The Fermi level tends to the height marked $E_i$, which can be calculated from eqn (3.4.6), which is applicable in this case.

Analogous arguments can be applied to $p$-type semiconductors with similar results except that the acceptor level $E_a$ replaces the donor level.

The foregoing arguments provide only a qualitative insight into $E_F(T)$, but indications have been given regarding the range of temperatures over which the formulae derived earlier can be applied to obtain quantitative results. The reader is invited to follow up these calculations as an exercise. To determine the position of the Fermi level at very low temperatures precisely, a much more general approach would be required than has been followed in the present treatment.

## 3.8  Variation of the Fermi Level with Impurity Density

A diagram illustrating qualitatively the variation of the Fermi level with impurity density is shown in Fig. 3.12. The horizontal axis of the graph represents the difference $2N = N_d - N_a$, the origin corresponding to the compensated case $N_d = N_a$ (see Section 3.6). Since the semiconductor is intrinsic in this case, $E_F = E_i$, the value given by eqn (3.4.6).

As long as impurity concentration remains within the limits $N_a \ll N_d \ll N_c$ in the case of $n$-type material, the Fermi level can be evaluated with the help of eqn (3.4.9). Similarly eqn (3.4.11) is applicable to a $p$-type sample provided that $N_d \ll N_a \ll N_v$. By these methods most of the graph of Fig. 3.12 can be evaluated, leaving a relatively small gap around $E_i$. The student will find it a worth-while exercise to go through the calculations for a typical case.

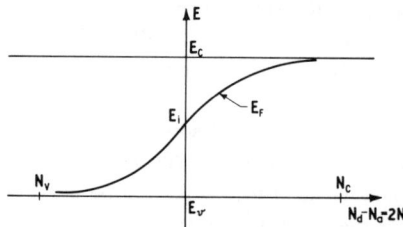

Fig. 3.12. Variation of Fermi level with impurity density.

In the case of high-impurity concentrations or *heavy doping* ($N_d \cong N_c$, $N_a \cong N_v$) the approximations used in the present treatment no longer apply. In particular the exponential or classical approximation to the Fermi-Dirac function becomes inadequate and the full function $f(E)$ must be used. In some materials, especially heavily doped for certain applications (tunnel diodes), the Fermi level may approach the impurity levels and indeed cross them.

## 3.9  Effect of Temperature on Conduction Processes in Semiconductors

As a first step towards a discussion of electrical conduction in semiconductors we consider the variation of charge carrier densities with temperature. A rough graph illustrating this relationship is presented in Fig. 3.13(a) for an $n$-type sample. At the absolute zero of temperature there are no free charge carriers since the electrons have settled in the available states in the valence band and on the donor atoms. As the temperature is increased slightly the donor electrons are gradually activated into the conduction band, thus providing a pool of free charge carriers. When all the donors have been ionized the saturated extrinsic condition has been reached. Over a range of temperature the density of free charge carriers now remains constant and equal to the impurity concentration (say $n = N_d$). At a temperature which depends on the value of the energy gap and the concentration of impurities, significant numbers of intrinsic electrons become excited into the conduction band. In due course they swamp the extrinsic electrons and come to dominate conduction of electrical currents.

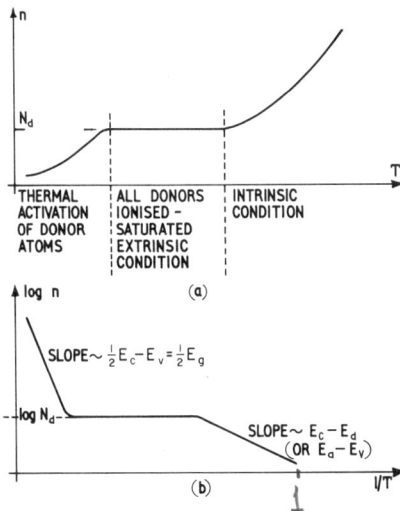

Fig. 3.13. Variation of charge carrier density with temperature.

The theoretical formulae derived earlier cannot be readily related to experimental findings on the basis of the graph of Fig. 3.13(a). For this reason it is customary to plot the logarithm of the carrier density against the reciprocal temperature as shown in Fig. 3.13(b). The thermal activation or excitation of intrinsic electrons at high temperatures appears as a straight line since it follows an exponential trend. Combining eqns (3.5.3) and (3.5.2) we find that the slope of the line is proportional to half the energy gap: $E_g/2k = (E_c - E_v)/2k$. Over the range of extrinsic temperatures the number of carriers remains constant, until a sufficiently low temperature is reached for some of the impurity atoms to retain their donor electron. The process of excitation of donor electrons also follows an exponential trend with temperature. Hence it will appear as another segment of a straight line in Fig. 3.13(b). Its slope will be proportional to the energy difference $E_c - E_d$.

Since electrical conduction in semiconductors largely follows the availability of free charge carriers (see below) we may assume that a graph of conductivity versus temperature will be similar in its principal features to Fig. 3.13. Some modifications must be expected, however, due to variations of mobility.

These are indicated in Fig. 3.14(a), which represents typical experimental data. It is beyond our scope to justify in any way the mobility curve but we can note that at high temperatures the mobility is affected primarily by the scattering of electrons by lattice vibrations (or phonons). As the temperature is decreased the mobility increases, but it does so only gradually. It hardly affects conduction over the intrinsic range but it makes a noticeable difference over the

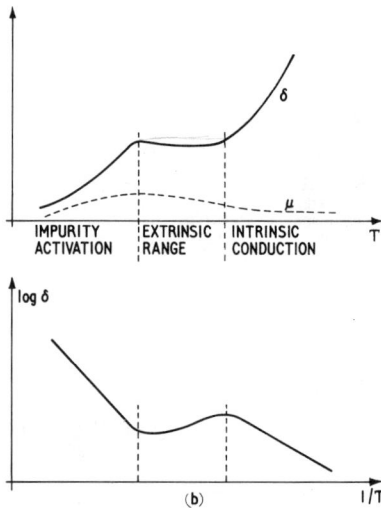

Fig. 3.14. Variation of conductivity with temperature.

extrinsic range, causing a dip in the conductivity curve which is absent from the carrier density curve.

It is instructive to contrast the conductivity of semiconductors with that of metals. The supply of conduction electrons in a metal is constant and abundant. The conductivity is limited by the scattering of electrons by phonons (or lattice vibrations) at high temperatures and by lattice defects and impurities at low temperatures. The latter mechanism is largely independent of temperature, hence the conductivity (and resistivity) remains constant over a limited range of low temperatures. As the temperature is increased lattice vibrations cause a steady decrease of conductivity (increase of resistivity).

## 3.10  Optical Phenomena in Semiconductors

It was explained in Section 3.5 that at the absolute zero of temperature all states in the valence band of a semiconductor are occupied by electrons, while all states in he conduction band are vacant. At higher temperatures some of the electrons are excited across the energy gap by thermal vibrations of the lattice which grow in intensity with rising temperature. In an intrinsic semiconductor, *to which the discussion of the present section is confined*, these electrons and their accompanying holes are the only possible carriers of electric currents. As their concentration increases with temperature the conductivity of the sample also increases.

The thermal activation process is not the only agency whereby electrons can be excited across the energy gap. Another possibility is provided by electromagnetic radiation or photons. If the semiconductor is illuminated by electromagnetic waves of frequency satisfying the relation $hf \geqslant E_c - E_v = E_g$ the incident radiation will be absorbed, while at the same time electrons will be excited into the conduction band. As a result the conductivity of the sample will be dramatically increased. This effect is usually referred to as *photoconductivity*. The frequency $f$ at which the onset of photoconductivity is observed is termed the *threshold frequency* and the corresponding wavelength, $\lambda = c/f$, is the threshold wavelength.

As the frequency of the incident radiation is varied across the threshold value the optical properties of the semiconductor change drastically. Below the threshold frequency the sample *transmits* radiation and is thus *transparent*. Above the threshold it *absorbs* radiation and is therefore *opaque*. The position is depicted diagrammatically in Fig. 3.15. Terminals A and B suggest the possibility of monitoring the resistance of the sample.

The foregoing optical properties can be deduced from the appearance of various semiconductors to the naked eye. Germanium and silicon are opaque because their every gap is small enough to permit the onset of absorption at infrared wavelengths, a fact which the student will no doubt verify by direct

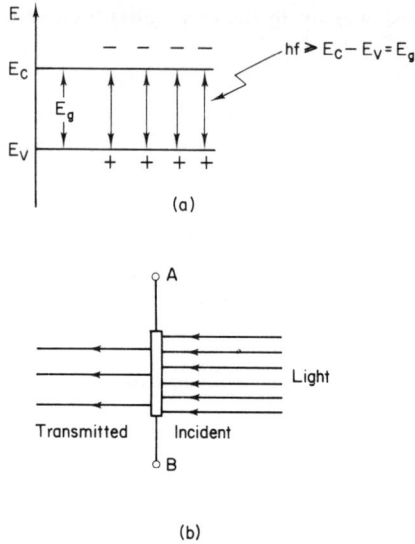

(a)

(b)

Fig. 3.15. Optical absorption and photoconductivity in semiconductors.

calculation. On the other hand cadmium sulphide has an energy gap large enough to absorb the blue part of the visible spectrum. Hence it is transparent and yellowish in colour.

Measurements of transmission and absorption provide perhaps the most direct determination of the energy gap of semiconductors. A typical but simplified graph of absorption against frequency is shown in Fig. 3.16. The onset of absorption is somewhat blurred by thermal effects and possibly the fact that the incident light is not strictly monochromatic. A true measure of the energy gap is provided by the dotted extrapolation of the principal *absorption edge*. This term is frequently used to describe the curve of Fig. 3.16.

The rate at which absorption increases with frequency is related to the increasing density of states as one moves away from the band edges. According

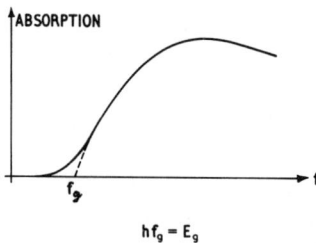

$hf_g = E_g$

Fig. 3.16. Absorption edge of a semiconductor.

to the discussion of Section 3.3 one would expect the absorption to reach its maximum value at 100% and remain at this value as the frequency is increased. However, the densities of states decrease as the conduction and valence band are penetrated beyond a certain depth, and absorption correspondingly also decreases.

### 3.11 The Hall Effect

Historically the Hall effect provided one of the decisive indications that semiconductors differ fundamentally from metals as regards the nature of the conduction mechanism. Measurements of the Hall coefficient make it possible to distinguish between negative and positive charge carriers. The discussion of the present section provides an outline of the Hall phenomenon as the experimental basis of the measurement of semiconductor parameters, and some useful devices. It is, perhaps, desirable to point out that it is quite possible to proceed with the study of subsequent chapters without any prior knowledge of the Hall effect.

The Hall effect is based on the Lorentz force which acts on charges moving in a magnetic field of induction **B**.

$$\mathbf{F} = q\mathbf{v} \times \mathbf{B} \qquad (3.11.1)$$

q is the charge of the current carrier and **v** its velocity. From the form of eqn (3.11.1) it follows that the vector product is equivalent to an electric field $\mathscr{E}_L$, sometimes referred to as the Lorentz field.

Let us consider more closely the geometrical relation of the vectors of eqn (3.11.1) with the help of Fig. 3.17. This represents a semiconducting slab carrying a transport current $J_x$ along the x-axis, in the presence of a magnetic field $B_z$ applied in a direction normal to one face of the slab. The current $J_x$ is in general made up of both holes and electrons, drifting with velocities $v_{px}$ and $v_{nx}$ in response to the externally applied field $\mathscr{E}_x$, in directions shown in Fig. 3.17.

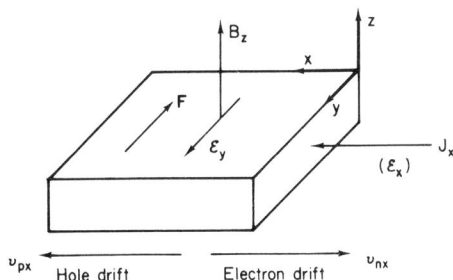

Fig. 3.17. Hall effect in a semiconductor.

To understand the Hall effect clearly it is best to consider one type of charge carrier at a time. Assuming that the sample is $p$-type under extrinsic conditions, the current $J_x$ consists almost exclusively of holes. Substitution of the relevant vectors into eqn (3.11.1) will disclose that the Lorentz force tends to deflect the holes in he *negative y*-direction. Under conditions of equilibrium the resulting charge accumulation sets up an electric field opposing the Lorentz force. This is the *Hall field* $\mathscr{E}_y$ in the *positive y*-direction in Fig. 3.17. In a *n*-type sample the current $J_x$ consists of electrons drifting in the negative *x*-direction. The effect of the Lorentz force is again to push them in the negative *y*-direction, giving rise to a Hall field which is now in the negative *y*-direction. Thus the reader will perceive that under certain conditions the Hall field will vanish, and there will be no Hall effect to observe.

The Hall effect is analysed in terms of the *Hall coefficient* $R_H$ defined by the relation

$$R_H = \frac{\mathscr{E}_y}{J_x B_z} \tag{3.11.2}$$

The advantage of the Hall coefficient is that it can be conveniently expressed in terms of charge carrier densities and mobilities. Measurements of the Hall coefficient as a function of temperature supply a considerable amount of information regarding these quantities.

To derive the requisite relationships we analyse the slab of Fig. 3.17 in terms of transverse currents set up by the Lorentz force and the Hall field. These must balance under conditions of equilibrium.

Current due to Hall field     $= \sigma \mathscr{E}_y = e(n\mu_n + p\mu_p)\mathscr{E}_y.$
(in positive *y*-direction)

Current due to Lorentz force     $= ep\mu_p v_{px} B_z - en\mu_n v_{nx} B_z$
(in negative *y*-direction)

where $v_{px} B_z$ and $v_{nx} B_z$ are the Lorentz force fields acting on the holes and electrons respectively. Equating the above currents yields an expression for the Hall field $\mathscr{E}_y$

$$\mathscr{E}_y = \frac{p\mu_p v_{px} - n\mu_n v_{nx}}{n\mu_n + p\mu_p}. \tag{3.11.3}$$

For substitution in eqn (3.11.2) an expression in terms of the transport current $J_x$ is required. Hence the carrier velocities are replaced by the relations

$$v_{px} = \mu_p \mathscr{E}_x = \mu_p \frac{J_x}{\sigma} = \frac{\mu_p}{e(n\mu_n + p\mu_p)} J_x$$

$$v_{nx} = \mu_n \mathscr{E}_x = \mu_n \frac{J_x}{\sigma} = \frac{\mu_n}{e(n\mu_n + p\mu_p)} J_x.$$

Substitution into eqn (3.11.3) yields the expression

$$\mathscr{E}_y = \frac{p\mu_p^2 - n\mu_n^2}{e(n\mu_n + p\mu_p)^2} J_x B_z. \tag{3.11.4}$$

Upon substitution of this result into the definition (3.11.2) we obtain the Hall coefficient in the following form

$$R_H = \frac{p\mu_p^2 - n\mu_n^2}{e(n\mu_n + p\mu_p)^2}. \tag{3.11.5}$$

Experimentally the Hall coefficient is obtained from measurements of the Hall voltage which appears across the sample in the presence of a given transport current $J_x$ and applied field $B_z$. In what follows we describe the typical variation of $R_H$ with temperature in $n$-type and $p$-type samples.

Let us first rewrite eqn (3.11.5) in terms of the *mobilities ratio* $c = \mu_n/\mu_p$. This parameter is invariably greater than unity since electrons are more mobile than holes.

$$R_H = \frac{p - c^2 n}{e(p + cn)^2} \tag{3.11.6}$$

For $n$-type samples the above expression reduces to

$$R_H = -\frac{1}{en} \tag{3.11.7}$$

since under extrinsic conditions $n \gg p$ and under intrinsic conditions $cn > p$ by virtue of $c > 1$. Under extrinsic conditions $n = N_d$ and

$$R_H = -\frac{1}{eN_d} = \text{const.} \tag{3.11.8}$$

This result is quite closely verified by experiment and can be used to measure the density of donor impurities. Under intrinsic conditions, at higher temperatures, $n \cong n_i = \sqrt{N_c N_v}\, e^{-E_g/2kT}$ by eqns (3.5.3) and (3.5.2). As an approximation the variation of the $N_c$ and $N_v$ numbers with temperature (see eqns (3.4.3) and (3.4.5)) can be neglected and we can write

$$R_H = -K e^{E_g/2kT} \tag{3.11.9}$$

where $K$ is a constant. Hence a plot of $\log R_H$ against $1/T$ will be a straight line of slope $E_g/2k$, which can be used to obtain a measure of the energy gap of the semiconductor under consideration. Figure 3.18 depicts such a graph and also shows the smooth transition from the extrinsic to the intrinsic condition, which is characterized by a more complicated relation between the impurity and intrinsic electrons.

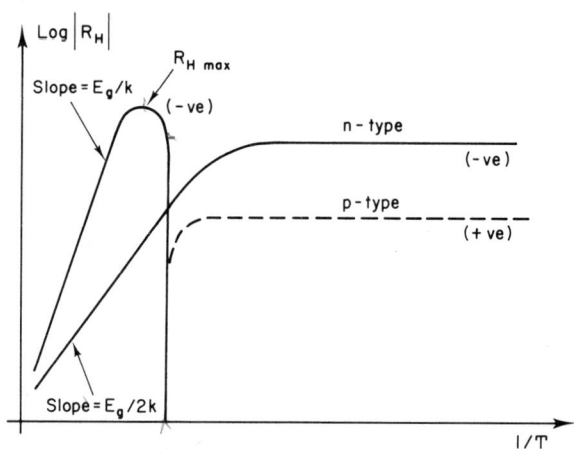

Fig. 3.18. Variation of the Hall coefficient with temperature.

The behaviour of $p$-type samples is rather more complicated except under strictly extrinsic conditions when $p \gg n$ and eqn (3.11.6) reduces to the constant (positive) value

$$R_H = \frac{1}{ep} = \frac{1}{eN_a} \qquad (3.11.10)$$

marked by the broken line in Fig. 3.18. As the temperature rises the value of $n$ increases in relation to $p$ and $R_H$ decreases until it vanishes, when the numerator of eqn (3.11.6) becomes zero. This happens when $p = c^2 n$. Hence the vanishing of the Hall coefficient in a $p$-type sample is affected by the mobilities ratio $c = \mu_n/\mu_p$ and does not occur when the hole and electron concentrations are equal. Having changed sign, $R_H$ reaches a maximum (negative) value when $p = cn$. It is useful to evaluate $R_{Hmax}$ in terms of the acceptor concentration $N_a$ by noting that $p = n + N_a$ and rewriting eqn (3.11.6) in the form

$$R_H = \frac{p - n}{N_a} \cdot \frac{p - c^2 n}{e(p + cn)^2}$$

Substitution of $p = cn$ then yields the following value

$$R_{H\,max} = -\frac{1}{eN_a} \cdot \frac{(c-1)^2}{4c} \qquad (3.11.11)$$

$R_{H\,max}$ is thus seen to depend both on the acceptor concentration $N_a$ and the mobilities ratio $c$ and the peak of the relevant curve in Fig. 3.18 can assume almost any position in relation to the remainder of the $p$-type graph. The illustration shows it above the value $1/eN_a$ for the sake of clarity.

It now remains to deal with the remaining section of the $p$-type graph and in particular to explain its steep slope. In this region $p < cn$ and the Hall coefficient is again approximated by eqn (3.11.7). However, the preponderance of electrons as transport-current carriers is assured in this case by the relatively high value of $c$, while the electron concentration is still much lower than the hole concentration given by $p = N_a + n$. Utilizing the $np$ product, eqn (3.5.3) we can write

$$n = \frac{n_i^2}{p} = \frac{n_i^2}{N_a + n} \cong \frac{n_i^2}{N_a} \sim e^{-Eg/kT} \qquad (3.11.12)$$

Hence $R_H \sim e^{Eg/kT}$ by eqn (3.11.7) and the slope of the relevant graph in Fig. 3.18 is seen to be $E_g/k$. This regime continues with increasing temperatures until the intrinsic condition is reached. Although even then the concentration of electrons in a $p$-type sample is slightly less than that of holes by the number $N_a$, activation across the gap $E_g$ predominates and the slope of the graph of $R_H$ assumes the value $E_g/2k$.

The foregoing outline of the Hall effect in semiconductors neglects the influence of a limited and unequal lifetime of individual hole-electron pairs. Although in some situations this influence must be taken into account, the specific results derived above remain substantially correct.

## Chapter 4

# Charge Continuity in Semiconductors

In preceding chapters the concentration of charge carriers in semiconductors was considered under conditions of equilibrium, that is in the absence of applied voltages, sudden changes in temperature or electromagnetic radiation. Although the question of current flow was discussed, the treatment was limited to homogeneous semiconductors under conditions which preserved the equilibrium state. Since the operation of most semiconductor devices requires the equilibrium state to be disrupted, it is now time to study some situations in which this happens. In this chapter we shall consider the possibility of setting up excess concentrations of minority carriers on the one hand. On the other we shall study the complete removal of minority carriers at specific locations within a semiconducting sample. Examples will be considered of the injection and subsequent motion of minority carriers, which are the central processes in transistor action. Specific results will be derived in a form to be used in subsequent chapters, dealing with *p-n* junctions and transistors.

## 4.1 Diffusion Currents in Semiconductors

In preceding chapters we have considered conduction currents in semiconductors which are visualized as drift currents superimposed on the random thermal motions of electrons, and caused by an applied electric field. There are other agencies which can cause current flows besides electric fields, notably *diffusion of charge carriers*. The reader is assumed to be familiar with the process of diffusion, which is applicable to a great variety of transport phenomena and is always associated with a *concentration gradient* in the quantity to be transported.

In semiconductors, in contrast to metals, excess local concentrations of free charge carriers can arise, or can be set up by design. As a result a gradient of charge density is present which causes a flow of current. The diffusion current, like any other diffusion flow, is proportional to the *negative density gradient* and the constant of proportionality is referred to as the *diffusion constant*. In mathematical terms one can write

$$\mathbf{J}_{\text{diff}} = -D \text{ grad } \rho \qquad (4.1.1)$$

where $\rho$ = charge density
$\quad\quad D$ = diffusion constant.

When the charge density varies in one dimension only, say along the $x$-direction, the above equation assumes the form

$$J_{x\,\text{diff}} = -D \frac{\partial \rho}{\partial x}. \qquad (4.1.2)$$

When applied to semiconductors the diffusion equation is best written in terms of the charge carrier concentrations $p$ and $n$, rather than in terms of the charge density.

$$J_{x\,\text{diff}} = -eD_p \frac{\partial p}{\partial x}$$

$$J_{x\,\text{diff}} = eD_n \frac{\partial n}{\partial x}, \qquad (4.1.3)$$

$D_p$ and $D_n$ are the diffusion constants for holes and electrons respectively. The electron charge is treated as a positive physical constant. Hence the sign of the second equation above is reversed to preserve the conventional direction of current flow.

## 4.2  Recombination of Charge Carriers

The distribution of charge carriers in a semiconductor may be affected by other processes besides diffusion. One of them is *recombination* of free electrons and holes. This is particularly likely to take place when an *excess concentration of minority carriers* is built up by some means, to be described at a later stage. Such an excess concentration of minority carriers constitutes a state of non-equilibrium and if left to itself will tend to disappear.

We shall denote the *rate of disappearance of excess minority charge carriers* by the symbol $r$. Hence the rate of disappearance of excess *minority charge* assumes the form $er$.

It is here accepted as intuitively clear that the *rate of recombination is proportional to the excess minority carrier concentration*. Hence it is possible to write for say, holes

$$r \sim p - p_0$$

where $p_0$ = equilibrium concentration of holes

$p$ = actual concentration of holes.

It is customary to set the constant of proportionality in the above relation equal to $1/\tau$. Hence we have

$$r = \frac{p - p_0}{\tau}. \qquad (4.2.1)$$

An analoguous relation applies to electrons. We shall see shortly that the constant $\tau$ is in the nature of a time constant or *lifetime* in an exponential process of decay of the excess charge concentration.

It is appropriate to mention at this point that there are several ways whereby excess concentrations of charge carriers can be created, either at well-defined loci in a semiconducting sample or uniformly throughout its volume. A great deal will be said about the former processes in the following sections, but it will be useful to realize now that the uniform generation of excess minority carriers throughout a sample is the exact opposite of recombination and in cases of equilibrium may be equated to it.

### 4.3  Sources and Sinks

The last possible method whereby charge can be generated or lost in a semiconductor is at localized sources or sinks. Theoretically these can have the form of points, lines, planes or other surfaces but in the elementary semi-conductor devices they usually take on the form of planes. In this form they may be treated as boundary conditions in analytical formulations of specific problems. Examples of such situations will be present shortly.

### 4.4  The Equation of Charge Continuity

All possible ways whereby the charge density at a point can vary are summarized in the equation of charge continuity. As derived in standard textbooks on electromagnetism it has the form

$$\frac{\partial \rho}{\partial t} = - \operatorname{div} \mathbf{J} = -\frac{\partial J_x}{\partial x} - \frac{\partial J_y}{\partial y} - \frac{\partial J_z}{\partial z} \qquad (4.4.1)$$

where the left-hand side gives the time rate of change of the charge density $\rho$ at a given point in space, while the right-hand side relates this rate to whatever currents may be flowing at that point. In the above form the equation is clearly inadequate for dealing with semiconductor problems, since it does not include the effects of charge recombination or its generation or loss at sources or sinks.

The deficiency is readily amended by including additional terms on the right side of eqn (4.4.1). Thus one can write

$$\frac{\partial \rho}{\partial t} = -\operatorname{div}\mathbf{J} - er + eg \tag{4.4.2}$$

where $er$ represents the time rate of loss of charge per unit volume due to recombination and $eg$ is the time rate of generation or loss of charge at a source or sink situated at the point under consideration. Examples will soon illustrate how these terms are to be handled in specific cases.

However, to begin with a few observations on the term $\operatorname{div}\mathbf{J}$ are necessary. The local current density $\mathbf{J}$ will in general contain two contributions: one due to an applied field $\mathscr{E}$, the other due to diffusion. The former can be expressed as follows, in the case of holes

$$\mathbf{J}_{\mathscr{E}} = e\mu_p p\mathscr{E}$$

$$\operatorname{div}\mathbf{J}_{\mathscr{E}} = e\mu_p \mathscr{E}.\operatorname{grad}p \tag{4.4.3}$$

where it is assumed that $\mathscr{E}$ is *constant*. In the case of current flow in one dimension the above expression simplifies to

$$\frac{\partial J_{x\mathscr{E}}}{\partial x} = e\mu_p \mathscr{E}_x \frac{\partial p}{\partial x}. \tag{4.4.4}$$

By eqn (4.1.1) the hole diffusion current is

$$\mathbf{J}_{\text{diff}} = -eD_p \operatorname{grad}p$$

$$\operatorname{div}\mathbf{J}_{\text{diff}} = -eD_p\nabla^2 p. \tag{4.4.5}$$

For one-dimensional current flow we have

$$\frac{\partial J_{x\,\text{diff}}}{\partial x} = -eD_p \frac{\partial^2 p}{\partial x^2}. \tag{4.4.6}$$

With the help of the above results the continuity equation (4.4.2) can be written in the following form (for holes)

$$\frac{\partial p}{\partial t} = D_p \nabla^2 p - \mu_p \mathscr{E}.\operatorname{grad}p + g - r. \tag{4.4.7}$$

For charge continuity in one dimension this reduces to the form

$$\frac{\partial p}{\partial t} = D_p \frac{\partial^2 p}{\partial x^2} - \mu_p \mathscr{E}_x \frac{\partial p}{\partial x} - \frac{p - p_0}{\tau} + g \tag{4.4.8}$$

where the recombination term has been written in the form of eqn (4.2.1).

To summarize let us note that diffusion currents are characterized by the second-order space derivative of carrier density in eqn (4.4.8). Likewise drift currents are characterized by the first-order space derivative. The term $g$, representing sources and sinks, will assume a variety of forms depending on circumstances.

## 4.5  Applications of the Equation of Charge Continuity

In the present section the continuity equation will be applied to several examples. The object will be in the first place to illustrate the processes of recombination and diffusion of charge carriers, and in the second place to derive some specific relations which will be applied later to the analysis of a *p-n* junction.

### (a)  Recombination of excess holes in an n-type semiconductor

A small sample of, say, *n*-type germanium is suddenly illuminated by a flash of radiation. This has the effect of exciting a number of electrons from the valence band into the conduction band in the manner described in Section 3.10. In the present context we are concerned with the fact that an *excess concentration* of holes, that is of minority carriers, is created in the sample. Although this excess number of holes may exceed the equilibrium value by orders of magnitude, it is still small compared with the density of impurity electrons which are the majority charge carriers. Hence the consequent change in the concentration of electrons is negligible and we are justified in considering the holes only. Before applying the continuity equation we make two assumptions: (a) the hole concentration is uniform throughout the sample; (b) since the sample is assumed to be isolated from any influence except the incident radiation there are no sources or sinks of charge inside it. The first assumption enables us to drop the space dependent terms in eqn (4.4.7), while the second eliminates the term $g$. Hence, in this case, the continuity of charge equation reduces to the simple form

$$\frac{dp}{dt} = -\frac{p - p_0}{\tau}$$

whose solution is

$$p = p_0 + Ce^{-t/\tau}$$

where $C$ is a constant of integration which represents the excess hole concentration immediately after the flash of radiation. The foregoing result shows that the excess concentration of holes dies away exponentially towards the equilibrium value $p_0$. The recombination process is characterized by the time constant $\tau$ which is the average *lifetime of excess minority carriers*. The process is

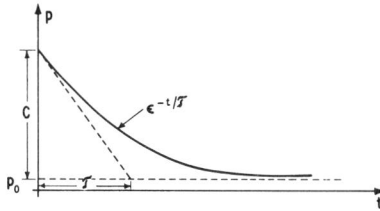

**Fig. 4.1.** Recombination of excess minority carriers.

illustrated graphically in Fig. 4.1. The exponential decay curve can be displayed experimentally on an oscilloscope screen with the help of a suitable test set up. A typical value of the lifetime is of the order of 100 $\mu$sec for transistor quality germanium or silicon.

*(b)  Simultaneous recombination and diffusion of excess minority carriers*

In the foregoing example there was no spatial variation of charge density, hence there could be no diffusion, only recombination. We now consider an idealized situation in which both processes take place simultaneously. A long bar of *n*-type semiconductor is connected to a circuit shown schematically in Fig. 4.2. The generator provides pulses of short duration which are used to inject short bursts or clouds of holes with the help of an emitter *p-n* junction, or some other method, details of which need not concern us at present. At first we assume that switch $S_1$ is closed, and $S_2$ is open, thus keeping the semiconductor rod free of any applied fields immediately after a pulse is applied. With these assumptions, and in the absence of sources or sinks, the one-dimensional equation of charge continuity reduces to the form

$$\frac{\partial p}{\partial t} = D_p \frac{\partial^2 p}{\partial x^2} - \frac{p - p_0}{\tau}.$$

We shall not carry out a detailed solution of this partial differential equation, but only state the final result and consider its physical meaning. The concentra-

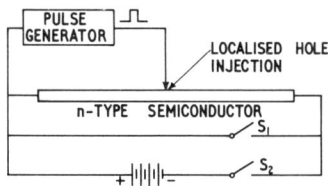

**Fig. 4.2.** Schematic diagram of apparatus for the injection of minority carriers.

tion of holes varies with position and time, after an injection pulse, according to the following relation

$$p = p_0 + \frac{N}{\sqrt{4\pi D_p t}} e^{(-x^2/4D_p t)-(t/\tau)}$$

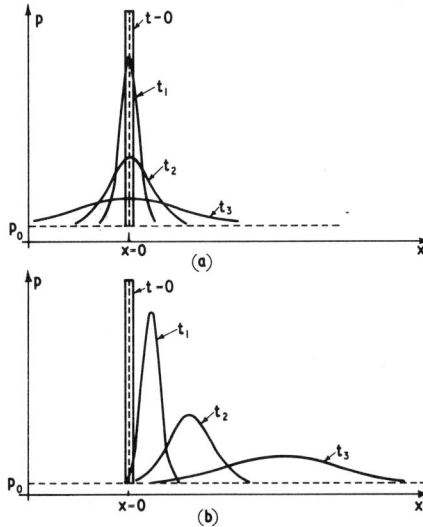

Fig. 4.3. Diffusion and recombination of a cloud of minority carriers.

where $N$ is the total number of holes contained in the injection cloud. The origin of the $x$ co-ordinate coincides with the point of injection. It is instructive to plot the above equation as a family of curves using time as parameter as shown diagrammatically in Fig. 4.3(a). At the instant $t = 0$ the hole distribution has the theoretical form of a $\delta$-function situated at $x = 0$ as suggested in the graph. After a while this collapses as a result of the combined effect of diffusion and recombination until the equilibrium concentraion $p_0$ is restored throughout the rod-shaped sample.

By closing switch $S_2$ while $S_1$ is open, an electric field is applied to the semiconductor sample. The effect of this field is to "blow" the cloud of holes along, as it gradually collapses due to recombination and diffusion. Fig. 4.3(b) shows the behaviour of minority carriers under such conditions.

### (c) Continuous injection of holes into n-type material

Once again we consider a rod-shaped sample of $n$-type material. A continuous supply of holes is maintained at one end of the bar, sufficient to establish a steady concentration of holes $p_1$, which substantially exceeds the equilibrium concentration $p_0$. The excess holes diffuse away from the source, along a charge concentration gradient, and at the same time recombine continuously with majority electrons. At a large distance from the end of the rod the hole concentration tends to the equilibrium value $p_0$ (Fig. 4.4). The semiconductor

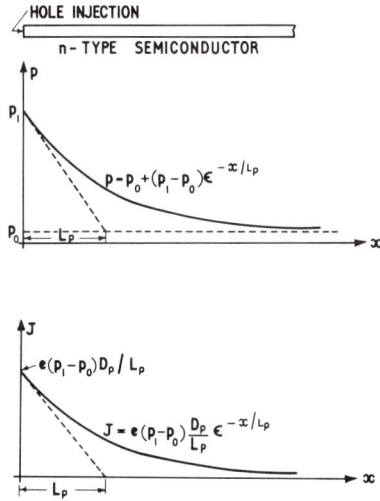

Fig. 4.4. Continuous injection of minority carriers.

bar is maintained in an electrically neutral state with the help of a closed circuit which returns the injected current to its source. Since a situation like this exists in a *p-n* junction, which will be discussed in the next chapter, it is desirable to obtain mathematical expressions for the hole distribution along the rod and to evaluate the diffusion current.

To this end the general equation of charge continuity eqn (4.4.8) is adapted to the present case and solved. In the first place it is emphasized that a steady state condition exists at every point along the rod. Hence the charge density does not vary with time and we have $\partial p / \partial t = 0$. Moreover, the field dependent term drops out since no voltage is applied to the specimen. Finally there are no sinks or sources *within* the sample. Hence $g = 0$. The effect of hole injection at the end of the bar will be taken into account through boundary conditions.

$$0 = D_p \frac{\partial^2 p}{\partial x^2} - \frac{p - p_0}{\tau_p}. \qquad (4.5.1)$$

The partial derivative may be replaced by an ordinary differential coefficient since only one independent variable is involved.

$$\frac{d^2(p-p_0)}{dx^2} - \frac{1}{D_p\tau_p}(p-p_0) = 0 \qquad (4.5.2)$$

where $p$ has been replaced by $p - p_0$ as the independent variable. This does not affect the equation but makes its solution simpler. The roots of the characteristic equation are

$$\lambda = \pm 1/\sqrt{D_p\tau_p} = \pm 1/L_p \qquad (4.5.3)$$

where $L_p$ is the *diffusion length* for holes. The solution of eqn (4.5.2) is

$$p - p_0 = A e^{-x/L_p} + B e^{x/L_p}. \qquad (4.5.4)$$

The constants of integration are determined with the help of the boundary conditions. As $x \to \infty$, $p = p_0$ or $p - p_0 = 0$; this is ensured only if $B = 0$. At $x = 0$, $p = p_1$; hence $A = p_1 - p_0$. Thus the hole distribution along the semiconducting rod is

$$p = p_0 + (p_1 - p_0)e^{-x/L_p}. \qquad (4.5.5)$$

This exponential variation is shown in Fig. 4.4 where the diffusion length $L_p$ is also clearly marked. The exponential curves of Figs 4.1 and 4.4 should be carefully compared. In the first case we have a *time decay* of excess minority carriers due to recombination only, characterized by the lifetime $\tau$. In the second case there is a *space variation* of the excess minority charge concentration associated with both recombination and diffusion, characterized by the diffusion length $L_p = \sqrt{D_p\tau_p}$, and independent of time at any location.

The diffusion current associated with the above hole distribution is given by the expression

$$J = -eD_p\frac{dp}{dx}$$

$$J = e(p_1 - p_0)\frac{D_p}{L_p}e^{-x/L_p} \qquad (4.5.6)$$

where the suffices on the current density symbol have been omitted for simplicity. The value of the current at the end of the rod is obtained by setting $x = 0$. Hence the injection current density is

$$J = e(p_1 - p_0)\frac{D_p}{L_p}. \qquad (4.5.7)$$

The foregoing results are summarized in Fig. 4.4.

*(d) Electron sink in p-type material*

As the next example of the continuity equation we discuss the effect of a sink on minority electrons in a *p*-type sample. As in the previous example we consider a rod-shaped specimen and we assume that complete extraction of electrons takes place at one end of the sample (see Fig. 4.5).

Since it is intended to deal with the electron sink as a boundary condition, the corresponding term drops out of the continuity equation. Moreover, there

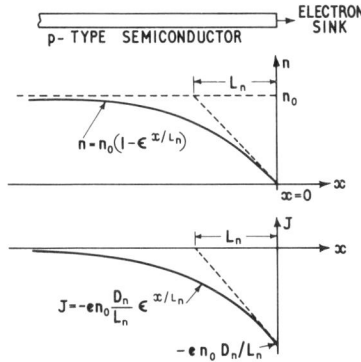

Fig. 4.5. Steady extraction of minority carriers at a sink.

are no applied fields and we assume a time independent steady state. Hence the continuity equation reduces to the form

$$0 = D_n \frac{\partial^2 n}{\partial x^2} - \frac{n - n_0}{\tau_n} \tag{4.5.8}$$

where $n_0$ is the equilibrium concentration of minority electrons in the sample and $\tau_n$ and $D_n$ are the lifetime and diffusion constant for electrons respectively. Dropping the partial derivative and writing $n - n_0$ instead of $n$, the above equation assumes the form

$$\frac{d^2(n - n_0)}{dx^2} - \frac{1}{D_n \tau_n} (n - n_0) = 0. \tag{4.5.9}$$

The roots of the characteristic equation and the final solution are

$$\lambda = \pm 1/\sqrt{D_n \tau_n} = \pm 1/L_n \tag{4.5.10}$$

$$n - n_0 = A e^{x/L_n} + B e^{-x/L_n}.$$

As shown in Fig. 4.5, the axes of co-ordinates in the present problem have been chosen in such a way that only negative values of $x$ are of interest. Since the second term of the above solution tends to infinity for large negative values of $x$, it is removed by setting $B = 0$. The other constant of integration is determined by the boundary condition at $x = 0$. Since the electron concentration $n$ is reduced to zero by the sink action at the end of the rod we find that $A = -n_0$. Hence the electron distribution in the sample has the form

$$n = n_0(1 - e^{x/L_n}) \qquad (4.5.11)$$

where the diffusion length $L_n$ for electrons is marked in Fig. 4.5.

The present example illustrates the deviation of the minority carrier concentration from the equilibrium value caused by the presence of a sink. The electron distribution away from the end of the sample is the result of the twin influences of diffusion due to a charge gradient and recombination. However, in this case "recombination" must be understood in a negative sense, since there is a deficiency of minority carriers near the sink. The "negative recombination" is in fact thermal excitation of electrons from the valence band, which tends to restore the equilibrium concentration $n_0$.

It will be useful for future reference to evaluate the diffusion current associated with the minority electron distribution of eqn (4.5.11).

According to the defining eqn (4.1.3) we write

$$J = eD_n \frac{dn}{dx}$$

$$J = -en_0 \frac{D_n}{L_n} e^{x/L_n} \qquad (4.5.12)$$

where the suffixes on the current density symbol have been omitted for simplicity. The diffusion current at the end of the sample, near the sink, is obtained on setting $x = 0$ in eqn (4.5.12).

$$J = -en_0 \frac{D_n}{L_n}. \qquad (4.5.13)$$

It should be noted that the sense of the current is correctly given by the above equations as negative, since the current is due to electrons flowing in the positive $x$ direction.

The foregoing results will be utilized in the analysis of the $p$-$n$ junction in the next chapter.

## 4.6 The Einstein Relation

When diffusion was first introduced in the opening section of this chapter no comment was made about the diffusion constant $D$ and its relation to the physical processes within a semiconductor crystal. However, it is clear that diffusion currents are influenced by other factors beside the charge density gradient and that these additional factors must be reflected in the magnitude of the diffusion constant. Foremost among these are the collisions which impede the movement of charge carriers through the lattice. Since this is the same agency which limits the mobility, it must be expected that the two are related, and we shall derive the necessary relation in the present section. The argument is based on a hypothetical sample of semiconductor which is non-uniformly doped by impurities. This novel situation has not been envisaged so far, but since it occurs frequently in practical devices it deserves close study.

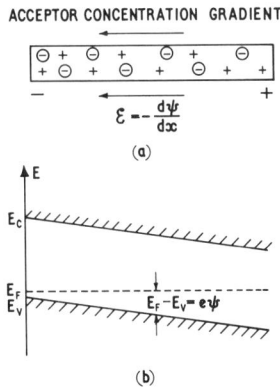

**Fig. 4.6.** Impurity gradient in a $p$-type semiconductor.

An idealized rod-shaped sample of $p$-type material is shown in Fig. 4.6(a). The concentration of acceptor impurities is assumed to increase uniformly from right to left of the sample. Under saturated extrinsic conditions, which are assumed to apply in the present case, there will be a corresponding gradient in the concentration of mobile holes, which will cause a diffusion current to flow to the right-hand side. This in turn will set up a net positive charge at that end of the bar, giving rise to an electric field as indicated in Fig. 4.6(a). The electric field can be expressed as the gradient of an electrostatic potential, to be denoted by $\psi$. As explained on p. 18 the influence of the potential is represented as tilting of the energy band scheme of the sample, indicated in Fig. 4.6(b).

An analysis of this case starts with the premise that the Fermi level $E_F$ is constant throughout the sample. This basic fact is derived from thermodynamics, where it is shown that any *thermodynamic potential is constant throughout a system in equilibrium*, and that the Fermi level or Fermi energy is a thermodynamic potential. Our sample is in equilibrium since it is isolated and outside the range of any external forces. In Fig. 4.6(b) the Fermi level is shown constant relative to the energy axis. Its position relative to the band edges is consistent with our knowledge of the variation of the Fermi level as a function acceptor density (see p. 42).

Since the tilting of the energy bands reflects the potential drop along the sample (see Fig. 1.11), the distance between the Fermi level and, say, $E_v$ can be written in the form

$$E_F - E_v = e\psi \qquad (4.6.1)$$

where the arbitrary constant implicit in any potential is assumed to be adjusted accordingly. The above relation can be differentiated with respect to distance along the sample to yield

$$\frac{d}{dx}(E_F - E_v) = e\frac{d\psi}{dx} = -e\mathscr{E}_x. \qquad (4.6.2)$$

The foregoing equation can be written in more general form as follows

$$\frac{1}{e}\,\text{grad}\,(E_F - E_v) = -\mathscr{E}. \qquad (4.6.3)$$

This emphasizes the general fact that variations of the Fermi level with respect to the energy bands in a semiconducting sample are always associated with the existence of a local electric field.

In the sample of Fig. 4.6(a) the diffusion current of holes from left to right is exactly balanced by the drift current in the opposite direction caused by the electric field. The balance is expressed by the equation

$$-eD_p\frac{dp}{dx} = -e\mu_p p\mathscr{E}_x. \qquad \text{should be.} \qquad (4.6.4)$$

Using the relations

$$\frac{dp}{dx} = \frac{dp}{d\psi}\frac{d\psi}{dx} = -\mathscr{E}_x\frac{dp}{d\psi} \quad \text{and} \quad \frac{1}{e}\frac{dp}{d\psi} = \frac{dp}{d(E_F - E_v)}$$

(the latter follows from eqn (4.6.1)) we can substitute in eqn (4.6.4), and with the help of eqn (3.4.10) we obtain

$$D_p = \frac{kT}{e}\mu_p. \qquad (4.6.5)$$

An analogous relation can be established for electrons.

$$D_n = \frac{kT}{e}\mu_n. \qquad (4.6.6)$$

Equations (4.6.5) and (4.6.6) are known as Einstein's relations. They show that the diffusion constants are indeed directly related to mobilities as might be expected. They also show that the diffusion constants increase with temperature since increasing thermal agitation of the lattice can be expected to speed the dissipation of excess charge concentrations.

*Chapter 5*

# The *p–n* Junction

## 5.1 Qualitative Charge and Energy Relations at a *p-n* Junction

We are now sufficiently equipped with basic concepts and methods to tackle the analysis of a *p-n junction*. Since the junction is the fundamental building block of all semiconductor devices, a careful study of its functioning and properties is an essential prerequisite for all electronic engineers. We start with a qualitative description of charge distributions and energy relations, following it up in subsequent sections with the derivation of mathematical relations leading up to the rectifier equation.

A *p-n* junction consists of two layers of material, *p*-type and *n*-type, joined intimately together. Indeed, the junction must be made so carefully that it is not possible simply to bring into contact two separate slices of semiconducting crystal, however carefully prepared and cleaned. The classical method of preparing a *p-n* junction starts with a single crystal having, say, *n*-type characteristics, and then to diffuse into it from one side *p*-type impurities at a sufficient concentration to over-ride the original *n*-type properties, according to the principles discussed in Section 3.6. This process is carried out at high temperatures in a specially designed furnace. Another method involves the deposition or growing of an epitaxial layer of, say, *n*-type material on a *p*-type crystal. The process is carried out in a vacuum chamber. The simplest method of forming a rectifying junction is the alloying of a small fragment of metal to a semiconducting crystal. However, the functioning of junctions prepared by this method is somewhat different and does not come within the scope of this chapter. It will be dealt with later under the heading of metal-to-semiconductor contacts.

A *p-n* junction is shown diagrammatically at the top of Fig. 5.1, where we assume that the *saturated extrinsic condition* applies—that is, all impurities are ionized. In fact, the analysis of the *p-n* junction throughout this chapter is based on this assumption. The region of prime interest is in the vicinity of the transition from *p*-type to *n*-type material. In a thin layer there are free, mobile electrons and holes in close proximity. These free charge carriers will recombine,

66

leaving behind immobile charges in the form of ionized donors and acceptors. The region containing this net remaining space charge is usually called the *depletion layer* or *barrier layer*. The thickness of the depletion layer is of the order of $10^{-6}$ to $10^{-4}$ cm. A diagram showing the charge distribution in the

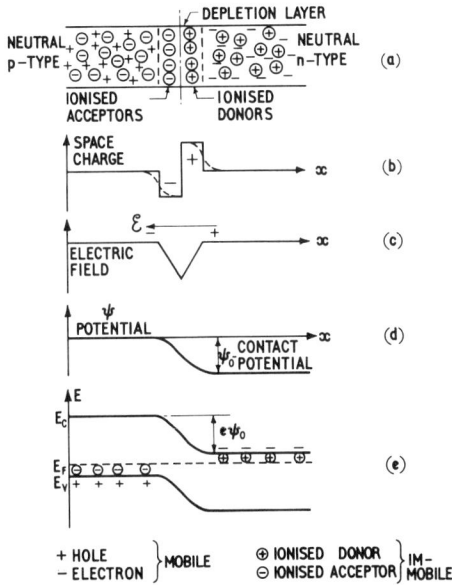

Fig. 5.1. Charge and energy relations at a *p-n* junction.

depletion layer is given in Fig. 5.1(b). The solid line indicates the ideal form of the space charge, while the dotted line suggests the actual charge distribution. The space charge gives rise to an electric field which is plotted in Fig. 5.1(c). The direction of the field is such as to *attract electrons to the right*. Hence the depletion layer will act as a *sink* for the very small number of thermal electrons present in the conduction band of the *p*-type material. On the other hand, the field repels majority holes to the left and majority electrons to the right.

Associated with the electric field is the electrostatic potential shown in Fig. 5.1(d). The potential is obtained by integration of the field according to electrostatic principles. However, as shown in the diagram, its sign has been reversed to conform with the energy level diagrams, in which we choose to plot increasing electron energies in the upward direction. The electric field and potential are essentially of the same type as those discussed in Section 4.6. They are caused by a nonuniformity in the distribution of impurities in the sample which, however, is much more drastic in the present case, changing not only in density but in character. The potential difference between the *p*-type and *n*-type sides of the junction is denoted by $\psi_0$. It will be later identified as the *contact potential* of the two materials.

Figure 5.1(e) depicts the energy level diagram of the junction. Since, at the present stage, we consider an isolated junction in equilibrium, the Fermi level is assumed constant throughout, as befits a thermodynamic potential (see also Section 4.6). The potential variation across the depletion layer causes the energy bands to be tilted as explained on p. 18, the overall difference in the height of the band edges being given by $e\psi_0$.

## 5.2  Calculation of the Contact Potential

As a preliminary step towards the evaluation of the various currents flowing across a *p-n* junction, we calculate the contact potential $\psi_0$. By inspection of Fig. 5.1(e) we can write

$$e\psi_0 = (E_c - E_v) - (E_F - E_v)_p - (E_c - E_F)_n \qquad (5.2.1)$$

where the subscripts *p* and *n* denote that the energy differences in brackets apply to the *p*-type and *n*-type material respectively. Since we are assuming the saturated extrinsic condition throughout this chapter, we can substitute for $(E_F-E_v)p$ and $(E_c-E_F)_n$ from eqns (3.4.11) and (3.4.9) respectively. Hence we find

$$e\psi_0 = (E_c - E_v) - kT \log \frac{N_v N_c}{N_a N_d}. \qquad (5.2.2)$$

The factor $N_v N_c$ is replaced with the help of the *np* product (eqn (3.5.2)), whence it follows that

$$e\psi_0 = kT \log \frac{N_a N_d}{n_i^2}$$

$$\psi_0 = \frac{kT}{e} \log \frac{N_a N_d}{n_i^2}. \qquad (5.2.3)$$

## 5.3  The *p-n* Junction in Equilibrium

Altogether, four different currents flow across the depletion layer of a *p-n* junction. In equilibrium, when no voltage is applied to the junction, these currents cancel. We shall now proceed to classify the currents and derive mathematical expressions for their magnitudes.

Figure 5.2 depicts a *p-n* junction in the form of a bar, with the energy level diagram shown below. The form of the latter has been explained in the preceding section.

### (i) *Saturation electron current $J_{sn}$*

Minority intrinsic or thermal electrons in *p*-type material are attracted by the electric field of the barrier to the right, and disappear down the potential gradient shown in Fig. 5.1(d). Hence, the depletion layer acts as a sink for minority electrons. The disappearance of the electrons at the sink sets up a concentration gradient of electrons in the *p*-type material, causing a *diffusion current* to flow towards the junction. The intrinsic electrons thus lost by the

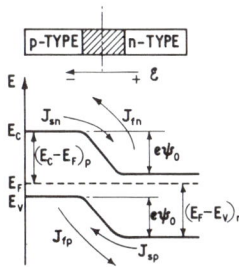

**Fig. 5.2.** Currents across a *p-n* junction in equilibrium.

*p*-type side are continually replenished by thermal excitation from the valance band.

The foregoing conditions are clearly analogous to example (d) discussed in Section 4.5 and the results obtained there can be applied immediately. In particular, the *magnitude* of the diffusion current $J_{sn}$ at the barrier is given by the expression

$$J_{sn} = e \frac{D_n}{L_n} n_0 = e \frac{D_n}{L_n} N_c e^{-(E_c - E_F)_p / kT} \tag{5.3.1}$$

where eqn (3.4.3) has been applied to substitute for the equilibrium concentration of minority electrons $n_0$. The above current depends only on the properties of the *p*-type material and will remain unchanged regardless of any bias voltage applied to the junction. For this reason, it is usually termed the *saturation current of electrons*. Once the electrons cross the barrier they join the pool of majority electrons in the *n*-type material. However, the numbers involved are extremely small compared with the density of majority electrons, and the latter remains substantially unchanged.

(ii) *Forward electron current $J_{fn}$*

The physical processes which enable some of the majority electrons to mount the potential step at the depletion layer and to find their way into the *p*-type material are more complicated than the sink effect which operates in the opposite direction. In the first place, only electrons which occupy levels above the height marked $e\psi_0$ in Fig. 5.2 will have enough energy to mount the barrier. However, many of them will be scattered on the way, while others will approach the barrier at an oblique angle, thus having effectively less kinetic energy associated with the component of their motion normal to the depletion layer.

For practical purposes, there is no need to know in detail what happens, because we require only the net current and this must be equal to $J_{sn}$. Otherwise, the equilibrium of the junction would be disturbed, contrary to our initial assumption. Hence the *magnitude* of the *forward electron* current must be

$$J_{fn} = e \frac{D_n}{L_n} N_c e^{-(E_c - E_F)_p/kT}. \qquad (5.3.2)$$

It is convenient to rewrite this expression so that the exponential factor contains the properties of *n*-type material instead of *p*-type. By inspection of Fig. 5.2, we can write

$$J_{fn} = e \frac{D_n}{L_n} N_c e^{-\{(E_c - E_F)_n + e\psi_0\}/kT}$$

$$J_{fn} = e \frac{D_n}{L_n} n_n e^{-e\psi_0/kT} \qquad (5.3.3)$$

where $n_n$ denotes the concentration of majority electrons in the *n*-type material from eqn (3.4.8). It is instructive to note that the exponential factor in eqn (5.3.3) reflects the fact that only electrons occupying levels above $e\psi_0$ contribute to the forward current. It is also useful to note for future reference that modifications of the potential step at the depletion layer will affect this current drastically through this exponential factor.

The electrons carried across the barrier by the forward current are said to be *injected* into the *p*-type material where they join the minority electrons. Hence the junction acts as a *source of electrons* and the situation is analogous to example (c) of Section 4.5, except that electrons must be substituted for holes. A comparison of eqns (5.3.1) and (5.3.3) shows that the density of injected electrons is

$$n_0 = n_n e^{-e\psi_0/kT}. \qquad (5.3.4)$$

These electrons will diffuse into the *p*-type material according to the law

$$n_p = n_0 \, e^{-x/L_n} \qquad (5.3.5)$$

where $n_p$ denotes the *density of injected electrons at a distance x* (positive) from the barrier. This is plotted in Fig. 5.3(a). It should be noted that the diagram *does not show* the intrinsic electrons "resident" in the *p*-type material. Their distribution follows an exponential of the form shown in Fig. 4.5 with a zero at the barrier. The sum of the two distributions adds up to $n_0$, the equilibrium concentration of intrinsic electrons in *p*-type material. However, this observation should not detract from the fact that there are two opposing currents of electrons in the vicinity of the depletion layer.

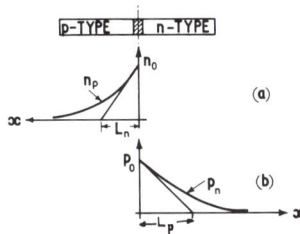

Fig. 5.3. Distribution of injected electrons and holes in the vicinity of the depletion layer.

### (iii) *Saturation current of holes* $J_{sp}$

This is analogous to the electron saturation current $J_{sn}$. The potential step at the barrier acts as a sink for holes, which drop "upwards" on the energy level diagram of Fig. 5.2, since the latter is constructed to show rising electron energies in the positive direction. By analogy with eqn (5.3.1) the *saturation current of holes* can be written in the form

$$J_{sp} = e \, \frac{D_p}{L_p} \, p_0 = e \, \frac{D_p}{L_p} \, N_v \, e^{-(E_F - E_v)_n/kT} \qquad (5.3.6)$$

where $p_0$ is the equilibrium concentration of intrinsic holes in *n*-type material. The above current depends only on the properties of the *n*-type material and remains unaffected by any bias voltage applied to the *p-n* junction, within limits. The holes carried by this current across the depletion layer make little difference to the concentration of extrinsic holes in the *p*-type material.

### (iv) *Forward current of holes* $J_{fp}$

This current is analogous to the forward current of electrons $J_{fn}$. Under equilibrium conditions it must be equal to the saturation current of holes $J_{sp}$.

Hence its *magnitude* is

$$J_{fp} = e \frac{D_p}{L_p} N_v e^{-(E_F - E_v)_n / kT}.$$ (5.3.7)

This expression is now rewritten in terms of the properties of the *p*-type material and the contact potential of the junction by reference to Fig. 5.2.

$$J_{fp} = e \frac{D_p}{L_p} N_v e^{-\{(E_F - E_v)_p + e\psi_0\} / kT}$$

$$J_{fp} = e \frac{D_p}{L_p} p_p e^{-e\psi_0 / kT}$$ (5.3.8)

where $p_p$ denotes the concentration of majority extrinsic holes in the *p*-type material.

The holes carried across the barrier by the forward current are said to be *injected* into the *n*-type material and the junction acts as a *source of holes*. The density of injected holes is obtained on comparing eqns (5.3.6) and (5.3.8) and is given by $p_0$.

$$p_0 = p_p e^{-e\psi_0 / kT}.$$ (5.3.9)

The injected holes spread into the *n*-type material by diffusion subject to recombination according to the relation

$$p_n = p_0 e^{-x / L_p}$$ (5.3.10)

where $p_n$ denotes the density of injected holes at a distance $x$ (positive) from the depletion layer. This is plotted in Fig. 5.3(b). The comments made above about injected electrons apply equally to injected holes.


## 5.4 The Reverse Biased *p-n* Junction

The application of a voltage to the *p-n* junction disturbs the equilibrium conditions assumed in the preceding section. The principal effect of this disturbance is to distort the Fermi level from its constant value. As a result, the Fermi level becomes "tilted" in the manner suggested in Fig. 5.4. Most of the tilting is confined to the depletion layer, because the applied voltage appears almost exclusively across it. This is due to the fact that the barrier is a region of very high resistance as compared with the *n*-type and *p*-type materials, because the populations of free charge carriers have been depleted. The height of the potential step at the barrier is thus enhanced by the applied voltage and assumes the value $\psi_0 + V_r$ shown in Fig. 5.4. $V_r$ denotes *reverse bias* and is here assumed to be a *positive number of volts*. At the same time, the depletion layer swells because the negative voltage applied to the $p$-type side attracts holes away from the barrier while the positive terminal on the $n$-type side attracts electrons.

Under these conditions, the reverse saturation currents $J_{sn}$ and $J_{sp}$ remain unchanged, as they depend only on the diffusion properties of the material adjoining the depletion layer. On the other hand, the forward currents $J_{fn}$ and $J_{fp}$ now have a higher potential step to surmount and are reduced to a negligible value by the application of a reverse bias of only a fraction of a volt.

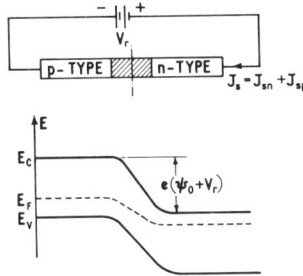

Fig. 5.4. Reverse-biased $p$-$n$ junction.

Expressions for the forward currents are obtained by a straightforward extension of eqns (5.3.3) and (5.3.8). It is only necessary to insert the value $\psi_0 + V_r$ into the index of the exponential factor to take account of the higher potential step at the barrier. Hence

$$J_{fn} = e \frac{D_n}{L_n} n_n e^{-e(\psi_0 + V_r)/kT} = J_{sn} e^{-eV_r/kT} \tag{5.4.1}$$

$$J_{fp} = e \frac{D_p}{L_p} p_p e^{-e(\psi_0 + V_r)/kT} = J_{sp} e^{-eV_r/kT}. \tag{5.4.2}$$

The expressions in terms of the saturation currents are obtained with the help of eqns (5.3.4) and (5.3.9). They make it clear that the forward currents are reduced to negligible values compared with the saturation currents through the agency of the exponential factor. What remains is the *reverse saturation current* of the junction made up of a sum of $J_{sn}$ and $J_{sp}$.

$$J_s = J_{sn} + J_{sp} = e \left( \frac{D_n}{L_n} n_n + \frac{D_p}{L_p} p_p \right) e^{-e\psi_0/kT}. \tag{5.4.3}$$

This current is marked in Fig. 5.4.

### 5.5 The Forward Biased *p-n* Junction and the Rectifier Equation

The application of a forward voltage to a *p-n* junction is illustrated diagrammatically in Fig. 5.5. The applied potential appears almost exclusively across the depletion layer because of its high resistance. The resulting "tilting" of the Fermi level reduces the potential step at the barrier. The depletion layer shrinks, since the applied electric field "pushes in" mobile charges from both sides.

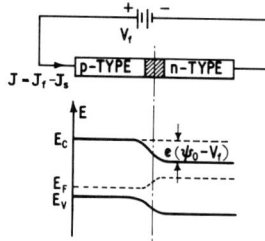

Fig. 5.5. Forward biased *p-n* junction.

The transport of charge across the junction is now dominated by the forward currents which can be written in the following form, by analogy with eqns (5.4.1) and (5.4.2).

$$J_{fn} = e \frac{D_n}{L_n} n_n e^{-e(\psi_0 - V_f)/kT} = J_{sn} e^{eV_f/kT} \tag{5.5.1}$$

$$J_{fp} = e \frac{D_p}{L_p} p_p e^{-e(\psi_0 - V_f)/kT} = J_{sp} e^{eV_f/kT} \tag{5.5.2}$$

where $V_f$ is again assumed to be a *positive number of volts*. It is clear from the above relations that the forward currents will swamp the reverse saturation currents on application of only a fraction of one volt. The rapidly increasing exponential factors ensure this state of affairs. The resulting *forward current* can be written in the form

$$J_{fn} + J_{fp} = J_f = J_s e^{eV_f/kT}. \tag{5.5.3}$$

We are now ready to formulate the *rectifier equation* of the p-n junction which describes the current/voltage relationship over all possible values of the applied bias. The *total current density* is always given by the difference of the forward and reverse saturation currents as follows

$$J = J_f - J_s = J_s(e^{eV/kT} - 1) \tag{5.5.4}$$

where $V$ has a *negative sign* when the bias is reverse and a *positive sign* when the bias is forward. It should also be noted that the above equation gives the total current density $J$, the correct sign depending on bias conditions.

A rectifier characteristic is shown schematically in Fig. 5.6. However, it should be noted that the vertical axis represents currents in amperes, not current

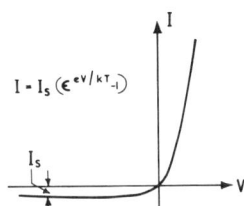

Fig. 5.6. Rectification characteristic of a *p-n* junction.

density. The rectifier equation of a specific junction is obtained on multiplying eqn (5.5.4) by the cross-sectional area of the junction and is written as follows

$$I = I_s(e^{eV/kT} - 1) \tag{5.5.5}$$

For positive values of $V$ exceeding a small fraction of one volt the negative term becomes negligible. Hence for most applications the junction current has the simpler form

$$I = I_s e^{eV/kT}. \tag{5.5.6}$$

### 5.6 Barrier Capacitance

The usefulness of the *p-n* junction is based mainly on its rectification properties. However, the *p-n* junction is also useful as a capacitor in some applications, while in others its capacitance assumes the role of a parasitic, undesirable circuit element.

The depletion layer consists of two layers of equal and opposite charge which are equivalent to a parallel plate capacitor. However, as pointed out in the preceding section, the thickness of the charge layers varies with the applied voltage. Hence it must be expected that the barrier capacitance will be a function of both the material properties of the junction and the applied bias.

The evaluation of the barrier capacitance presents a more difficult mathematical problem than a parallel plate condenser because the charge layers are *distributed* over a finite thickness. This makes it necessary to apply Poisson's equation to calculate the thickness from a knowledge of the space charge densities in the *p*-type and *n*-type materials and the voltage across the barrier. The calculations constitute an exercise in electrostatics and will not be reproduced in the present context. We shall only state the final results and explain their significance.

The barrier thickness is given by the expression

$$t = 1{\cdot}05 \times 10^{-6} \left\{ \frac{\epsilon_0\,\epsilon_r(\psi_0 - V)}{2\pi e(N_a + N_d)} \right\}^{1/2} \left\{ \left(\frac{N_a}{N_d}\right)^{1/2} + \left(\frac{N_d}{N_a}\right)^{1/2} \right\} \text{cm}$$

$$(5.6.1)$$

where $\epsilon_r$ is the relative permittivity of the semiconductor and $V$ is the voltage applied to the junction (negative for reverse bias). The above formula applies only when all impurities are ionized.

Equation (5.6.1) reduces to simpler forms in some cases of practical importance. Thus, for a *symmetrical junction* for which $N_a = N_d = N$, we obtain

$$t = 1{\cdot}05 \times 10^{-6} \left\{ \frac{\epsilon_0\,\epsilon_r(\psi_0 - V)}{\pi e N} \right\}^{1/2} \text{cm.} \qquad (5.6.2)$$

A highly *unsymmetrical junction* satisfies the condition $N_a \gg N_d$. Hence the barrier thickness is

$$t = 1{\cdot}05 \times 10^{-6} \left\{ \frac{\epsilon_0\,\epsilon_r(\psi_0 - V)}{2\pi e N_d} \right\}^{1/2} \text{cm.} \qquad (5.6.3)$$

It should be noted that in this case, the thickness of the depletion layer in *n*-type material is very much greater than in *p*-type.

The junction capacitance is evaluated from a knowledge of the space charge densities and the barrier thicknesses given above. The following are typical results.

Symmetrical junction:

$$C = 1{\cdot}05 \left\{ \frac{\epsilon_0\,\epsilon_r\,eN}{16\pi(\psi_0 - V)} \right\}^{1/2} \mu\text{Farad/cm}^2. \qquad (5.6.4)$$

Unsymmetrical junction:

$$C = 1{\cdot}05 \left\{ \frac{\epsilon_0\,\epsilon_r\,eN_d}{8\pi(\psi_0 - V)} \right\}^{1/2} \mu\text{Farad/cm}^2. \qquad (5.6.5)$$

The last result also applies to a metal-to-semiconductor junction. As will be explained in a later chapter, a very thin layer of charge is present on the metallic side of the contact while the depletion layer in the semiconductor is relatively thick. Moreover, the semiconductor may be either $n$-type or $p$type. In such cases, it is customary to replace the charge density $eN_d$ in eqn (5.6.5) in terms of the resistivity of the material through the relation $1/\rho = eN_d\mu$. Hence the capacitance of the metal-to-semiconductor contact is

$$C = 1\cdot05 \left\{ \frac{\epsilon_0 \epsilon_r}{8\pi\mu\rho(\psi_0 - V)} \right\}^{1/2} \mu\text{Farad/cm}^2 \qquad (5.6.6)$$

where $\mu$ and $\rho$ may represent the properties of either $n$-type or $p$-type material. For comparison with experimental data the foregoing result is usually expressed in the form

$$C = k(\psi_0 - V)^{-1/2}. \qquad (5.6.7)$$

The principal applications of the junction capacitance are in *parametric amplifiers* and as capacitors in *integrated microcircuits*. In the first case, the dependence of the capacitance on the applied voltage is utilized. The capacitance is varied with time in response to an applied signal. This process enables desirable energy exchanges to take place, which result in the amplification of another signal. When used as a capacitor in a microcircuit the $p$-$n$ junction is operated under conditions of fixed reverse bias. The bias can be used to adjust the value of the capacitance.

## 5.7 The Photovoltaic Effect

Before parting company with the $p$-$n$ junction as such, we introduce a phenomenon which forms the basis of many important applications.

A $p$-$n$ junction is exposed to electromagnetic waves of sufficiently high frequency to excite intrinsic electrons from the valance band to the conduction band. According to Section 3.10, the frequency must satisfy the relation $hf \geqslant E_c - E_v$. The position is suggested by the diagram of Fig. 5.7, in which the densities of minority carriers on both sides of the junction are indicated by the symbols $n_p$ and $p_n$.

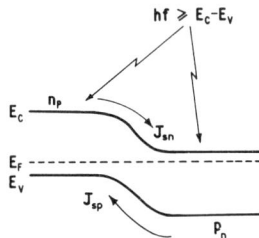

Fig. 5.7. The photovoltaic effect.

The effect of the incident radiation is to increase the densities of the minority charge carriers to values very much in excess of the equilibrium concentrations $n_0$ and $p_0$. As a result, the reverse saturation currents $J_{sn}$ and $J_{sp}$ are also increased to values very much in excess of the equilibrium conditions discussed in Section 5.3. At the same time, the forward currents $J_{fn}$ and $J_{fp}$ remain unchanged, since the potential barrier at the junction remains unchanged. The imbalance between the currents crossing the depletion layer causes a net accumulation of charge at the terminals of the junction. The resulting potential difference can be utilized to supply power to an external load.

Two types of devices of great practical importance are based on the foregoing effects: the *photodiode* and the *solar cell*. Photodiodes can be used as indicators of light intensity in that they provide a current which is related to the incident radiation. The radiation wavelength may be in the infrared band, in which case the diode acts as an infrared detector. Solar cells are used extensively to convert energy from solar radiation into electrical power in communications satellites and space vehicles in general.

## 5.8  Breakdown of a *p-n* Junction

The rectification characteristic of Fig. 5.6 is subject to limitations of two kinds. In the first place the validity of eqn (5.5.5) along the voltage axis is restricted by the onset of breakdown processes in the depletion layer, to be discussed in the present section. Secondly the actual current values at given junction voltages will be affected by temperature variations which change carrier concentrations.

As explained in Sections 5.4 and 5.5 a voltage applied to a *p-n* junction is dropped almost entirely across the depletion layer. The application of an increasing reverse bias will therefore produce an increasingly intense electric field in this layer. To some extent this tendency will be counteracted by the widening of the barrier resulting from a higher bias, but the net effect will still be an increased field. This field will accelerate the electrons crossing the depletion layer, imparting to them ever-increasing amounts of kinetic energy, which the electrons may give up to lattice atoms with which they collide. At some value of the field, that is of the applied voltage, the electrons gain sufficient energy between successive collisions to *ionize* lattice atoms *by impact*. The additional electrons which are produced, immediately contribute to the ionizing process, thus building up to an *avalanche breakdown* of the junction. The outward symptom of breakdown is a steep increase of the junction current at the breakdown voltage $V_B$, as depicted in Fig. 5.8. Avalanche breakdown of a *p-n* junction is entirely analogous to the breakdown in a gas. The latter is discussed at length in Chapter 9. The method of analysis used in Chapter 9 has been applied to semiconductors to calculate breakdown conditions. Values of the breakdown voltage vary widely with doping concentrations.

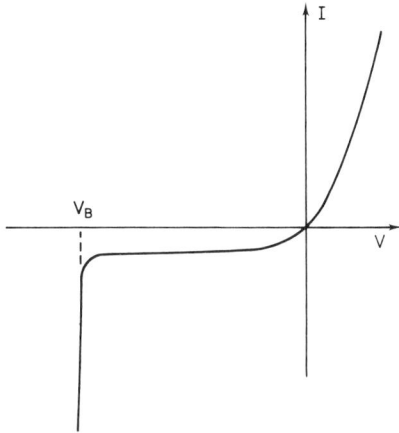

Fig. 5.8. Breakdown in a *p-n* junction.

Break down of a *p-n* junction can also be caused by a different mechanism usually called *Zener breakdown*. A reference to Fig. 5.9 will disclose that there are numerous vacant electron states in the conduction band of the *n*-type material lying at the same energy level as the electrons in the valance band of the *p*-type material under conditions of reverse bias implied by the figure. Under certain conditions these electrons can cross the distorted energy gap by the process of *tunnelling* and thus add to the reverse saturation current of the junction by becoming conduction electrons.

The tunnelling current can only set in if the depletion layer is very short, that is the forbidden gap is very narrow as measured along the horizontal axis of Fig. 5.9. Once this point is reached, a further narrowing of the gap, caused by a slight increase of the applied bias, will swell the current very rapidly giving rise to an

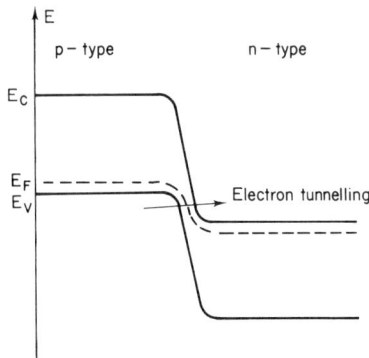

Fig. 5.9. Principle of Zener breakdown.

*I*/*V* characteristic similar to Fig. 5.8. Tunnelling of electrons through space which is effectively insulating is a quantum mechanical phenomenon. It is dealt with in Chapter 18 on the basis of quantum theory and readers interested in the physical basis of the process are referred to that discussion.

## Chapter 6

# The Work Function and Electron Emission Processes

## 6.1 The Work Function

The energy gap between the valence and conduction bands of solids has been found to be of crucial importance to the discussion of electrical conduction in semiconductors. There is another important concept, the work function, which relates to other electrical characteristics, notably the ability of metals to emit electrons.

We define the *work function* as the energy difference of an electron just outside the solid, and at the Fermi level. It is equal to the quantity $E_0 - E_F = e\phi$ marked in Fig. 6.1, where $E_0$ denotes the energy level just outside the metal and

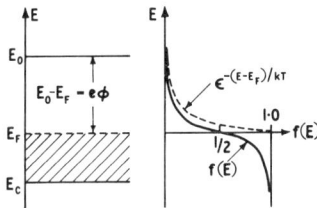

Fig. 6.1. The work function of a metal.

$e\phi$ is the work function measured in electron volts. The symbol $\phi$ represents the equivalent potential difference in volts and is itself frequently referred to as the work function.

Typical values of the work function of metals are several electron volts, e.g. gold, 4·8 eV and tungsten, 4·5 eV. According to the discussions of Section 3.2, the fraction of states just below $E_0$ occupied by electrons at normal temperatures is given by the exponential factor $e^{-e\phi/kT}$. The reader will no doubt satisfy himself by direct substitution that this is minute. The number of electrons per unit volume of metal actually occupying these high lying levels is obtained by multiplying the Fermi-Dirac function by the density of levels,

81

following the method established in Section 3.3. Under normal conditions these electrons are very few indeed, and since they are the only charge carriers capable of leaving the host metal, special measures must be taken to increase their numbers if it is desired to extract a significant current from the metal. The most obvious of these special measures is to raise the temperature of the metal and thus increase the thermal factor $e^{-e\phi/kT}$. This gives rise to *thermionic emission,* to be discussed in the next section.

The *work function of a semiconductor* is defined by analogy with a metal as the energy interval between the Fermi level $E_F$ and the level $E_0$ just outside the solid. The significance of this concept will become apparent in connection with the contact potential to be discussed in a subsequent section.

## 6.2 Thermionic Emission and its Applications

When a metal is heated to a sufficiently high temperature the "tail" of the Fermi-Dirac distribution at the height of the work function, $e\phi$, above the Fermi level assumes a significant percentage value and a measurable number of electrons can escape from the metal. The electrons thus emitted constitute the *thermionic current* which can be attracted by a positive electrode placed in the vicinity. The positive electrode is referred to as the *anode* while the emitting surface is called the *cathode.* Together they form a *vacuum diode* shown schematically in Fig. 6.2.

As explained in the preceding section the number of electrons available for thermionic emission is proportional to $e^{-e\phi/kT}$. Hence it can be expected that the thermionic current will itself be proportional to this thermal factor. It will

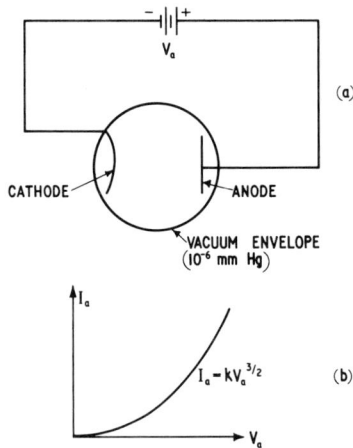

Fig. 6.2. Thermionic vacuum diode.

be shown at a later stage, on the basis of quantum mechanics, that the *current density* is in fact given by the expression:

$$J = AT^2 e^{-e\phi/kT} \qquad (6.2.1)$$

where $A$ is theoretically a universal constant of the order of 100 amps/cm$^2$deg$^2$, but varying widely from one cathode material to another under actual experimental conditions. The reader will satisfy himself by direct substitution that the temperature must be raised to very high values ($\cong 2000°C$) for the thermionic current to become significant in metals having typical work functions, i.e. 4 to 5 eV. Hence only metals with a high melting point can be used as cathodes, tungsten being most common.

Since the high temperatures required with metallic cathodes present various technical difficulties, efforts were made at an early stage in the development of thermionic valves to find materials having lower work functions. These were found among oxides of the rare earths and gave rise to *oxide-coated cathodes*. The latter have work functions ranging down to 1 eV and are used in all small-size thermionic valves.

The characteristics depicted schematically in Fig. 6.3 apply only to a vacuum diode, one in which electrons do not collide with gas molecules, while crossing the cathode-to-anode space. Figure 6.3(a) shows how the emitted current varies with increasing cathode temperature while the anode potential is fixed. At low temperatures there is no measurable current. Then, as the exponential factor in eqn (6.2.1) assumes significant values, the current rises rapidly. Under these

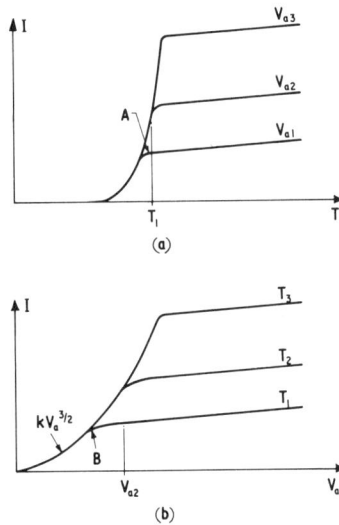

Fig. 6.3. Thermionic current as a function of temperature and anode potential.

conditions every electron emitted by the cathode is immediately attracted to the anode and there is no significant space charge. The current is said to be *temperature limited*. However, when the temperature is sufficiently increased the number of emitted electrons grows to a point when they form a *space charge* which effectively screens the cathode from the influence of the positive anode. Electrons appearing at the surface of the cathode are no longer attracted away and the current *saturates*. If the anode potential is, say, $V_{a1}$, this condition is reached at point $A$ of Fig. 6.3(a). Under these conditions the current is said to be *space charge limited*. At higher applied voltages a greater number of electrons is required to form a screening space charge and the *saturation current* is correspondingly greater. It should be noted that the current continues to increase slightly beyond point $A$. This is due to the fact that many electrons leave the cathode with a certain initial velocity, or kinetic energy, and that this increases on average as the cathode temperature rises.

Figure 6.3(b) shows the variation of the current with the applied anode potential, after the cathode has been heated to a point when it emits significant numbers of electrons. The cathode temperature is then kept constant at some value $T_1$, say, marked in Fig. 6.3(a). As the anode voltage is increased from zero up to point $B$ we have a space charge limited current. Anode potentials beyond this value "sweep" the cathode-to-anode space clear of charge and set up the temperature limited condition represented in Fig. 6.3(b) by the nearly flat section of the current curves. The current continues to increase slightly, however, due to the *Schottky effect*. This term denotes a slight lowering of the energy level $E_0$ at the surface of the metal, caused by externally applied anode potential.

Thermionic vacuum valves are almost universally operated under conditions of space charge limited currents, that is below point $B$ of Fig. 6.3(b). The current/voltage characteristic of a thermionic vacuum diode operated under these conditions can be derived by application of Poisson's equation to the cathode-to-anode space. It has the form

$$I_a = k V_a^{3/2} \qquad (6.2.2)$$

where $k$ is a constant. This equation applies only for positive values of the anode voltage $V_a$, as defined in Fig. 6.2(a). When $V_a$ is negative no current flows through the diode, as indicated in Fig. 6.2(b). This result should be compared with the rectifier equation of a semiconductor diode, Fig. 5.6.

The outstanding importance of the vacuum diode is to be found in developments which followed it, namely the introduction of a *control grid* to make a *triode*, followed by multigrid valves such as tetrodes, pentodes etc. Referring to Fig. 6.2 it is readily seen that a third meshlike electrode can be interposed between the cathode and anode. If this electrode, called the control grid, is given a potential above or below that normally prevailing in its neighbourhood, the flow of electrons across the vacuum space will be enhanced or depressed. The resulting current fluctuations can be utilized to produce voltage variations in a suitable anode circuit which are an *amplified* replica of the signal applied to the control grid. Thus an *amplifier* can be obtained.

Although small-size thermionic valves have been largely superseded by transistors and solid state microcircuits in circuit applications, thermionic emitters in cathode ray tubes and high-power valves still retain their importance in the absence of early replacement for these devices.

## 6.3  Contact Potential of Two Solids

If two metals having different work functions are placed side by side but are maintained in *electrical isolation,* their energy level diagrams align themselves as shown in Fig. 6.4(a). Since the only area of "contact" between the samples is the free space separating them it must be assumed that they share the energy level $E_0$ of an electron which has just managed to escape. Consequently the Fermi levels $E_{F1}$ and $E_{F2}$ of the two metals are distinct.

If the metals are next joined by an electric contact as suggested in Fig. 6.4(b) some electrons will flow from metal 1 into metal 2 to occupy the lower energy vacant states available there. As a result a potential difference, the *contact potential,* is established between the two samples while the Fermi levels align themselves, as they must in a *single* system in equilibrium, to which the metals have now been reduced.

The number of electrons required to set up the contact potential is very small compared with the total populations of conduction electrons. Hence their transfer from sample 1 into sample 2 will not affect the position of the Fermi levels relative to $E_0$, leaving the values of the work functions $e\phi_1$ and $e\phi_2$

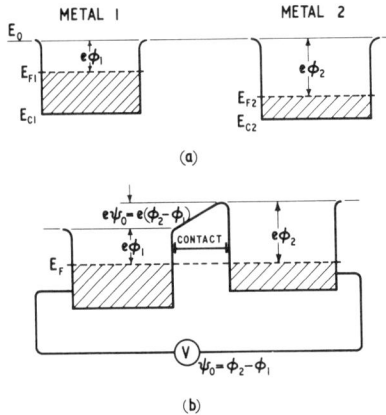

(a)

(b)

Fig. 6.4. Contact potential of two metals.

unchanged by the contact. In view of this observation we can give the contact potential a definite value as suggested by the energy level diagram of Fig. 6.4(b).

$$\psi_0 = \phi_2 - \phi_1. \qquad (6.3.1)$$

Hence the *contact potential* $\psi_0$ *equals the difference of the work functions of the two metals* expressed in volts. It can be measured directly by a voltmeter shown schematically in Fig. 6.4(b).

The above arguments apply to any solids, not just metals. In particular they apply to semiconductors, as implied by the discussion of the *p-n* junction in Section 5.1. In that case we have the interesting example of a contact within a single sample of the same material. The contact potential is due to the different positions of the Fermi level relative to the energy band scheme caused by the *n*-type and *p*-type impurities. The work functions of the two types of material are given by the expression

$$e\phi_p = (E_0 - E_F)_p > (E_0 - E_F)_n = e\phi_n \qquad (6.3.2)$$

where the symbols have obvious meanings with reference to Fig. 5.2. Hence the contact potential of the junction has the form

$$\psi_0 = \phi_p - \phi_n \qquad (6.3.3)$$

also given by eqn (5.2.3).

A different possibility of great practical importance arises in the case of contacts between a metal and a semiconductor, which may act as rectifiers or ohmic connections. These will be discussed in a later section.

Other practical applications of the contact potential are found in thermo-couples and in the generation of thermoelectric power.


## 6.4 Photoelectric Emission

In thermionic cathodes the emission of electrons was secured by raising the temperature of the conduction electrons to impart to them sufficient kinetic energy to enable them to leave the parent metal. The same end result can be achieved at normal temperatures by irradiating a *photocathode* with electro-magnetic waves of a sufficiently high frequency (or short wavelength). The quanta of such radiation can be absorbed by conduction electrons occupying states around the Fermi level. As a result the electrons are excited to higher energy levels and in some cases this extra energy is sufficient to cause *photoemission* (emission due to photons).

The diagram of Fig. 6.5 shows the energy relations involved. Photoemission is likely to take place if the photon energy is at least equal to the work function of the cathode

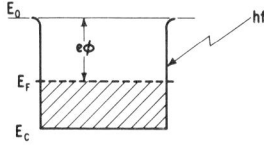

**Fig. 6.5.** Photoemission.

$$hf_s = e\phi \tag{6.4.1}$$

where $f_s$ is termed the *threshold frequency* of the incident radiation. If the frequency of the incident radiation exceeds the threshold, electrons will be emitted by the photocathode with initial kinetic energies given by the relation

$$\tfrac{1}{2}mv^2 \leqslant hf - e\phi \tag{6.4.2}$$

where $f$ is the photon frequency.

### (i) *Photoemission between two metals*

The emitted electrons will constitute a *photocurrent* between the photocathode and an anode placed in the vicinity, provided certain energy relations are satisfied. One possible arrangement, involving an anode having a greater work function than the cathode, is depicted in Fig. 6.6. The diagram allows for the possibility that a potential may be applied between the cathode and anode, but to begin with we consider the condition of equilibrium when $V_a = 0$.

It will be clear from the energy level diagram that it is not enough for the photons incident on the cathode to satisfy eqn (6.4.1), applied to the cathode, in order to cause a photocurrent to flow to the anode. The photons must satisfy the more stringent condition

$$hf = e\phi_2 \tag{6.4.3}$$

where $\phi_2$ is the work function of the anode. Photons having quanta of energy greater than $hf$ will secure a flow of photocurrent.

The application of a potential between the anode and cathode will displace the Fermi levels of the two electrodes relative to each other and will either

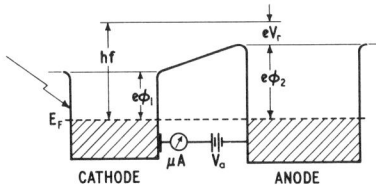

**Fig. 6.6.** Photoemission between two metals.

facilitate or hinder the flow of photoelectrons depending on its polarity. A negative potential will naturally hinder the flow of electrons by raising the Fermi level of the anode. When the potential reaches the value $V_r$ given by the relation

$$eV_r = hf - e\phi_2 \qquad (6.4.4)$$

the flow of photoelectrons will be completely stopped. $V_r$ is called the *retarding potential*. The negative magnitude of the energy quantities shown in Fig. 6.6 implies that $V_r$ is negative. However, this need not always be the case as the reader will satisfy himself by assuming that $hf < e\phi_2$. However, even in the presence of positive anode voltages photoemission will only take place if the photon frequency exceeds the threshold value given by eqn (6.4.1) as applied to the cathode.

Once the anode voltage is increased above $V_r$ the photocurrent rises steadily. This is because more and more electrons occupying states below the Fermi level receive sufficient energy from the incident photons to enable them to mount the energy step $e\phi_2 - eV_a$. Moreover, since the density of states in a close neighbourhood of the Fermi level can be assumed to be constant, the probability that an incident photon excites an electron grows linearly with the depth to which the photon can penetrate the conduction band. Hence the photocurrent rises approximately linearly with $V_a$. The process continues until the applied voltage reaches a value satisfying the condition

$$e\phi_2 - eV_a = e\phi_1 \qquad (6.4.5)$$

when the excited electrons do not have to rise in energy above $e\phi_1$ to reach the anode. After that the photocurrent remains constant or saturates. The current/voltage relationship for a *photocell* of this type is depicted in Fig. 6.7. The onset of photocurrent at $V_r$ and the saturation condition are smeared out at temperatures above absolute zero, due to the gradual transition from fully occupied states below the Fermi level to vacant states above it. Moreover, a

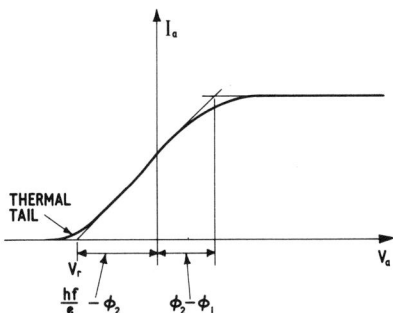

Fig. 6.7. Current/voltage graph of a photocell.

photon striking the metal may be absorbed some distance below the surface and the excited electron may lose some energy in collisions before it leaves the metal.

### (ii) *Photoemission between a semiconductor and a metal*

The energy relations explained above must be modified if the photocell consists of a semiconducting cathode and a metallic anode as shown in Fig. 6.8(a). Since

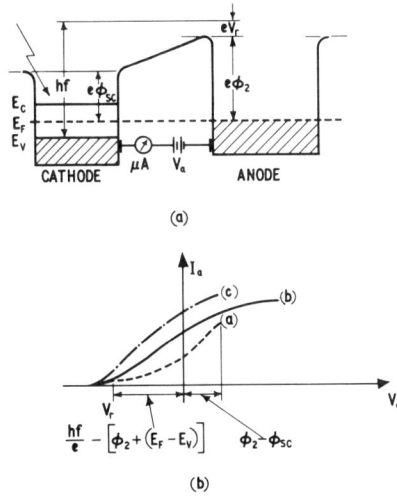

Fig. 6.8. Photoemission from a semiconductor to a metal.

the Fermi level is situated somewhere in the energy gap and there are no electron states in its vicinity, the photons must have sufficient energy to excite electrons in the *valence band* to cause photoemission to the metallic anode. Electron densities in the conduction band are insufficient under normal conditions to yield a significant photocurrent.

The current voltage graph for a photocell of this type is shown in Fig. 6.8(b). Its general form is the same as in Fig. 6.7 but the value of the retarding potential allows for the position of the Fermi level in the energy gap. Hence a cell containing a semiconducting photocathode can be used to determine experimentally with the help of a voltmeter the height of the Fermi level above the top of the valence band. The frequency of the incident radiation must, of course, be known in an experiment of this type, and it must exceed the threshold value given by eqn (6.4.1), as applied to the semiconductor cathode, i.e.

$$hf_s = e\phi_{sc} + (E_F - E_v) \tag{6.4.5}$$

It was emphasized in connection with the metal cathode that the photo-current is expected to increase linearly, because the density of states is constant in the vicinity of the Fermi level. Now, we know (see Section 3.3) that the density of states is low near the band edges of a semiconductor and increases deeper in an allowed band. Hence the photocurrent shown in Fig. 6.8(b) should be expected to follow the lower dotted line, increasing slowly compared with the straight line.

It is instructive to continue this argument at this juncture and to consider a hypothetical material in which the density of states *decreases* as one moves below the edge of the allowed band. The photocurrent would then increase at a higher rate than linear, as suggested by the chain line of Fig. 6.8(c).

The three types of density-of-levels functions are shown schematically in Fig. 6.9 and labelled to correspond with the current characteristics of Fig. 6.8(b).

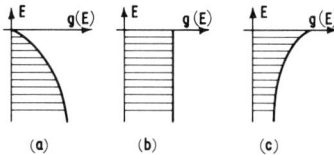

Fig. 6.9. Various types of density-of-levels functions.

The foregoing discussion illustrates the importance of the density of levels functions in relation to the performance of actual devices. On the other hand Fig. 6.8 shows how photoemissive phenomena can be used to secure experimental data regarding the density-of-levels functions.

### 6.5 Metal-to-Semiconductor Contacts

The properties of contact surfaces between metals and semiconductors are of importance for two reasons. In the first place they act as rectifiers in certain cases in a manner similar to *p-n* junctions. Secondly they provide non-rectifying or ohmic contacts between semiconductors and wires used as interconnections in electrical circuits. The distinction between the two cases depends on the relative configurations of the energy bands of the metal and semiconductor in a manner to be discussed below.

Before embarking on this subject, however, it should be noted that the contacts under discussion cannot be obtained simply by touching a semiconducting sample with a piece of wire. Alloying processes must be used and a variety of precautions must be taken to minimize contamination of the surfaces of contact since it is known that the performance of junctions is greatly affected by surface states, which in turn are modified by impurities.

(i) *Rectifying contact between a metal and an n-type semiconductor*

In the discussion of metal-to-semiconductor contacts we introduce the concept of the *electron affinity* of the semiconductor. This is defined by the distance between the "vacuum level" $E_0$ and the bottom of the conduction band $E_c$, and is denoted by $\chi$

$$e\chi = E_0 - E_c. \tag{6.5.1}$$

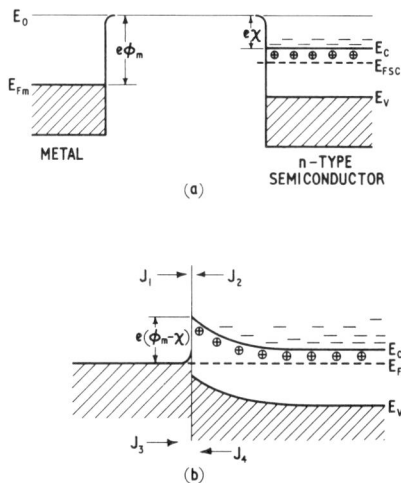

Fig. 6.10. Rectifying metal-to-semiconductor contact.

The electron affinity is marked in Fig. 6.10(a); it is the energy required by a conduction electron to be emitted from the semiconductor.

We assume that the work function of the metal is greater than the electron affinity of the semiconductor, $\phi_m > \chi$, as shown in Fig. 6.10(a), which depicts the two materials *in isolation*. It should be particularly noted that there is a copious supply of free donor electrons in the semiconductor, occupying states relatively high above the Fermi level $E_{Fm}$. As the two materials are brought into contact the donor electrons in the vicinity of the interface drop in energy to occupy the lower levels available in the metal. As a result a depletion layer is formed in the semiconductor and $E_{Fsc}$ is shifted locally downwards in relation to the energy gap, to coincide with $E_{Fm}$. The extra electrons which entered the metal distribute themselves as a charge layer over the surface of contact to balance the positive charge of the depletion layer in the semiconductor. The energy relations applicable to the completed contact are shown in Fig. 6.10(b). It should be noted that the position of the Fermi level in the semiconductor near the surface of contact adjusts itself to leave a barrier equal to the energy difference $e(\phi_m - \chi)$. The barrier is usually referred to as a *Schottky barrier*. It is an obstacle in the way of electrons trying to cross from the metal to the semiconductor and vice versa.

As in the case of the *p-n* junction four distinct currents can be identified across the contact surface.

$J_1$: This is an electron current flowing from the metal to the semiconductor, composed of electrons which are energetic enough to surmount the contact potential barrier. It is in effect a thermionic current emitted by the metal against the potential $\phi_m - \chi$. Since this barrier is less than $e\phi_m$, the current can be expected to be higher than thermionic emission into free space, discussed in Section (6.2).

$J_2$: This again is an electron current consisting of conduction electrons from the semiconductor which occupy levels above the barrier. Under equilibrium conditions (no applied voltage) $J_2 = J_1$.

$J_3$: Current of holes from the metal to the semiconductor having enough energy to mount the energy step into the valance band of the semiconductor.

$J_4$: Current of intrinsic holes generated thermally in the semiconductor and diffusing towards the metal which acts as a sink. Only holes generated within one diffusion length of the surface contribute to this current. In equilibrium $J_3 = J_4$.

In the absence of externally applied voltages the four currents cancel each other to yield no net flow, as in a *p-n* junction. The cases of practical interest arise when voltages are applied.

Let us first consider what happens when a *reverse bias* is applied to a metal-to-semiconductor contact of the above type (negative battery terminal applied to the metal). Its effect is to depress the Fermi level in the semiconductor in relation to the metal, the distortion being confined to the depletion layer as in a *p-n* junction. As a result the conduction electrons in the semiconductor have a much higher barrier to surmount before they can be emitted into the metal. Hence the current $J_2$ is rapidly reduced to zero. A similar argument applies to the current of holes $J_3$. However the currents $J_1$ and $J_4$ remain unaffected by the reverse bias and their sum forms a small reverse saturation current $J_s = J_1 + J_2$ in analogy to the *p-n* junction.

The application of a forward bias to the metal-to-semiconductor junction lifts the Fermi level in the semiconductor relative to the metal. This effectively lowers the contact barrier to the motion of donor electrons from the semiconductor to the metal and allows the current $J_2$ to rise rapidly. By a similar argument the hole current $J_3$ increases rapidly under the influence of a forward bias. The sum of the currents $J_2 + J_3$ soon swamps the saturation current and constitutes a substantial forward current in analogy to the *p-n* junction.

On the basis of the foregoing arguments it is possible to write down quantitative expressions for the current components using a form of thermionic emission equation, eqn (6.2.1) and to assemble these into a rectifier equation of the same form as applies to a $p$-$n$ junction. The conclusion would then be that a metal-to-semiconductor contact has the same rectification characteristics as a $p$-$n$ junction. Although contacts of the type described above have rectifying properties, the detailed theoretical results are not confirmed by experimental data.

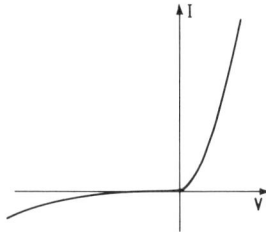

Fig. 6.11. Current/voltage curve of a copper-oxide rectifier.

In the first place the current/voltage curve, Fig. 6.11, of a metal-to-semiconductor junction differs significantly from that of a $p$-$n$ junction. In particular the reverse current is not constant, but increases steadily with increasing reverse bias. Moreover, since according to the theory currents across the junction depend on the relative magnitudes of the work functions of the metal and semiconductor, this should appear in experiments when different metals are used with the same semiconductor. Although such experiments do show the expected trend, the actual magnitudes of currents measured do not agree with predictions. On these grounds it is concluded that the theory outlined above does not take into account all the phenomena which influence the properties of a metal-to-semiconductor junction, although the general principle is valid.

Some of the discrepancy can be accounted for if allowance is made for the effect of the *image charge* of electrons just outside the metal. An electrostatic calculation shows that the effect of the image field in the presence of a reverse bias is to lower the barrier $\phi_m - \chi$ and hence to increase the thermionic currents. Moreover, the effect is accentuated by an increasing bias. Hence the growing reverse current shown in Fig. 6.11 is at least qualitatively accounted for. However, there are other properties of the metal-to-semiconductor contact, e.g. breakdown conditions, which cannot be satisfactorily explained in such terms. There is, moreover, both theoretical and experimental evidence which suggests that surface states have a profound influence on the height of the barrier, and it is generally assumed that an accurate theory of the metal-to-semiconductor contact must take them into account. The general conclusion must be, therefore, that the functioning of these devices is not yet properly understood, although various techniques for their preparation are well known.

### (ii) *Ohmic contact between a metal and an n-type semiconductor*

A non-rectifying or ohmic contact is obtained if the electron affinity of the $n$-type semiconductor is greater than the work function of the metal, as indicated in Fig. 6.12(a). The diagram makes it clear that there are many vacant

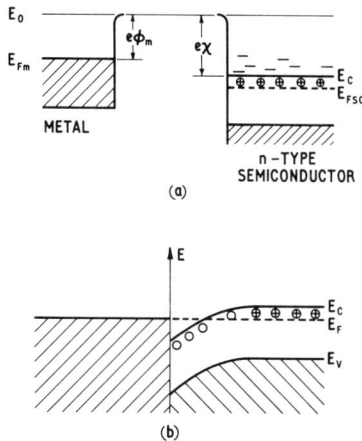

Fig. 6.12. Ohmic contact between a semiconductor and a metal.

electron states in the conduction band of the semiconductor, at energies well below the Fermi level of the metal. When contact is established between the two materials electrons move from the metal to the semiconductor, settling at first on the ionized donors and then occupying many of the vacant states in the conduction band. The migration of electrons sets up a potential difference between the metal and semiconductor which confines the extra electrons to a thin layer adjoining the metal. The concentration of electrons in this region is sufficient to distort the energy bands to such an extent that the Fermi level is moved up into the conduction band of the semiconductor as shown in Fig. 6.12(b). As a result of these modifications of the energy band scheme the conduction bands are at similar heights on the energy axis and electron transport across the metal-to-semiconductor interface is unimpeded in either direction. Ideally the resistance of a contact of this type would be the sum of the bulk resistances of the metal and semiconductor. In fact the actual contact surface adds to the overall resistance but it does so in a linear, ohmic fashion.

(iii) *Contacts between p-type semiconductors and metals*

The principles outlined above apply to *p*-type semiconductors with some fairly obvious modifications. Thus a rectifying contact is obtained if the electron affinity of the *p*-type material is greater than that of the metal, and conversely for an ohmic contact. Moreover, the polarity of the applied voltage is reversed for the corresponding bias. Thus a reverse bias is obtained on application of the positive battery terminal to the metal and vice versa for a forward bias.

## 6.6  Other Emission Phenomena

There are other processes which can give rise to electron emission besides thermionic and photoemission. Of considerable technological importance are secondary emission and field or Schottky emission. In the present section we state briefly what comes under these headings but to begin with we mention the method used in the production of x-rays.

A stream of *x-rays* is obtained when a target electrode is bombarded by electrons having energies of the order of 20 KeV. The target is usually rhodium and the electrons are in the form of a beam accelerated by an electron gun. The phenomenon is the inverse of photoelectric emission in that photons (of x-ray wavelength) are obtained in exchange for electrons incident on a solid. The x-rays thus obtained have many medical and industrial applications by virtue of their ability to penetrate tissues and other materials which are opaque to visible light. Since the wavelengths are comparable in magnitude with interatomic spacings, x-rays are widely used for the analysis of crystal structures based on the principle of Bragg diffraction.

Let us now return to electron emission from metals.

(i) *Secondary emission*

Electron emission from metals can be caused by the impact of primary electrons or other particles, e.g. gas ions. The precise physical mechanisms whereby the kinetic energy of the impinging particles is imparted to the metallic electrons and enables them to escape are not properly understood at present. There is, however, a body of empirical knowledge which makes it possible to predict under what conditions this *secondary emission* is strong or weak. To quantify experimental data it is customary to use the *secondary emission coefficient* which is defined as follows.

$$\gamma = \frac{\text{number of emitted electrons}}{\text{number of incident electrons or other particles}}$$

$\gamma$ varies over a very wide range—from small values of $10^{-2}$ to really large values such as $10^2$.

Secondary emission appears as an undesirable, parasitic phenomenon in some cases while it is useful in others. Thus the appearance of secondary electrons at the anode of a thermionic vacuum tube is unwanted and measures must be taken to suppress them. On the other hand, secondary emission is utilized in electron multipliers, devices specifically designed to produce copious secondary electrons. Some phenomena depend on secondary emission for their existence. Thus the glow discharge in a gas (see Chapter 9) which forms the basis of fluorescent lighting, is sustained by the emission of relatively few secondary electrons by the cathode due to positive ion impact.

### (ii) *Field emission*

The presence of an intense electric field adjacent to the surface of a metal can cause electron emission from a *cold cathode*. It can be shown by an electrostatic calculation that the effect of an applied field is to lower the potential barrier at the surface of the metal and thus to enable more electrons to escape. This is the cause of the Schottky effect mentioned in Section 6.2 in connection with thermionic emission from heated cathodes. The effect is, however, insufficient to produce significant numbers of electrons at normal temperatures and some other mechanism must be found to account for emission from cold cathodes. This is now understood to be electron tunnelling. It is shown in Section 18.3 on the basis of quantum mechanics that some electrons can penetrate through a potential barrier if the latter is sufficiently thin. The application of a strong electric field to a metal changes the step-like increase in potential at the metal surface into a barrier of finite thickness. Theoretical estimates show that the thickness is small enough at applied fields of the order of $10^2$ KV/cm to allow large numbers of electrons to escape.

*Chapter 7*

# The Transistor

## 7.1 The *p-n* Junction as an Emitter of Charge Carriers

In the preceding chapter various methods of electron emission were discussed. Only emission from a metallic source was envisaged and in most cases charge carriers entered a vacuum region in which they were exposed to the action of electric fields originating at a nearby positive electrode. The metal-to-semi-conductor contacts were an exception in that the emitted electrons entered another solid placed in intimate contact with the metal.

In the present section we go on to discuss the charge emitting properties of suitably designed *p-n* junctions. We shall find that they are more versatile than metallic electron emitters in the sense that both negative and positive charge carriers can be emitted. On the other hand emission can only take place into the crystal lattice of another solid. In a way this is a limitation because it is not possible to form and accelerate well-defined electron beams. However, the process forms the basis of transistors which in addition to an emitting junction contain a means of collecting the emitted charge carriers, be they electrons or holes.

To maintain an analogy with the electron emitters of the preceding chapter we discuss first an electron emitting *n-p* junction. We refer to Fig. 7.1 in which the sequence of materials has been deliberately interchanged as compared with Fig. 5.1, because we shall want to add another *n*-type section to the right-hand side of this combination. The discussion of Chapter 5 was formulated in general terms, allowing for the *n*-type and *p*-type sides of a junction to be either equally or unequally doped. The reader should note from Fig. 7.1 that now we only consider *unequally doped* juctions and specifically that the *n*-type side is heavily doped, while the *p*-type side is lightly doped. The difference in the impurity densities must be sufficiently great to cause a significant difference in the position of the Fermi level relative to the energy gap. Thus in the *p*-type material the Fermi level is closer to the mid-gap position than in the *n*-type material which means that $(E_F - E_v)_n > (E_c - E_F)_p$. This reflects the fact that there are

97

far more electrons in the $n$-type material than holes in the $p$-type. A further consequence of the unequal doping is an unsymmetrical depletion layer. The space charge layer of immobile ionized acceptors is much thicker on the $p$-type side of the junction to maintain the requirement of overall space charge balance. Finally, the contact potential drop is confined almost entirely to the lightly doped side of the unsymmetrical junction.

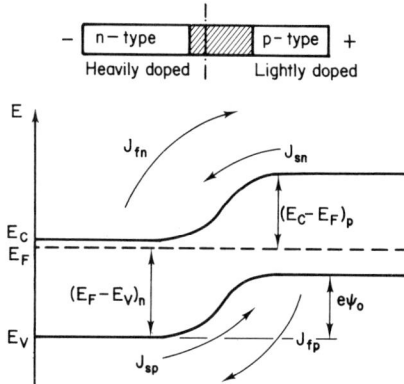

Fig. 7.1. Electron emitter junction.

The emitter junction operates under conditions of small forward bias, sufficient for the reverse saturation currents $J_{sn}$ and $J_{sp}$ to be negligible compared with the forward currents. A fraction of one volt is enough to ensure this condition. We now turn our attention to the forward currents which have been evaluated in Chapter 5. According to eqns (5.5.1) and (5.5.2) the forward currents are directly proportional to the concentrations of majority carriers $n_n$ and $p_p$ which in turn are equal to the impurity concentrations $N_d$ and $N_a$ under conditions of extrinsic saturation, assumed throughout this chapter. Hence the numbers of electrons injected into the $p$-type material will greatly exceed the number of holes crossing into the $n$-type material. The doping of the two sections of the junction can be easily arranged to differ by a factor of 100 which means that electrons will constitute some 99% of the current crossing the barrier. Thus the junction serves as an effective emitter of electrons.

The relative balance of the electron and hole currents can also be quantified in terms of the position of the Fermi level. It is readily verified, using the methods of Sections 3.1 and 3.2, that a difference in the position of the Fermi level of about 0·1 eV in the sense shown in Fig. 7.1, implies that the density of holes in $p$-type material is only a few per cent of the density of electrons in $n$-type material. A shift in the position of the Fermi level of this magnitude is readily accommodated within the typical semiconductor energy gap of 1 eV without affecting its character.

The electrons injected into the p-type side of the junction assume the character of excess minority carriers. They are transported away from the barrier by diffusion, subject to the concentration gradient and recombination with majority holes. This process has been discussed in Section 4.5 and represented diagrammatically in Fig. 4.4 on the example of holes. An analogous description applies to electrons which concern us here. There are also holes injected into the n-type material but in the present context it is emphasized that they are much less numerous.

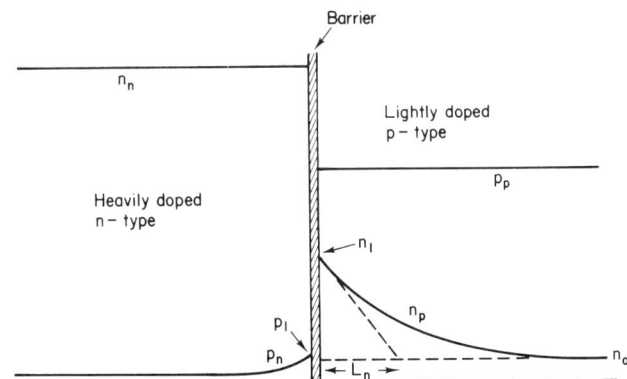

Fig. 7.2. Charge carrier distribution in the vicinity of an electron emitter junction.

The relative distribution of the various species of charge carriers on both sides of the electron emitter junction is summarized in Fig. 7.2. However, the reader should note that the vertical axis does not indicate relative densities to a linear scale. Thus $n_n \gg p_p$ than suggested by the graph. Similarly $n_1 \gg p_1$.

It is important to note for future reference that current transport is effected by different types of carrier at different points of the junction. Thus, well to the left of the barrier majority electrons are practically the sole current carriers, the intrinsic holes $p_0$ being negligible. Within a diffusion length of the barrier, injected minority holes contribute a little to the current but this is very small in a highly unsymmetrical junction like the one considered here. Immediately to the right of the barrier the current is largely carried by injected minority electrons $n_1$. The flow of majority holes is still small, being just enough to supply the injected holes $p_1$. As one moves further into the p-type material, the excess minority electrons are gradually removed by recombination and the transport of current is taken over by majority holes. On the extreme right of Fig. 7.2 the current is carried entirely by majority holes drifting to the left.

A quantitative measure of the effectiveness of an emitter junction is provided by the *minority carrier emitting efficiency*, usually denoted by $\gamma$ and defined by the relation

$$\gamma = \frac{J_{fn}}{J_{fn} + J_{fp}} = \frac{1}{1 + \dfrac{\sigma_p L_n}{\sigma_n L_n}} = \frac{1}{1 + \dfrac{\rho_n L_n}{\rho_p L_p}} \qquad (7.1.1)$$

where the last expression has been obtained on substitution from eqns (5.5.1) and (5.5.2) and by using Einstein's relations eqns (4.6.5) and (4.6.6).

A discussion entirely analogous to the foregoing can be applied to an unsymmetrical junction with a heavily doped $p$-type side which acts as an emitter of holes. An emitting efficiency is defined for this junction by an obvious modification of eqn (7.1.1).

If it is assumed as a first approximation that the diffusion lengths for electrons and holes are roughly equal, the emitting efficiency of a junction depends only on the relative resistivities of the two materials. The $n$-type material, being more highly doped, has a lower resistivity than the $p$-type material. The emitting efficiency is brought closer to unity by making this unsymmetry more pronounced. As pointed out above it is quite feasible to obtain forward current ratios of 100 thus securing an efficiency of some 99%.

## 7.2 The Junction Triode

After the discussion of the preceding section we are in a position to assemble a three layer junction triode and explain its operation. The supply of emitted minority electrons on the $p$-type side of the junction of Fig. 7.1 is readily controlled by varying the forward bias applied to the junction. If a sink for electrons is placed in the close vicinity of this junction to *collect* the emitted electrons and to pass them through a suitable circuit, a replica of the emitter current variations will be obtained in that circuit. Under suitable conditions this replica will be an amplified version of the original.

At this point we recall from Section 5.1 that a reverse biased $p$-$n$ junction acts as a sink for the minority electrons present in the $p$-type material. Hence if a reverse biased $p$-$n$ junction is placed close to the right of the emitter of Fig. 7.1 it will serve as a *collector* of the injected minority electrons. The desired configuration is depicted diagrammatically in Fig. 7.3. The three layer device constitutes an *n-p-n junction transistor*. The central region between the two junctions is called the *base* for historical reasons, because it formed a kind of solid base in the early transistors.

To ensure efficient operation the base region must be kept thin, allowing the collector junction to be situated well within the electron diffusion length $L_n$, in order to intercept as many of the minority electrons as possible before they recombine.

To appreciate how the collector current can be controlled by the emitter current we go back to the rectification characteristic of a *p-n* junction (Fig. 5.6). With a reverse bias the junction carries only the reverse saturation current $I_s$,

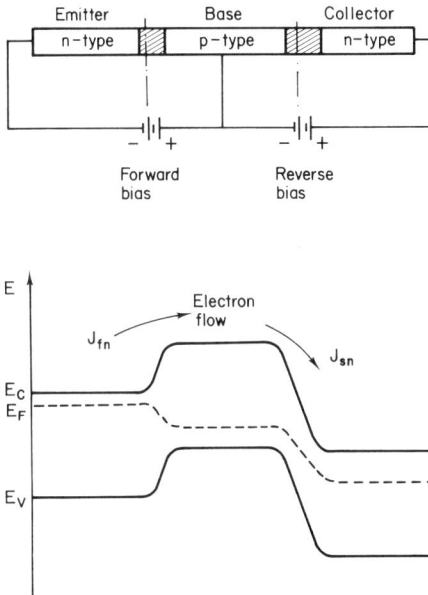

Fig. 7.3. *n-p-n* junction transistor.

made up of minority electrons and holes for which the junction acts as a sink. The magnitude of this current depends solely on the concentrations of intrinsic electrons and holes in a specific junction. In the absence of any voltage on the emitter this is the current that flows across the collector junction of Fig. 7.3. It is shown in Fig. 7.4 by the line marked $I_e = 0$.

On application of a fixed forward bias to the emitter junction the supply of minority electrons to the *p*-type base region is increased above the equilibrium intrinsic concentration $n_0$. Hence the electron current flowing to the collector is correspondingly increased. Since this is accomplished by the sink action of the collector junction, which is largely independent of the bias voltage within breakdown limits, the collector current does not change substantially with the voltage. Thus the collector behaves very much like a constant current source. As the forward bias on the emitter is increased to higher values, correspondingly higher currents flow to the collector. Figure 7.4 shows schematically a family of collector current/voltage curves with the emitter current as parameter.

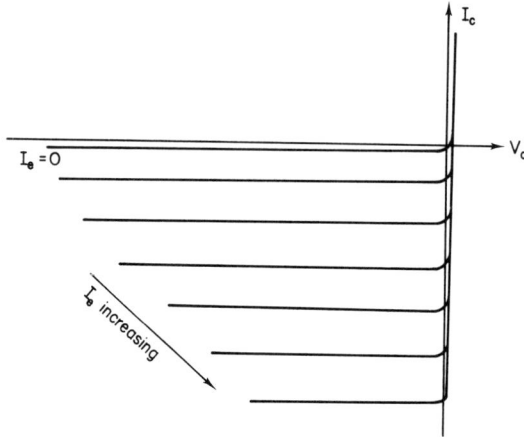

**Fig. 7.4.** I/V collector characteristics of an *n-p-n* transistor with the emitter current as parameter.

The method of presenting the characteristics of Fig. 7.4 has been dictated by the arguments which led to their development. Practical data sheets published by manufacturers are arranged somewhat differently to facilitate their use in circuit design. Thus the sense of the current and voltage axis is reversed and the graphs are shown in the first quadrant as depicted in Fig. 7.5. This presentation corresponds with the fact that transistors are usually operated with either their

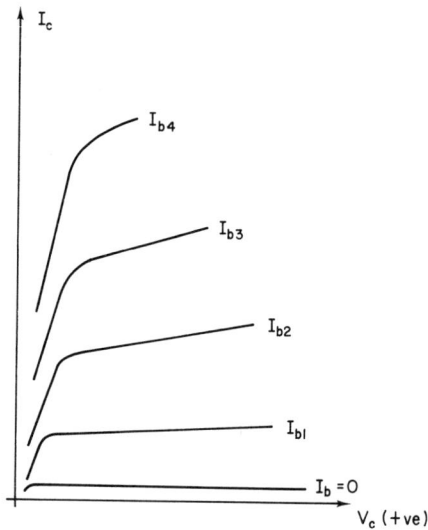

**Fig. 7.5.** Practical transistor characteristics with the base current as parameter.

emitter or base earthed and a *positive* voltage applied to the collector of an *n-p-n* device. Moreover the base current is used as parameter for the set of curves. It is a small fraction of the emitter current as will be explained below.

The use of the curves in circuit design is a subject appropriate to texts dealing with transistor circuits, but in the present context it is desirable to reflect on the physical aspects of amplification which can be obtained with the help of a transistor triode. The biasing arrangement shown in Fig. 7.3 can only be used to measure the d.c. characteristics. To obtain amplification a signal source must be inserted in the emitter circuit and a load circuit, a resistor in the simplest case, must be connected to the collector circuit. The signal source causes the forward emitter current to fluctuate around an average value set by the d.c. bias. These current fluctuations are intercepted by the collector and passed through the load resistor across which they develop a voltage. Since not quite all the emitter current reaches the collector due to recombination and other causes it may seem strange how amplification, and more especially power amplification is realized. The answer lies principally in the fact that much higher voltages are applied to the collector than to the emitter. In fact the desire to operate the collector circuit at the highest possible voltages makes it necessary to assure against reverse breakdown by making the depletion layer as wide as possible. This is achieved by light doping of the collector section, lighter even than the base section. The actual doping figures, as measured by resistivities, vary between various types of transistor. For germanium junctions they are typically 10 ohm-cm, 1 ohm-cm and 0·01 ohm-cm for collector, base and emitter layers respectively.

The foregoing outline of the functioning of an *n-p-n* transistor is exactly paralleled by a *p-n-p* junction with holes assuming the role of charge carriers emitted into the *n*-type base region and then collected by a reverse biased junction. It is an excellent exercise to write out the reasoning in full and to sketch the relevant graphs, noting the polarities of the applied voltages.

## 7.3 Properties of the Base Layer

From the foregoing qualitative description of the junction transistor it will be clear that the properties of the base are crucial to the operation of the device. In the present section we shall discuss somewhat more quantitatively how the juxtaposition of the two junctions affects the properties of the emitter, and how effective is the base in transmitting the minority carriers to the collector.

The proximity of the collector means that the distribution of minority electrons no longer follows the exponential curve of Fig. 7.2 and that their density drops to zero rather than the equilibrium value $n_0$. The actual distribution is suggested in Fig. 7.6 by the solid line, while the exponential distribution is shown dotted for comparison. To evaluate the actual distribution we use the equation of charge continuity, which is applicable to the present case in the form of eqn (4.5.1), adapted to electrons.

$$0 = D_n \frac{\partial^2 n}{\partial x^2} - \frac{n - n_0}{\tau_n}. \tag{7.3.1}$$

The general solution is analogous to eqn (4.5.4)

$$n_p - n_0 = A\,e^{-x/L_n} + B\,e^{x/L_n} \tag{7.3.2}$$

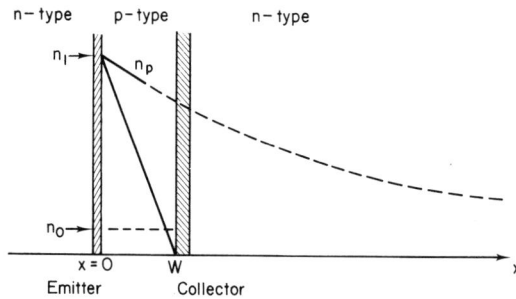

Fig. 7.6. Minority charge carrier distribution in the base layer of a junction transistor.

but the constants of integration must be disposed of differently because of different boundary conditions which are $n = n_1$ at $x = 0$ (see Fig. 7.2) and $n = 0$ at $x = W$. Rather than write out the result in full now, we shall utilize it as the need arises.

An analysis of eqn (7.3.2) shows that it represents a slightly concave line with a somewhat greater gradient at the emitter than at the collector. In some situations it is perfectly adequate to assume that the injected minority carrier distribution in the base follows a straight line.

Such a case arises when it comes to evaluating the efficiency of the emitter. Since the gradient of the linear electron distribution is $n_1/W$ the resulting diffusion current is $J_{fn} = -D_n\,n_1/W$ which should be compared with the value of $J_{fn} = -D_n(n_1 - n_0)/L_n$ for the exponential distribution. Noting that $n_1 \gg n_0$ in actual emitters we find that the proximity of the collector has increased the injected minority current by the factor $L_n/W$. Hence the minority carrier emitting efficiency, eqn (7.1.1), must be modified to read

$$\gamma = \frac{1}{1 + \dfrac{\rho_n}{\rho_p}\dfrac{W}{L_p}} \tag{7.3.3}$$

which represents a significant improvement caused by the sink action of the collector. It is emphasized that the initial density of injected electrons is the same in both cases, being given by $n_1 = n_n \, e^{-e(\psi_0 - Ve)/kT}$ according to eqn (5.5.1).

Although the linear approximation to the minority carrier distribution in the base forms an adequate basis for the calculation of the emitter efficiency it cannot be used to evaluate the *emitter-to-collector transport efficiency*. This is defined by the ratio of the diffusion current at the collector to the diffusion current at the emitter and is usually denoted by $\beta$.

$$\beta = \frac{\left(\dfrac{dn_p}{dx}\right)_{x=W}}{\left(\dfrac{dn_p}{dx}\right)_{x=0}}. \tag{7.3.4}$$

Since a linear distribution of $n_p$ would yield a value of unity for $p$, the more accurate concave distribution of eqn (7.3.2) must be used. The constants of integration must now be evaluated from the boundary conditions and the final expression for the electron distribution substituted in the above definition.

After a certain amount of manipulation the following expression can be obtained

$$\beta = \frac{2(n_1 - n_0) + n_0(e^{W/L_n} + e^{-W/L_n})}{2n_0 + (n_1 - n_0)(e^{W/L_n} + e^{-W/L_n})}. \tag{7.3.5}$$

Since in an actual base layer $n_1 \gg n_0$ and $L_n \gg W$ the following approximations apply

$$\beta \cong \mathrm{sech}\left(\frac{W}{L_n}\right) \cong 1 - \tfrac{1}{2}\left(\frac{W}{L_n}\right)^2. \tag{7.3.6}$$

The reader will readily verify that if $W = 0\cdot3 \, L_n$ the base layer transport efficiency exceeds 90%. Conditions better than this can be realized in practical materials for which the diffusion length is a fraction of a millimetre.

It would follow from the foregoing results and the discussion at the end of the preceding section that the current flowing around the collector circuit can never exceed the emitter current. This is not strictly the case, due to the intervention of a charge multiplying process in the collector depletion layer. The electric fields present there when a reverse bias is applied may be intense enough to impart sufficient energy to the free electrons to enable them to excite additional electrons into the conduction band by the process of avalanche breakdown discussed in Section 5.8. This process of *collector multiplication* may increase the collector current substantially. In transistors using metal-to-semiconductor contacts the multiplication factor may be 2 or 3. In junction transistors the effect is much less pronounced but may still significantly affect their operation.

The above definition of the emitter-to-collector transport efficiency is only applicable under d.c. or quasi-static conditions, that is as long as the period of an a.c. signal applied to the emitter is much longer than the *transit time of the minority charge carriers* crossing the base layer. If the frequency of the applied signal is increased to a point when the period becomes comparable with the transit time the base transport efficiency decreases and a condition is reached in which no amplification can be obtained from the transistor. It can be shown by a theoretical analysis that this cutoff frequency is related to base parameters by the relation

$$f \cong \frac{D_n}{\pi W^2}.$$
(7.3.7)

As might be expected this shows that the frequency range of a transistor is increased by making the base layer thinner as well as modifying its diffusive properties. A great deal of development has been applied to this problem with the result that transistor devices are now as versatile as thermionic vacuum valves, their operation extending into the u.h.f. range.

A transistor parameter of great practical importance is the impedance presented to an a.c. signal source, inserted into the emitter circuit in addition to any forward bias dictated by operational requirements. As we are interested in variations of the emitter current in response to an *incremental* voltage the *emitter input impedance* is obtained by differentiation of the forward emitter voltage with respect to the current.

We start with eqn (5.5.6) which gives the current of a forward biased *p-n* junction under the conditions envisaged here. Substituting for the reverse saturation current from eqn (5.4.3) (which has been converted from current densities to total currents through multiplication by the junction area) we find

$$I = e \left( \frac{D_n}{L_n} n_n + \frac{D_p}{L_p} p_p \right) e^{-e\Psi_0/kT} e^{eV/kT}.$$
(7.3.8)

Remembering that the emitter current is of one species only the above equation simplifies to the following form for an electron emitter

$$I = e \frac{D_n}{L_n} n_n e^{-e\Psi_0/kT} e^{eV/kT}.$$
(7.3.9)

The proximity of the collector further modifies the current, on the basis of the argument which led up to eqn (7.3.3), by the factor $L_n/W$. Hence

$$I_e = e\,\frac{D_n}{W}\,n_n\,e^{-e\Psi_0/kT}\,e^{eV_e/kT} \tag{7.3.10}$$

where $I_e$ and $V_e$ denote actual d.c. emitter current and voltage.

After the above preliminaries which are intended to emphasize the differences between the forward current of an emitter junction and a general isolated $p\text{-}n$ junction, we proceed to evaluate the differential coefficient $dV_e/dI_e$. After some manipulation we find

$$r_e = \frac{dV_e}{dI_e} = \frac{kT}{e}\,\frac{1}{I_e} \text{ Ohms.} \tag{7.3.11}$$

This interesting relation between the emitter input impedance and the reciprocal of the emitter current is a direct consequence of the exponential relation between the current and the emitter voltage. It shows that the incremental impedance can be varied over a wide range through bias conditions.

It is useful to reflect at this point on the limitations of the current controlling function of the emitter. In the first place increasing the forward bias on the emitter will increase indefinitely both the emitter and collector currents. The operation of the transistor is limited in this direction by resistive heating, which will ultimately destroy the device. On the other hand device currents are reduced by decreasing the emitter bias but can never be completely shut off because the reverse biased collector will still pass its reverse saturation current (see Fig. 7.4). In applications of transistors to switching circuits this is significant and efforts are then made to design devices having a low *collector saturation current*. It must not be assumed that this current is identical with the reverse saturation current of an isolated $p\text{-}n$ junction. The proximity of the emitter junction has a reducing effect on it.

### 7.4 Current Gain of a Transistor

In the present section we shall deal with a parameter which is universally used to quantify the properties of transistors used in amplifying circuits. It is the *current gain*, usually denoted by $\alpha$, and defined as the ratio of the incremental current flowing at the collector terminal to the incremental current at the emitter terminal. In symbols

$$\alpha = \left|\frac{\Delta I_c}{\Delta I_e}\right|. \tag{7.4.1}$$

Since in most practical cases the incremental currents are small a.c. currents superimposed on d.c. bias currents, it is customary to denote them by lower-case letters. Thus

$$\alpha = \left|\frac{i_c}{i_e}\right|. \tag{7.4.2}$$

Our object is to express the current gain $\alpha$ in terms of the several current flows dealt with in preceding sections.

We refer to Fig. 7.7(a) which depicts an *n-p-n* transistor, like the one of Fig. 7.3, but with emphasis on those features which concern us here. The arrows show the actual direction of flow of each species of charge carrier, which means that electron flow is indicated on the metallic leads. The conventional current direction in the leads is therefore opposite to the arrows.

Having entered the emitter terminal the current $i_e$ is still carried by majority electrons in the *n*-type material until it reaches the vicinity of the emitter junction. Here a minute fraction of the current is taken over by injected holes, within one diffusion length of the barrier (see Fig. 7.2). The bulk of the current, given by $\gamma i_e$, is emitted into the base region as injected minority electrons. The minority electrons are transported through the base, with efficiency $\beta$, to the collector. Hence the electron current reaching the collector is given by $\beta \gamma i_e$. On passing the collector depletion layer the electron current undergoes multiplication (see p. 105) by some factor which we shall denote by $M$. Hence the electron current reaching the collector terminal has the magnitude $M\beta\gamma i_e$. As this is the current which leaves the collector we find

$$i_c = M\beta\gamma i_e = \alpha i_e.$$

$$\alpha = M\beta\gamma. \tag{7.4.3}$$

Meanwhile the holes produced as a result of collector multiplication enter the base region as a current of magnitude $(M-1)\,\beta\gamma i_e$ and become available for recombination with the excess minority electrons present. If $\alpha > 1$ the number of holes in the base is too great to be used up in the recombination process with minority electrons. Extra electrons, which must be supplied to neutralize these surplus holes, enter the base lead. The magnitude of the base current is then given by

$$i_b = (\alpha - 1)\,i_e. \tag{7.4.4}$$

If $\alpha < 1$ some electrons are removed from the base, the current still being given by eqn (7.4.4). It will be noted that the requirements of the external circuit are met by the currents marked on the individual branches.

The currents entering and leaving the transistor terminals are marked as electron flows in Fig. 7.7(a) to facilitate the foregoing physical explanations. For purposes of circuit applications the currents must be indicated in accordance with conventional directions which are opposite to electron flow but agree with hole flow. Figure 7.7(b) summarizes accepted conventions as applicable to both *n-p-n* and *p-n-p* transistors.

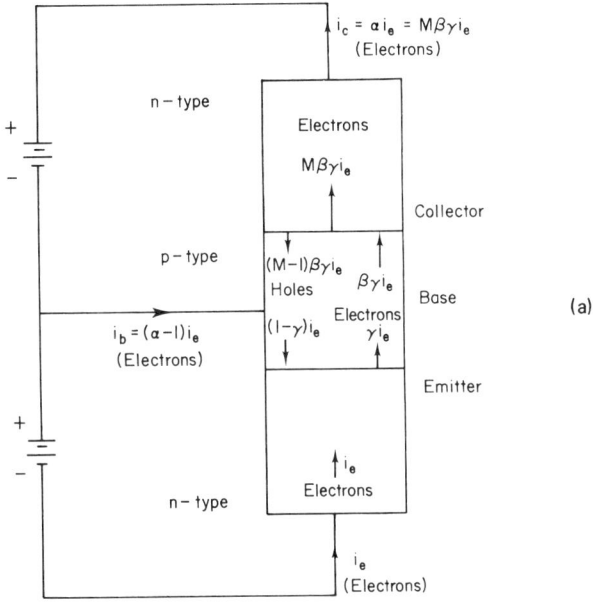

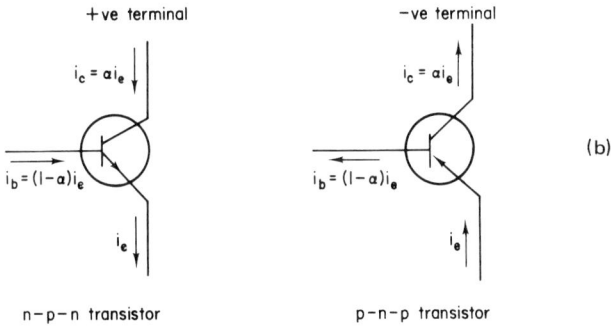

Fig. 7.7. (a) Current gain of an *n-p-n* transistor. (b) Circuit symbols and conventional current directions in transistors.

## 7.5 Alloyed Junctions

The *n-p-n* transistor discussed in preceding sections was tacitly assumed to be a single crystal device cut from a grown ingot of, say, germanium of silicon, the three regions suitably doped in the process of growing or by selective diffusions at a later stage. However, as explained in Section 6.5 metal-to-semiconductor contacts can fulfil the rectifying function of a *p-n* junction and consequently should be capable of acting as emitters and collectors in transistor devices.

At this point a distinction should be made between metal-to-semiconductor contacts of the type discussed in Section 6.5 and alloyed junctions. The latter are more complicated in structure and function in that they contain both a *p-n* junction and a metal-to-semiconductor contact. The position is best explained by reference to Fig. 7.8, which represents the end result of an alloying process,

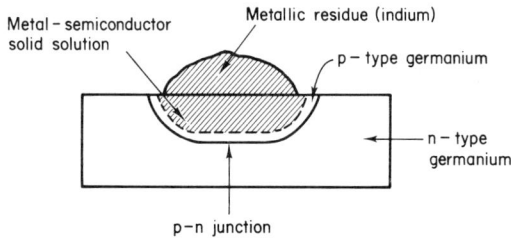

Fig. 7.8. Alloyed *p-n* junction.

which started with a blob of metal ($\cong 1$ mm dia.) placed on a slab of semiconductor and heated to a sufficient temperature for the metal to be melted. Perhaps the simplest example is provided by indium alloyed with *n*-type germanium. The region just below the metal fragment is a solid solution of the metal and semiconductor. At its lower periphery is a thin layer of semiconductor into which indium has diffused at the peak of the heating cycle as *p*-type impurity in sufficient numbers to render the semiconductor *p*-type (see Section 3.6). This *p-n* junction has been formed with an ohmic metal-to-semiconductor contact automatically included to make connections to outside circuitry.

The foregoing example was particularly simple because the indium served both as a *p*-type trace impurity and a metallic connection element. A somewhat more complicated situation arises when gold with a small alloy fraction of gallium is used. In this case gallium serves as the *p*-type impurity while gold provides the metallic connection. Another possibility is lead with an admixture of antimony. The latter is an *n*-type impurity to be used with *p*-type germanium, the lead serving as the metallic connection.

A junction of the type shown in Fig. 7.8 can, of course be made on both sides of the semiconductor slab. An *n-p-n* transistor is then obtained with the *p*-type semiconductor slab functioning as its base. This technique formed the basis of the earlier transistor manufacturing processes.

## 7.6 The Field Effect Transistor

The principle of transistor action as explained in foregoing sections was based on the controlling function of the emitter which regulated the amount of current reaching the collector. The charge emitting properties of the forward biased *p-n*

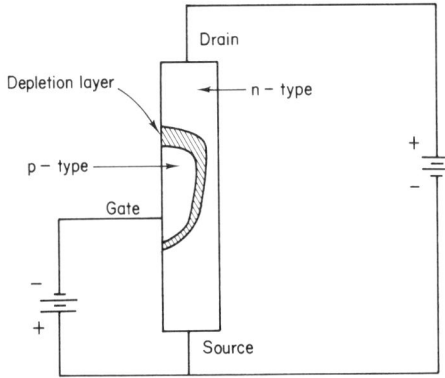

Fig. 7.9. Principle of the field effect transistor.

junction were thus crucial to the operation of the device. The current controlling action can be secured in other ways, without the use of an emitting junction. In the present section we describe one example of such possibilities called the *field effect transistor* or FET for short.

The device depicted diagrammatically in Fig. 7.9 consists of an *n*-type conduction channel with a *p*-type region encroaching on it partially. The conduction channel is relatively long and narrow, its width being comparable with the thickness of the depletion layer. The functioning of the device depends on the variation of the barrier thickness with the voltage applied to the *p-n* junction as explained on p. 72.

On application of a positive voltage to the *drain* terminal of the device a flow of electrons, supplied at the *source* terminal, is established. The main obstacle to this current is provided by the ohmic resistance of the constricted part of the *n*-type channel. The *p-n* junction is reverse biased by the negative voltage applied to the *gate* electrode. Hence it passes only a fixed current, the reverse saturation current, which is small compared with the main channel current. However, as the voltage applied to the gate is varied, the depletion layer alternately swells and shrinks, thus varying the thickness and hence the resistance of the constricted electron channel. As a result the drain current fluctuates and develops a replica of the gate voltage across a load resistor, if one is inserted in the drain circuit. Under suitable conditions amplification is obtained. Readers familiar with the properties of thermionic vacuum triodes will note that the field-effect transistor is very similar, the gate playing the part of the control grid.

A typical set of $I/V$ curves of a FET is shown in Fig. 7.10. Typical drain voltages are of the order of tens of volts while drain currents are tens of

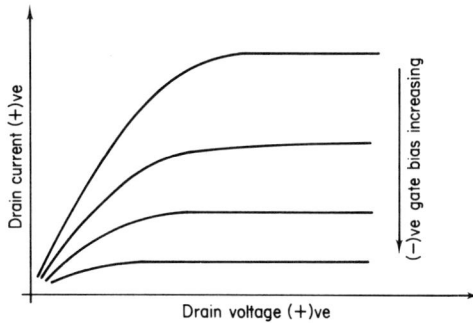

Fig. 7.10. $I/V$ characteristics of a field effect transistor.

milliamperes. A sufficiently high negative gate voltage may cause the swelling depletion layer to block or "pinch off" the conduction channel completely. The drain current is not completely cut off under these conditions, because the reverse saturation current will continue to flow from the gate.

The thickness of the depletion layer is shown to vary from point to point in Fig. 7.9. This is due to the fact that the $IR$ voltage drop along the main channel has to be progressively added to the gate voltage, resulting in a higher reverse bias on the p-n junction at the drain end of the device.

A similar field effect can be obtained without the use of a p-n junction. The gate voltage can be applied via a flat metallic electrode, separated from the main channel by a thin insulating layer. A negative gate voltage then sets up electric fields within the semiconductor channel which restrict the flow of electrons in n-type material, very much like the effect of a control grid in a thermionic triode. A device of this type can be called a *metal-insulator-semiconductor* triode or briefly MIS. The first examples of these transistors were made of silicon, the insulating layer being silicon oxide. Hence they were called *metal-oxide-silicon* devices or MOS devices for short.

One of the principal advantages of field effect transistors is the immunity of their $I/V$ characteristics to temperature variations over a wide range. This is because the current flow depends entirely on the concentration of majority carriers, which is largely insensitive to temperature changes in the extrinsic region. Against this the junction transistor relies for its operation on the injection and transport of minority carriers, processes which are much more temperature sensitive.

## Chapter 8

# Semiconductor Devices

## 8.1 Thermistors

As discussed in Section 3.9 the conductivity or resistivity of semiconductors depends largely on the availability of charge carriers, be they positive or negative. The charge carrier concentrations in turn are functions of temperature, varying rapidly under intrinsic conditions or in the range of thermal activation of impurities (see Fig. 3.13). This suggests the possibility of constructing resistors, whose resistance varies with temperature in a predictable manner. Such resistors are termed *thermistors*. The resistance of a thermistor may change either due to changes of ambient temperature, or due to temperature changes induced by power dissipation in the device itself. Thermistors are usually applied under the latter conditions and can therefore be used for switching or power-measuring purposes.

Thermistors are usually operated in the intrinsic range where the carrier concentration varies according to the exponential factor $\exp(-B/T)$ where $B$ is constant (see Chapter 3). Using this factor the resistance of a thermistor can be expressed in the form (see eqn (2.4.1))

$$R = R_0 \, e^{B(1/T - 1/T_0)} = V/I \tag{8.1.1}$$

where $R_0$ is the resistance at the ambient temperature $T_0$ (usually around room temperature$-300$ °K). The power dissipated in a thermistor can be expressed in the form

$$W = C(T - T_0) = VI. \tag{8.1.2}$$

Differentiating the above equations with respect to $I$ and setting the derivative equal to zero, values of temperature are obtained at which the $I/V$ characteristic of the thermistor has maximum or minimum values. After some manipulation a quadratic relation is obtained.

$$T_m{}^2 = B(T_m - T_0). \tag{8.1.3}$$

The solutions are

$$T_m = \frac{B}{2}(1 \pm \sqrt{1 - 4T_0/B}). \tag{8.1.4}$$

Figure 8.1 depicts diagrammatically the voltampere characteristic in the form in which it is usually presented for thermistors. The two values of $T_m$ correspond to the minimum and maximum of this curve respectively. It should be noted

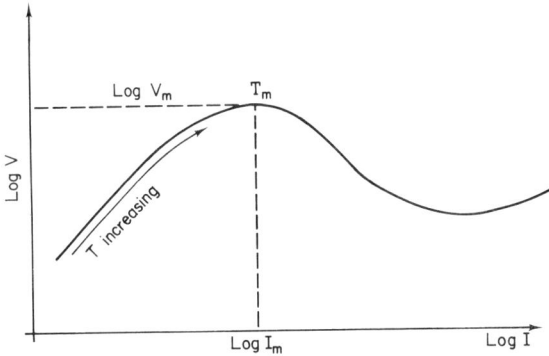

Fig. 8.1. Typical voltage/current curve of a thermistor.

that the current and voltage values of practical thermistors cover a vast range, depending on material parameters such as the energy gap and doping of the semiconductor used. Moreover, the curve of Fig. 8.1 need not have a maximum or minimum at all, a case which arises when the quantity under the square root of eqn (8.1.4) is negative. This happens when a semiconductor with a small band gap is operated at a high ambient temperature. However, the typical applications of thermistors are based on the characteristic of Fig. 8.1.

Thermistors can be used to compensate for the increasing resistance of a metallic conductor with temperature. The temperature coefficient of a thermistor is

$$\alpha = \frac{1}{R}\frac{dR}{dT} = -\frac{B}{T^2} \tag{8.1.5}$$

a negative quantity, consistent with the qualitative form of Fig. 3.14 in the intrinsic range. A carefully designed thermistor, connected in series with a metal, will yield a conduction path with a resistance independent of temperature over a wide range.

High-frequency power can be measured with the help of a thermistor mounted as the termination of a transmission line or waveguide and arranged to absorb either a known fraction or all of the incident energy. In the latter case a d.c. biasing current is applied and adjusted to match the thermistor resistance to the waveguide. A measure of the power is obtained by measuring the thermistor resistance in a bridge circuit and relating it to the temperature.

In the above application the thermistor is effectively a thermometer indicating a rise in its own temperature caused by the dissipation of energy in itself. A thermistor can, of course, be used as a conventional thermometer measuring changes in ambient temperature via resistance changes. The latter are monitored by passing through the thermistor a current small enough not to affect its resistance significantly, by comparison with the effect of the ambient temperature.

A thermistor responding to ambient temperatures can be used as a fire alarm. It is biased to operate well to the left of the maximum in Fig. 8.1 at normal temperatures. When a fire raises the ambient temperature sufficiently the operating point is shifted to the right of the voltage maximum. The consequent increase in current is designed to be sufficient to operate a relay which in turn actuates an alarm bell.

The thermistor characteristic has a range of negative incremental resistance to the right of the maximum. This can be utilized to obtain amplification. However, the usefulness of such amplifiers is limited by the slowness of thermistor response. This limitation is absent in a tunnel diode, to be discussed in a subsequent section.

## 8.2 Hall Effect Devices

The Hall effect discussed in Section 3.11 lends itself to some useful applications, two of which will be described briefly in the present section.

From eqn (3.11.4) it is apparent that the Hall field is directly proportional to the normal applied magnetic field $B_z$ provided the transport current $J_x$ is kept constant. Thus a thin slab of semiconductor can be used to measure magnetic fields. In this application the slab usually takes the form of a thin film deposited on an insulating non-magnetic substrate. Two pairs of ohmic contacts are attached to opposite edges of the slab as depicted diagrammatically in Fig. 8.2. One pair of terminals is used to apply a preset transport current to the slab while the other is connected to a sensitive voltmeter. The latter will indicate a voltage $V_y = W\mathscr{E}_y$, where $W$ is the width of the slab. $V_y$ is a direct measure of the magnetic field in which the slab is immersed. The ancillary circuitry is usually housed in a compact box with a meter calibrated to indicate the magnetic induction directly. A suitable control enables the value of the transport current density to be adjusted in steps, to secure the desired degree of sensitivity of the instrument.

Since the Lorentz force and hence the Hall voltage depends on the vector product of the field and current (see eqn (3.11.1)), the field value indicated by a Hall probe will vary with the relative orientation of the semiconducting slab and field, reaching a maximum when the field is normal to the plane of the slab. In practice this is easily ascertained by manipulation of the probe, thus yielding a unique value of the field.

A Hall probe can also be used as a simple multiplier to obtain the product of two quantities. In this application a small coil is wound around the plate of Fig. 8.2 in such a way that its field is represented by the vector $B$. One of the

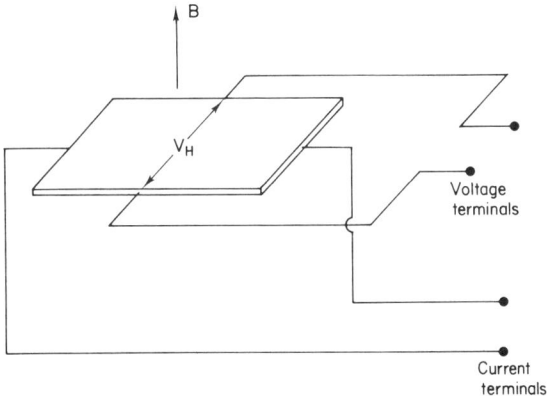

Fig. 8.2. Magnetic field measurement by Hall probe.

quantities to be multiplied is translated into a current applied to this coil. The transport current $J_x$ is arranged to be proportional to the other quantity. The resulting Hall voltage is proportional to the product of the two quantities according to eqn (3.11.4).

## 8.3 Tunnel Diodes

All the discussion of the *p-n* junction contained in earlier chapters was based on the assumption that the Fermi level is well within the energy gap of the semiconductors under consideration, and that the simplified forms of the Fermi-Dirac distribution (Section 3.2) are applicable. The term "non-degenerate" is sometimes used to classify semiconductors in this category.

In the present section we propose to describe a device whose properties and operation lie outside these limitations. The *tunnel diode* can function as it does because the Fermi level lies within the allowed bands: in the conduction band of *n*-type material and in the valence band of *p*-type material. Semiconductors having this property are sometimes referred to as "degenerate".

The positioning of the Fermi level within the allowed bands is achieved through heavy doping of the material. In fact, the concentration of impurities must exceed the numbers $N_c$ and $N_v$ of equivalent states in the conduction and valence bands respectively. Since situations of this kind were specifically excluded from the scope of the mathematical analysis of earlier chapters we can only formulate the present discussion in qualitative terms. This provides a sufficient basis for an understanding of tunnel diodes but is inadequate for analysis and design purposes.

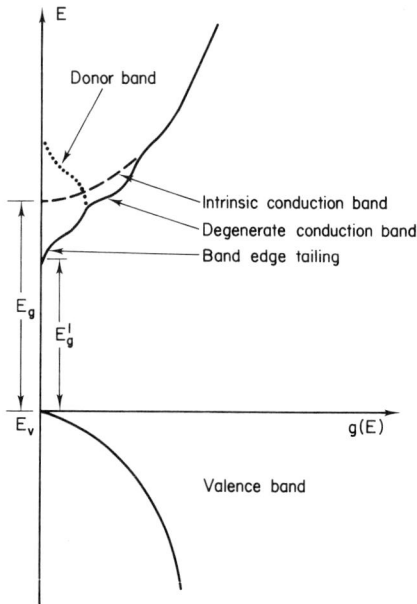

Fig. 8.3. Energy band structure of a heavily doped semiconductor.

As a first step we consider more closely the effects of heavy doping on the energy band structure of semiconductors. A high concentration means that impurity atoms are within reach of mutal interaction. According to the principles laid down in Chapter 1 this means that their states must spread over a band of energies. Thus in the case of $n$-type material there is no longer a single donor level $E_D$ but a band of donor states. If the doping is heavy enough this band may overlap with the original conduction band. The resulting band structure is shown diagrammatically in Fig. 8.3. Two features should be particularly noted. (a) The energy gap $E_g$ has been effectively reduced in width to the value $E_g'$. (b) A region of rather sparse density of states appears near the bottom of the modified conduction band, referred to as "band edge tailing". Similar effects of energy band modification appear at the top of the valence band of heavily doped $p$-type semiconductors.

Let us now consider a *p-n* junction sufficiently heavily doped to have the Fermi level situated within the valence and conduction bands respectively. The equilibrium configuration is depicted diagrammatically in Fig. 8.4(b). It should be noted that the energy gap configuration resembles that of Fig. 5.9, applicable to conditions of Zener breakdown in the presence of a reverse bias. As in that case, electron tunnelling can take place across the barrier, if empty states are

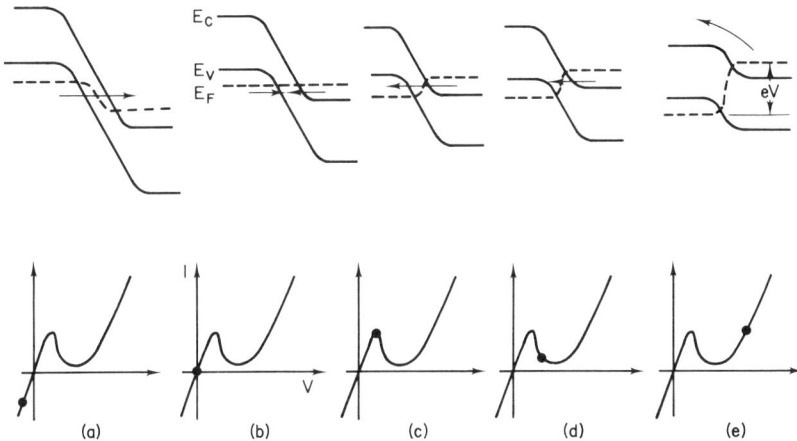

**Fig. 8.4.** Energy relations and current flow in a tunnel diode. The arrows indicate electron currents.

available on the other side. At temperatures above absolute zero there will be some empty states near the Fermi level on both sides of the junction. Hence some electrons will tunnel, but since the movement is in both directions no net current flow will be observed. This is indicated by the dot at the origin of the *I/V* coordinates of Fig. 8.4(b).

To simplify the discussion of current flow in the presence of applied voltages we assume that the junction temperature is $0\,^{\circ}$K. The application of a reverse bias is represented in Fig. 8.4(a). As there are numerous empty states on he *n*-type side of the junction, lying opposite occupied states on the *p*-type side, a tunnel current will flow. Since increased reverse bias exposes more and more states to tunnelling electrons, the current will increase steeply. The conditions are, in fact, entirely analogous to Zener breakdown discussed in Section 5.8. The reverse saturation current $I_s$ of both electrons and holes continues to flow across the junction, but it is swamped by the tunnel current. In any case its value is very small, the densities of intrinsic electrons and holes being negligible by virtue of the remote position of the Fermi level from the respective allowed bands.

The application of a forward bias exposes empty states in the valence band of p-type material to the numerous electrons resident in the conduction band of n-type material. An increasing tunnel current will flow from n-type to p-type, reaching a maximum value under the conditions of Fig. 8.4(c). This shows the empty states on the p-type side lying exactly opposite the occupied states on the n-type side. A further increase in the forward voltage restricts the number of empty states lying at the same level as occupied states and the tunnelling current drops, reaching a minimum value corresponding to the valley of Fig. 8.4. Over this range of applied voltages the tunnel diode displays *negative resistance*. A continued increase of voltage finally brings into play the normal forward current of the *p-n* junction composed of electrons thermally excited to states above the Fermi level of the n-type material. This happens only at temperatures above $0\,°K$ and follows the trend discussed in Chapter 5. There is, of course a corresponding current of holes flowing from p-type to n-type material and not shown in Fig. 8.4(e).

The precise shape of the tunnelling characteristic depends on the material parameters of the junction. In particular, the minimum "valley current" is greatly affected by the band edge tailing mentioned above. Without it the tunnel current drops to zero before the thermal current reaches a significant value.

The negative resistance characteristic of the tunnel diode is utilized in amplifier and oscillator circuits. In this connection it is important to note that the frequency range of such amplifiers is not limited by electron transit time across the barrier as discussed on p. 106. The tunnelling effect relies on what can be described as *simultaneous* presence of an electron on both sides of the insulating barrier. This being so no time delay is involved in crossing the barrier. Hence tunnel diodes can be used even at microwave frequencies.

## 8.4  Opto-Electronic Devices

Semiconductors are subject to various optical effects some of which have already been briefly described in Sections 3.10 and 5.7. In the present section we give an outline of the functioning of some devices based on these effects and others, to be explained as we go along. Devices which operate by virtue of an interaction between electromagnetic radiation and the device material are sometimes referred to collectively as *opto-electronic devices*. The present section does not, however, include any discussion of semiconductor lasers. This topic is reserved for Chapter 26 which deals exclusively with lasers.

The effects described in Sections 3.10 and 5.7 amount to the possibility that electrons can make transitions from one energy level to another under the influence of electromagnetic radiation. In that context only upward transitions were implied. These are associated with *absorption* of photons or quanta of the incident radiation. There is also the possibility that downward transitions may occur resulting in the *emission* of photons. Apart from the laser this is the case in *electroluminescent* devices which will be described later in this section. All these possibilities are usually referred to as *radiative transitions.*

Radiative processes in semiconductors are not confined to transitions across the forbidden gap, but may include transitions to levels within the gap. Before embarking on the desription of specific devices we summarize the various possibilities with the help Fig. 8.5.

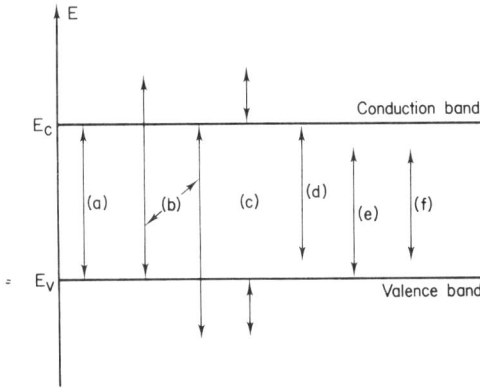

**Fig. 8.5.** Possible transitions in a semiconductor.

(i) *Interband transitions* are most probable when they take the form of recombination of electron-hole pairs occupying energy levels close to the band edges as suggested by arrow (a). Such transitions are frequently referred to as *intrinsic or edge emission.* Less likely are transitions of type (b) since they involve high energy or *hot electrons and holes.* They may take place when a semiconductor is on the verge of avalanche breakdown (see Section 5.8).

(ii) *Intraband transitions* do not involve recombination in the usual sense, only the transfer of electrons or holes from one level to another. They also require hot carriers (c).

(iii) Transitions involving crystal defects present the greatest diversity of possibilities. They include transitions between substitutional impurities such as (d) conduction band to acceptor, (e) donor to valence band and (f) donor to acceptor. In all these cases recombination of mobile and immobile charges is involved. Transitions of these types are sometimes called *extrinsic.* Apart from impurities other crystal defects may give rise to discrete energy levels situated within the forbidden gap. Such levels are collectively called *trapping levels* or *traps* for short. All traps can participate in transitions, the probabilities depending on the nature of the trapping centres and their density.

In addition to radiative processes the transitions of electrons from one energy level to another may take the form of an *Auger process*. In this the surplus energy is given up to free electrons or holes as kinetic energy.

Many transitions are only possible if certain momentum conservation laws are observed. Thus in an Auger process the free electron which absorbs energy must also change its momentum in such a way that the sum of its momentum and the momentum of the other electron is the same before and after the event. The momentum conservation laws have a generally restricting effect on transition probabilities and have a significant effect on the operation of devices.

A figure of merit frequently used to judge the effectiveness of radiative processes is the *quantum efficiency*. This is defined as the ratio of hole-electron pairs generated per incident photon. Only under optimum conditions does this quantity rise to 50%.

An opto-electronic device based on the absorption process described in Section 3.10 is called a *photoconductor*. It consists of a slab of semiconducting material provided with ohmic contacts and mounted for easy exposure to light as suggested in Fig. 8.6. The increase in conductivity of this sample when irradiated is arranged to enhance the flow of current in the external circuit. This in turn is displayed on a meter or recorded on a chart. Both intrinsic and extrinsic transitions may be used, the range of detectable light frequencies being correspondingly limited on the lower side by the minimum energy difference between levels involved in transitions. On the high frequency side sensitivity is limited by other effects, including diminishing densities of levels deep in the allowed bands. Some photoconductors are cooled to reduce the concentration of thermally excited carriers.

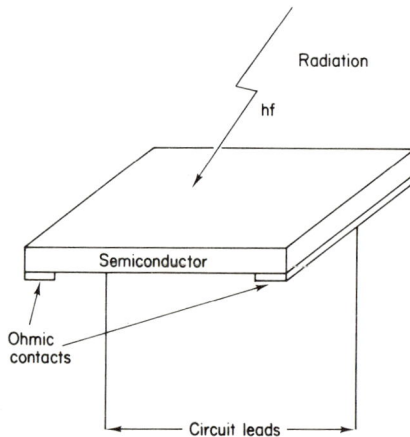

**Fig. 8.6.** Basic elements of a photoconductor.

The performance of photodetectors is measured by various figures of merit, the principal ones being the *photoconductive gain* and the *response time*. The gain is defined as follows

$$G = \frac{\Delta I}{eg} \qquad (8.4.1)$$

where $g$ is the number of hole-electron pairs generated by the radiation per second, and $\Delta I$ is the consequent increase in the current passing through the semiconductor. Photoconductors have a high gain according to this measure, $10^5$ being a typical figure. The response time of a photoconductor is clearly related to the transit time of charge carriers between the electrodes. In this respect the photoconductor compares unfavourably with other types of photodetector, yielding figures of the order of $10^{-3}$ sec. Against this the response time of *p-n* junction photodetectors, to be discussed next, is about $10^{-10}$ sec while their gain is 1. The wide variation of the figures of merit of different types of detector provides ample scope for choice in particular applications.

As mentioned in Section 5.7 a *p-n* junction can be used as a photodetector and several possibilities of detailed arrangement arise. Perhaps the simplest design uses a reverse biased diode, with only one side of the *p-n* junction exposed to the incident radiation. Thus the latter has the effect of modulating the reverse saturation current of, say, electrons by affecting the concentration of minority intrinsic electrons in the *p*-type material. The modulated electron current is then processed by external circuitry and displayed as information. It follows from the discussion of the *p-n* junction in Chapter 5 that only electrons produced by radiation within a diffusion length of the barrier can contribute towards the photodetecting function.

A refinement on the above principle is obtained if the optically active side of the junction is processed to have a non-uniform concentration of impurities or to be *graded*. According to the principle discussed in Section 4.6 such material has a built-in potential gradient which helps to sweep the minority carriers towards the junction. In this way an effectively greater volume of material can contribute towards the photocurrent.

A photodector which relies very much on the sweeping effect of a built in electric field is the *depletion layer photodiode*. As explained in Section 5.1 the depletion layer is a region of intense electric fields, in which free mobile charge carriers are absent except for those in transit. Production of electron-hole pairs by irradiation of the depletion layer results in a modulation of the reverse saturation current. To allow this the device is operated with a reverse bias which also serves to widen the depletion layer and thus to expose more material to the incident radiation. Another measure, taken to increase and otherwise tailor the depletion layer to requirements, is the insertion of an intrinsic layer between the *p*-type and *n*-type sections resulting in a *p-i-n photodiode*. The values of reverse bias applied to depletion layer photodiodes are well below avalanche breakdown conditions. Consequently there is no photoconductive gain but response times of the order of $10^{-11}$ sec are available.

The opposite conditions apply in *avalanche photodiodes*. These devices are biased to the verge of breakdown. Electron-hole pairs created by radiation are immediately accelerated by the barrier field to energies sufficient to cause impact ionization. The resulting avalanche contributes towards an internal current gain which constitutes photoconductive gain as defined in eqn (8.4.1). The fact that avalanche photodiodes have gains of the order of $10^4$ together with response times of $10^{-10}$ sec makes them very promising devices.

As an opto-electronic device the *solar cell* is assuming increasing importance with the growth of space exploration and satellite communications by virtue of the fact that it is the principal source of energy in space. Its mode of operation has been explained briefly in Section 5.7. Beyond this there is a host of technological problems which interested readers should follow up in specialist sources.

*Electroluminescence* is the reverse of the photovoltaic effect: optical radiation is obtained in return for an input of electrical energy. The electrical input is designed to generate a non-equilibrium, excess concentration of charge carriers, whose recombination is accompanied by the emission of radiation.

Several methods of electrical excitation exist. The *intrinsic* method is perhaps the simplest, amounting to the application of a sufficiently strong electric field to a volume of semiconductor to cause impact ionization by accelerated electrons. The electron-hole pairs thus created recombine via an interband radiative process (see Fig. 8.5). To obtain visible radiation a semiconductor with a relatively large gap must be used (ZnS; $Eg = 3 \cdot 6$ eV). The best results are obtained from powdered semiconductors embedded in a dielectric such as glass. The applied electric field can be obtained from the a.c. mains supply.

Basically the same excitation mechanism is used when a *p-n* junction is reverse biased into the avalanche breakdown condition. The electron-hole pairs generated by impact ionization recombine via interband or intraband radiative transitions. This method of exciting electroluminescence requires a d.c. voltage.

Tunnelling across suitable *p-n* junctions or metal-to-semiconductor contacts may also give rise to electroluminescence. Electrons tunnelling from, say, the valence band of the *p*-type material may recombine in the *n*-type material via a radiative process.

Perhaps the most important method of exciting electroluminescence used at present is by *injection of minority* carriers under forward bias conditions in a *p-n* junction. As will be apparent from the discussion of Chapter 5 there is a substantial excess minority carrier concentration within a diffusion length of the barrier when a forward current is flowing across the junction. The excess carriers are removed by recombination. It is beyond the scope of the present discussion to go into the question why in some semiconductors the recombination process is more likely to be radiative than in others. Suffice it to say that gallium arsenide (GaAs) is in this category. This fact, coupled with the highly advanced technology of GaAs, has resulted in the development of efficient lamps having, if required, a directional radiation pattern. On a frequency scale the radiation is concentrated just below the visible range, as is to be expected from the GaAs band gap. The radiation peak rises in frequency as the temperature of the device is lowered, as expected from the variation of the gap width with temperature. The efficiency of GaAs lamps varies widely with temperature reaching values of 40% at 20 °K and dropping to less than 10% as room temperature is approached. The response time is of the order of $10^{-9}$ sec by virtue of the very short lifetime of minority carriers in GaAs.

## 8.5 Semiconductor Controlled Rectifiers

The use of *p-n* junctions and metal-to-semiconductor contacts as rectifiers will not be discussed in the present book beyond the statement of the physical principles of operation contained in Chapters 5 and 6. The characteristics of devices and their use in specific applications are best covered in texts dealing with circuits, and interested readers are referred to such sources.

Apart from the single junction there are four layer *p-n-p-n devices* which can be used as rectifiers with a built in control facility. Although *p-n-p-n* devices are currently used to fulfil a multitude of functions, the description of the present section is intended to provide a background for their application as *semiconductor controlled rectifiers.*

A typical *p-n-p-n* device is depicted diagrammatically in Fig. 8.7. The terminal markings of *A* and *K* stand for "anode" and "cathode" to indicate that the application of a positive voltage to the anode is equivalent to a "forward" bias. The upper part of the figure shows a typical doping profile while the lower part gives an energy level diagram applicable under equilibrium conditions. The doping contrast between the inner and outer sections of the device should be particularly noted.

The application of a forward voltage to the *p-n-p-n* junction is represented by the diagrams of Fig. 8.8. The effect of a positive voltage at the anode is to push in holes towards the first junction *J*1. The resulting electrical imbalance of the barrier exerts an attractive force on majority electrons on the other side. The net result is a narrowing of the depletion layer which is symptomatic of the presence of a forward bias on junction *J*1. At the same time the attraction of electrons away from junction *J*2 has a widening effect on the depletion layer. This indicates that there is a reverse bias on *J*2. By an analogous argument the application of a negative voltage to the cathode is equivalent to a forward bias on junction *J*3. It also has a reinforcing effect on the reverse bias across *J*2. The

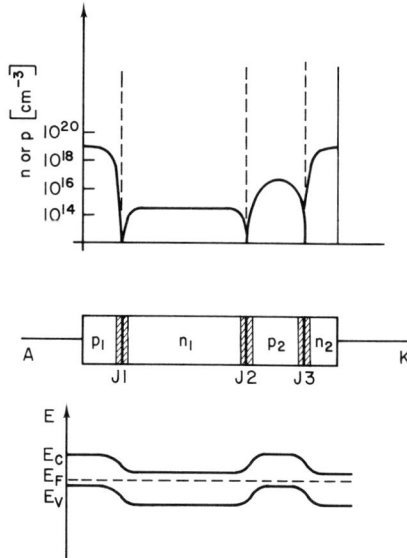

Fig. 8.7. A *p-n-p-n* junction in equilibrium and its doping profile.

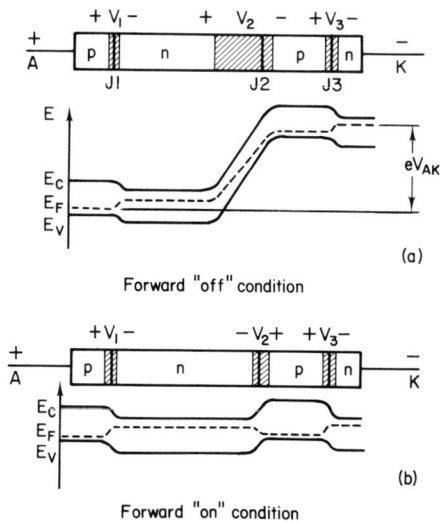

Forward "off" condition

(a)

Forward "on" condition

Fig. 8.8. Forward voltage on a *p-n-p-n* junction.

energy level diagram of the *p-n-p-n* junction in this condition is shown in Fig. 8.8(a). It should be noted that the middle junction acts as a collector of holes from *J*1 and of electrons from *J*3. The reverse bias on the middle junction, together with the designed properties of the device ensure that the current flowing under these conditions remains small. This corresponds to the segment $O - V_s$ of the $I/V$ characteristic of Fig. 8.9, which is referred to as the

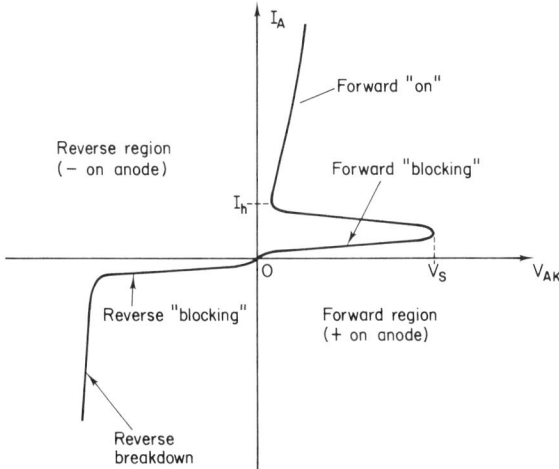

**Fig. 8.9.** $I/V$ characteristic of a *p-n-p-n* junction.

"forward blocking" or "off" state, $V_s$ being the "switching" voltage. The "off" state is dominated by the high impedance of the reverse biased middle junction which takes up most of the applied voltage. The latter is a sum of the three voltage drops as indicated in Fig. 8.8(a)

$$V_{ak} = V_1 + V_2 + V_3. \tag{8.5.1}$$

Increasing the applied voltage does increase the current slightly, by virtue of the increased emission of holes and electrons by the emitter junctions. The increasing rate of flow of electrons originating in *J*3 tends to compensate the depletion layer of *J*2 for the loss of electrons attracted to *J*1. At a crucial point the flow is sufficient to cause a slight contraction of this depletion layer. The result is a slightly diminished loss of electrons to *J*1 which in turn emphasizes the narrowing of the depletion layer. These mutually reinforcing tendencies swiftly convert the reverse biased middle junction into a forward biased junction as suggested in Fig. 8.8(b). The current flowing through the device rises suddenly, while the voltage decreases assuming the value

$$V_{ak} = V_1 - |V_2| + V_3. \qquad (8.5.1)$$

This is roughly equivalent to the voltage drop across a single *p-n* junction. The catastrophic switching change into a new forward "on" condition is clearly the result of an internal positive feed-back process built into the device. Although the process has been explained in terms of electrons emitted by *J*3, the holes emitted by *J*1 clearly contribute towards the switching action in an analogous way. In an actual experiment the current must be limited by external circuitry, otherwise the *p-n-p-n* junction will be destroyed. Using a suitable arrangement the "on" part of the forward *I/V* characteristic above the "holding current" $I_h$ can be plotted.

The application of a negative voltage to the anode causes the outer junctions to become reverse biased. The middle junction is forward biased, and since it allows a free passage for both types of charge carriers along the central region of the device, it has little effect on its operation in this condition. As there are effectively two reverse biased *p-n* junctions connected in series, the *I/V* relationship is that of a *p-n* junction with a reverse voltage. As shown in Fig. 8.9 the current is low and the device is in the "reverse blocking" condition. In a device having the doping profile of Fig. 8.7 most of the applied voltage is dropped across junction *J*1. It is also this junction which is instrumental in precipitating the "reverse breakdown" condition and a sudden increase in current, shown in Fig. 8.9.

To summarize the *p-n-p-n* junction has a reverse *I/V* characteristic resembling that of a single *p-n* junction. However, in the forward direction it has the properties of a bistable device capable of operating either in a low current or high current state. The changeover from one condition to the other is accomplished through regenerative feedback inherent in the structure of the device. Details of the switching action, which have been explained above in physical terms, can also be analysed in terms of an equivalent circuit of two transistors. Interested readers are referred to specialist sources for particulars of this approach. In the present context we shall continue to rely on the physical argument to show that the *p-n-p-n* device can be used as a controlled rectifier.

The key to this application of the device lies in the provision of a third electrode, the *gate* attached to the segment $p_2$ via an ohmic contact, as shown diagrammatically in Fig. 8.10. When a positive voltage is applied to the gate, while the device is in the forward "blocking" condition, a current of holes is forced into segment $p_2$. These extra holes are attracted to the reverse biased junction *J*2 and tend to diminish the width of the depletion layer. In this way the gate current tends to precipitate the transition to the forward "on" condition which now takes place at a lower value of the forward voltage. A

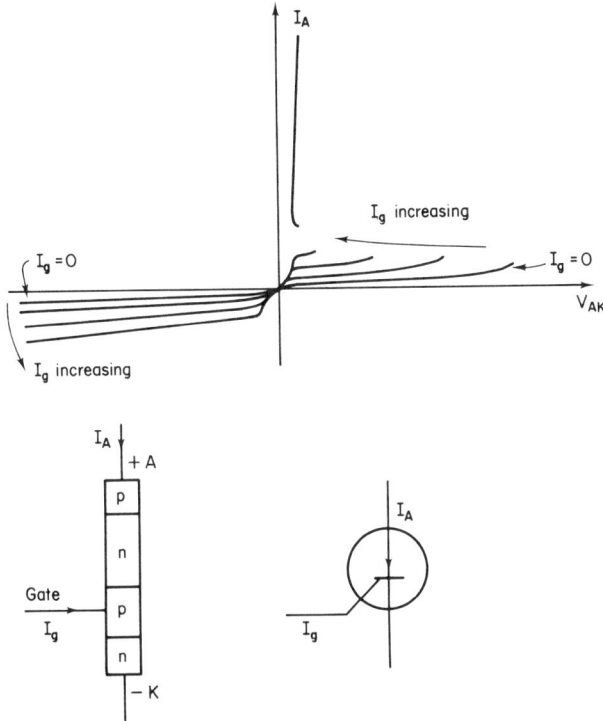

**Fig. 8.10.** Gate control of a semiconductor controlled rectifier.

family of $I/V$ curves is shown in Fig. 8.10, with the gate current $I_g$ as parameter. From this it is seen that the forward switching voltage is drastically reduced, while the reverse current is increased but still remains small on the scale of the forward current.

A *p-n-p-n* junction with a gate is usually referred to as a *thyristor. Controlled rectification* is obtained with the help of a thyristor by applying to the gate a fraction of the a.c. voltage to be rectified, through an adjustable phase shifting network. By adjusting the phase of the gate signal, the main cathode-to-anode path is switched into the "on" condition for any desired fraction of the a.c. cycle. The average value of the rectified d.c. current can thus be controlled as required.

*Chapter 9*

# Electrical Properties of Gases

The conduction of electricity by gases takes place through the medium of charged particles as it does in solids. However, by virtue of the different structure of a gas, or lack of it, the laws of conduction are quite different. To emphasize some of the differences from the outset let it be mentioned that the structure of a gas has no periodic properties and the motion of gas molecules is not confined to vibrations about an equilibrium point as is the case with lattice ions. Moreover, there is a greater variety of charge carriers, which includes ionized atoms and molecules in addition to free electrons. In what follows we survey various processes whereby charged particles can be created and annihilated and then derive some of the basic laws of conduction of electricity through gases.

Any treatment of the properties of gases must draw inevitably on the background of the kinetic theory. Although few specific mathematical results are invoked in the present chapter the reader is assumed to be familiar with basic concepts such as the mean free path.

### 9.1  Collisions in a Gas

The principal agency whereby charge carriers are created in a gas are collisions between the molecules, atoms and electrons which make up a volume of gas. Other agencies are present in many situations, e.g., thermionic or photoelectric emission, or ionization by energetic particles such as x-rays. These phenomena will be touched on in later sections as the need arises, but we open with a discussion of collision processes, as causes of ionization.

As in a solid, ionization takes place as a result of energy transfers which must be related to the energy levels of the ionized particles. In this connection we note that a volume of gas possesses additional features not found in individual atoms or solids. These are due to the additional dynamical degrees of freedom possessed by molecules.

130

(a) A molecule can rotate about its axis in space and thus may have up to three additional degrees of freedom depending on its symmetry.

(b) The constituent atoms may vibrate about the centre of mass of the molecule. The number of extra degrees of freedom will depend on the molecular structure, being one for the simplest diatomic molecules.

Energies associated with motion within these extra degrees of freedom are quantized into discrete levels just like the levels of individual atoms. Such states can be investigated theoretically by methods of quantum mechanics, or experimentally by methods of microwave spectroscopy, to be described in later chapters.

Collisions are classified into two distinct types according to the nature of energy exchanges which occur between the colliding particles.

## (i) *Elastic collisions*

In this kind of collision only *translational kinetic energy may be exchanged* between the particles, while the sum of the kinetic energies of the particles remains the same before and after the collision. This implies that the internal energy of the colliding atoms or molecules remains unchanged, that is the structure of their energy levels remains unaffected. Although the energy of individual particles may be conserved in an elastic collision, a change in their (vectorial) momenta may appear as a symptom of a collision.

Although elastic collisions cannot by definition affect the electrical charge of an atom or molecule, they influence energy interchanges between the constituents of a gas.

## (ii) *Inelastic collisions*

In this case the energy transfer will include *changes of the internal energies* of one or both colliding particles, as well as of the translational kinetic energies. The internal configuration of the particles may be modified as a result of excitation to a higher quantum state or as a result of ionization.

Let us consider more closely some examples of the internal changes which may be caused by this type of collision. One or both of a pair of colliding atoms or molecules may be excited to a higher energy state or even ionized. As a result the sum of the translational kinetic energy of the particles will be less after the event. A collision between an excited atom and, say, an electron may cause the atom to drop to a lower energy state. The internal energy thus released may go to ·increase the translational kinetic energy of one or both of the colliding particles after the event. Alternatively the energy may be emitted in the form of electromagnetic radiation.

The foregoing description of collisions implies that the particles were in some kind of intimate short-range contact for a short though finite duration of time. However, there are situations where the motion of particles is modified at long range as a result of mutual interaction but without obvious contact. For example in a highly ionized gas at low pressure (long mean free path) the electrostatic forces between particles may modify their paths at relatively great distances. To cover events of this kind by the concept of collisions a different definition is required. However, since such events have no bearing on ionization processes in gases, the matter will not be pursued any further in the present context.

### 9.2 Collision Probability and Cross-section

It will be shown at length in Part II of the present book that results of atomic scale events can only be predicted in terms of probabilities and averages. Thus the outcome of collisions between particles of a gas, must be formulated in terms of probabilities. From a practical point of view this is not really a limitation since one is concerned with quantities of gas containing very large numbers of atoms and electrons and only their average properties are of consequence.

We start with a definition of the probability $P$ that a particle will be involved in a collision. An intelligible measure of this quantity will be obtained by noting that it must be proportional to the number of collisions the particle undergoes in unit distance of its path. We therefore adopt as the definition of *the collision probability the number of collisions per unit length of path.*

$$P = \text{number of collisions per unit length of path}$$

$$P = \frac{1}{\text{mean free path}} = \frac{1}{\lambda}. \qquad (9.2.1)$$

The collision probability is readily evaluated for an idealized gas consisting of hard spheres of equal diameter. Figure 9.1 depicts the path of such a sphere. Moving along the dotted line the sphere will collide with all other spheres whose centres lie within a cylindrical volume of diameter $2d$. Hence, if the number of particles per unit volume is $n$, the number of collisions will be

$$P = n\pi d^2. \qquad (9.2.2)$$

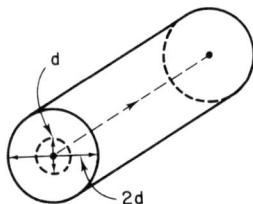

Fig. 9.1. Collision probability within a cylindrical volume of gas.

This result is consistent with the value of the mean free path in this type of gas, which is shown in any text dealing with the kinetic theory to be $\lambda = 1/(n\pi d^2)$.

The above calculation might appear to be remote from reality as it assumes that all molecules within the cylindrical volume of Fig. 9.1 are stationary and remain undisturbed by the moving molecule which itself continues along a straight line. The kinetic theory makes it possible to allow for the random motion of individual molecules and predicts the mean free path to have the value $\lambda = 1/(\sqrt{2}n\pi d^2)$. As can be seen the correction is of a relatively minor nature and for many practical purposes eqn (9.2.2) provides a perfectly adequate estimate. It may be noted in passing that allowance can also be made for the non-spherical shape of individual molecules and the random distribution of their velocities. In each case the resulting correction of the mean free path is found to be equally insignificant.

Although the above definition provides a sensible measure of the collision probability within a volume of gas at a given pressure, it is desirable to go beyond it and seek a quantity which would characterize the collision properties of an individual particle, regardless of the pressure. Noting that a particle is more likely to be involved in a collision if its cross-sectional area is greater, we are led to drop the particle density $n$ from eqn (9.2.2) and retain only the quantity

$$q = \pi d^2 \qquad (9.2.3)$$

which is the cross-sectional area of a spherical particle. In general terms the *effective cross-section for collision*, $q$, of a particle is defined in terms of the collision probability by the relation

$$P = nq; \quad q = \frac{P}{n} = \frac{1}{n\lambda}. \qquad (9.2.4)$$

In other words the effective cross-section for collisions is defined as the area by which the particle density must be multiplied to obtain the collision probability.

The above definition of the collision cross-section is only applicable to events involving identical particles, be they hard spheres or otherwise. To treat collisions between particles of different types it is necessary to make distinctions between the target particle and the other particle involved in the collision and between possible types of collision, since the collision cross-sections for, say, elastic or inelastic collisions will, in general, differ. Thus for particles of types 1 and 2 involved in a collision of type $x$ the effective collision cross-section is defined as follows

$$q_{2x} = \frac{1}{n_2 \lambda_{1x}} \qquad (9.2.5)$$

where $x$ = type of collision: excitation, ionization, elastic etc.

     2 = target particle

     1 = other particle involved in the collision

     $\lambda_{1x}$ = mean free path of type 1 particle between collisions of type $x$.

Collisions between electrons and other, heavier particles provide a good example of the above definition. In this case the other particle is always treated as a target of diameter $d$. It is now established that the diameter, as well as mass, of an electron is vanishingly small in collision events. Hence by eqn (9.2.2): the collision probability is $p = n_2 \pi d^2/4$ where $n_2$ is the density of target particles. The effective cross-section for collision of type $x$ between the electron and target particle is $q_{2x} = \pi d^2/4$.

Equation (9.2.5) lends itself to further extension to include the possibility of collisions of various types with particles of different kinds. However, it is unnecessary to pursue the matter further in the present context and readers interested in this topic are referred to specialist texts dealing with ionized gases.

Quantitative data regarding collision cross-sections can be obtained theoretically by methods of quantum mechanics. However, since the necessary calculations are complicated and lengthy, the bulk of available data has been secured by experimental measurements of collision probabilities. As the latter depend on the density $n$, it is customary to refer everything to the standard conditions of 1 mm $H_g$, 0 °C and $n = 3 \cdot 56 \times 10^{16}$ cm$^{-3}$.

In the context of the electrical properties of gases the most important are both elastic and inelastic collisions between electrons and neutral atoms or ions. The elastic collisions have probabilities of the order of 100 cm$^2$/cm$^3$, while inelastic collisions are about 10 times less probable.

### 9.3 Energy Transfer in Elastic Collisions

Since elastic collisions are relatively probable compared with others it is desirable to have some idea regarding energy transfer between various particles of a gas involved in such collisions. As a starting point we consider the model of two hard spheres one of which, the target, is assumed to be stationary, while the other moves along a line which does not coincide with the line of centres of the spheres. The arrangement is illustrated in Fig. 9.2.

The problem of evaluating the fraction of kinetic energy $\Delta E$ transferred to the stationary target particle is treated by classical dynamics and the result is

$$\Delta E = E_1 \frac{4 m_1 m_2}{(m_1 + m_2)^2} \cos^2 \theta \qquad (9.3.1)$$

where: $m_2$ = mass of stationary target particle

        $m_1$ = mass of moving particle

$E_1$ = kinetic energy of moving particle

$\theta$ = angle between velocity vector and line of centres.

As we are only interested in averages over a large number of collisions we must evaluate the average value of the above fractional energy transfer over the range from 0 to $\pi/2$ of the angle of incidence $\theta$. This means finding the average value of $\cos^2 \theta$ with respect to the probability distribution of collisions over $0 < \theta < \pi/2$.

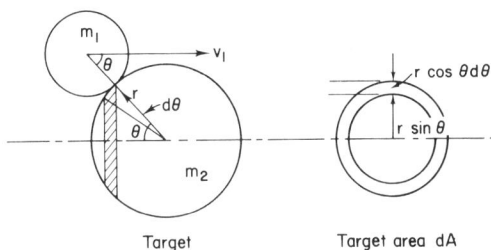

Fig. 9.2. Energy transfer in elastic collisions.

To find the latter we note that the probability of collision at an angle between $\theta$ and $\theta + d\theta$ is proportional to the area of the zone surface within these angles, projected on a plane normal to the velocity vector of the impinging particle. This is indicated by the ring on the right-hand side of Fig. 9.2, and denoted by $dA$

$$dA = \pi r^2 \sin 2\theta \, d\theta. \qquad (9.3.2)$$

Thus the desired probability distribution, normalized to unity over the range $0 < \theta < \pi/2$, is given by $2 \sin 2\theta$.

The average value of $\cos^2 \theta$ is now readily evaluated.

$$\langle \cos^2 \theta \rangle = \frac{2}{\pi} \int_0^{\pi/2} \cos^2 \theta \sin 2\theta \, d\theta = \tfrac{1}{2}. \qquad (9.3.3)$$

Substitution of this result into eqn (9.3.1) yields the average energy transfer in elastic collisions

$$\langle \Delta E \rangle = \frac{2m_1 m_2}{(m_1 + m_2)^2} = \tfrac{1}{2}(\Delta E)_{\max} \qquad (9.3.4)$$

where $(\Delta E)_{\max} = (4m_1 m_2)/(m_1 + m_2)^2$ is the maximum energy transfer possible in a head-on collision.

A special case of practical importance is provided by electrons incident on atoms. As their kinetic energies are generally comparable, the electrons are very much faster by virtue of their small mass, and it is fair to assume that the target particle is stationary. Hence for $m_1 \ll m_2$

$$\langle \Delta E \rangle \cong \frac{2m_1}{m_2} \ll 1.$$

Thus electrons loose very little energy in elastic collisions with atoms.

## 9.4 Inelastic Collisions

Although inelastic collisions in a gas are much less probable than elastic collisions, they are nevertheless of greater practical relevance to the electrical properties of gases, since they are the principal means of ionization. In the present section we classify and briefly describe the principal types of inelastic collisions.

At the outset we state the *principle of detailed balancing:* this postulates that (a) every collision is reversible and (b) in a volume of gas in equilibrium each type of collision and its reverse occur at the same rate. For if this were not the case the state of the gas would change with time, contradicting the assumption that it is in equilibrium.

One way of categorizing inelastic collisions is according to which of the two particles involved gains or gives up internal energy.

### (i) *Inelastic collisions of the first kind*

The impinging particle, usually a light and fast electron, looses kinetic energy which goes to increase the internal energy of the target particle, usually a slow and heavy atom or molecule. The increased internal energy appears in he form of excitation or ionization.

### (ii) *Inelastic collisions of the second kind*

The impinging particle gains kinetic energy at the expense of the internal energy of the target particle which reverts to a lower energy state. This type of collision can only happen if the target particle is above its lowest possible energy state, called the *ground state.*

Inelastic collisions can also be subdivided into two categories according to whether they provide energy for ionization or not. Several distinct processes can be listed under each heading.

### 1. *Types of Ionizing Collisions*

(a) *Thermal ionization* takes place in a gas sufficiently hot for some of the particles to have enough kinetic energy to cause ionization on impact. This process requires temperatures well above normal.

(b) *Cumulative ionization* may result from successive collisions each causing excitation to a higher energy level in relatively low energy steps.

(c) *Autoionization* happens in an excited atom when one electron drops back to the ground state while another is ejected. This process may be induced by a collision, but since much if not all ionizing energy is provided internally within the target particle, the impinging particle may not have to give up any kinetic energy.

2. *Other Types of Collision Processes*

(a) *Charge transfer* sometimes takes place between colliding atoms and/or ions resulting in the exchange of an electron.

(b) A *negative* ion is formed as a result of *attachment* following a collision between an electron and neutral atom. It is instructive to observe that the ionization of acceptor impurities in semiconductors is based on this process.

(c) Some collisions between electrons and positive ions, or negative and positive ions, result in *recombination.* In these processes both kinetic energy and ionization energy must be given up in the form of radiation or to third bodies. Recombination of electrons and holes in semiconductors comes under this heading, the holes being the positive ions.

(d) *Penning effect* is the name given to a process which involves the transfer of excitation energy from one neutral atom to another. This may result in ionization, especially if one of the atoms is in a metastable excited state, that is one which has a long lifetime—of the order of $10^{-3}$ sec.

## 9.5  Interaction Between Gas Particles and Electro-magnetic Radiation

It is not necessary for gas atoms or molecules to be involved in collisions in order to change their electrical and energetic state. Excitation to a higher energy state or indeed ionization and vice versa may be caused by electromagnetic radiation or photons. Such processes can take place in any volume of gas and are of crucial importance to the operation of lasers. Several types of processes can be distinguished.

### (i) *Photoexcitation*

Photons of the right wavelength or frequency can be absorbed by atoms which become excited. This process can be used as a source of "pump" energy in lasers.

### (ii) *Photoionization*

The absorption of photons of sufficient energy, may cause ionization. The quanta of radiation required for this purpose are in the x-ray or higher range. This process allows fluorescent lighting to function. The presence of a few atoms ionized by cosmic rays in a tube enables a glow discharge to be started on application of a voltage to the electrodes. This matter will be dealt with at greater length in subsequent sections.

The above two processes involve the *absorption* of radiation by atoms. *Emission* can also take place through the medium of the following processes.

### (iii) *Stimulated emission*

Photons incident on excited atoms may *stimulate* transitions to lower energy levels. The energy freed by the transition is emitted as radiation of the same frequency as the stimulating photons. The probability of such transitions is proportional to the intensity of the stimulating radiation. This process forms the basis of maser and laser devices and will be discussed more thoroughly in Part III of the present book.

### (iv) *Spontaneous emission*

Excited atoms or ions can revert *spontaneously* to a lower energy level. The energy balance is emitted in the form of electromagnetic radiation. This tendency to drop back to the ground state is always present and it means that the lifetime of excited states is strictly limited. Typical lifetimes are of the order of $10^{-8}$ sec. Some states, called *metastable*, can last up to $10^{-3}$ sec. They play a crucial part in the operation of gas lasers.

## 9.6 The Function of Electrodes in Gas Discharges

To utilize the electrical properties of gases in circuit applications it is necessary to connect a volume of gas, contained within a suitable envelope, via metallic contacts called *electrodes*, to external apparatus. The function of electrodes is best viewed in terms of sources and sinks of charge carriers, specially inserted into a gas to maintain and control desired electrical conditions. The charge carriers originating at electrodes are almost exclusively electrons. They are liberated by the mechanisms of electron emission discussed in Chapter 6.

### (i) *Thermionic emission*

Thermionic emission may be introduced deliberately by the provision of heated cathodes which may be oxide coated. This is done in gas-filled rectifying diodes and thyratrons. On the other hand an electrode may be heated by positive ion bombardment to a temperature at which it emits significant numbers of thermionic electrons. These electrons then enter the gas and modify its electrical properties by their presence. The arc discharge, to be discussed in a later section, is caused by thermionic emission of this type.

(ii) *Photoelectric emission*

Photoelectric emission plays a similar part to thermionic emission in that it is applied by design in some devices while it appears as a parasitic phenomenon, albeit a useful one, in others. Thus the start of the glow discharge in a fluorescent tube can be attributed in part to the presence of some electrons photoemitted through the agency of the odd cosmic ray present in he atmosphere.

(iii) *Secondary emission*

Secondary emission phenomena are present in most gas-filled devices to a greater or lesser extent. This may involve electron emission by a positive electrode due to *electron impact.* Alternatively additional electrons may be emitted by the cathode due to *positive ion impact.* Finally *neutral atoms impinging* on any electrode may cause a few secondary electrons to be emitted. However, the last two processes have low probability.

(iv) *Field emission*

Field emission takes place where very intense electric fields are present at the surface of the cathode. This mechanism is responsible for the emission of high density currents from the liquid mercury pool of mercury arc rectifiers. As explained in Section 6.6 field emission is an example of electron tunnelling which will be discussed in Part II of the present book.

　　Sometimes a cathode fulfils a function quite distinct from a source of electrical charge and become a source of the material of which it is made, in atomized form. Under suitable conditions a signficant amount of material is dislodged from the cathode in the form of individual atoms, by positive ion bombardment. The metallic atoms thus liberated mix with the gaseous atmosphere by diffusion and deposit themselves on neighbouring cold objects in the form of a thin layer or film. The process is referred to as *sputtering* and is used as a method of preparing thin films. The metallic ions mixing with the gas may modify its composition by first reacting with some component of the gas and then depositing themselves as films of the resulting chemical compound. A desirable outcome of this operation may be either the film itself or the resulting purification of the gas mixture. Thus niobium and tantalum may be used to remove traces of oxygen because they have a strong chemical affinity to this element. A purifying process of this type is frequently referred to as *gettering.*

## 9.7  Basic Concepts of Charge Transport in Gases

The application of an electric field to a volume of ionized gas will cause a flow of current by virtue of the forces acting on individual charges. The mobility of charge carriers in a gas is defined by the same relation as was formulated for semiconductors in Section 2.4. Thus we can write

$$\mathbf{J} = e(n\mu_n + p\mu_p)\mathscr{E} \tag{9.7.1}$$

where $p$ represents the density of positive ions which are mobile in a gas. This is an important distinction between ionized gases and semiconductors. In the latter the ions are themselves immobile, the transport of positive charge being effected through the medium of electrons drifting from one ion to another, thus giving rise to the concept of a moving hole.

Since electrons are very much more agile than heavy ions, we conclude that their mobility is much greater: $\mu_n \gg \mu_p$. Hence in lightly ionized gases, in which ions and electrons are evenly distributed and $n = p$, eqn (9.7.1) assumes the form

$$\mathbf{J} = en\mu_n\mathscr{E} \tag{9.7.2}$$

indicating that the current is carried entirely by electrons.

Under other conditions there may be regions of appreciable net space charge. Hence $n \neq p$ and eqn (9.7.2) does not necessarily hold. Situations of this kind will be described in a later section in connection with the glow discharge.

Excess concentrations of electrical charge may exist locally in gases as they do in semiconductors. These give rise to diffusion currents entirely analogous to those discussed in Section 4.1. For electrons the diffusion current assumes the form

$$\mathbf{J}_{\mathrm{diff}} = eD_n\,\mathrm{grad}\,n \tag{9.7.3}$$

and similar equations apply to other species of charged particles with obvious modifications. The diffusion constants naturally differ from one particle species to another.

It should be borne in mind that in a gas concentrations of *uncharged particles* are also subject to diffusion processes, the flow of particles being proportional to the pressure gradient. In this connection partial pressures of the gas species concerned are implied.

## 9.8 Ohmic Conduction and Saturation in a Weakly Ionized Gas

Gas samples under normal conditions contain a small number of ions and electrons produced by the action of ultraviolet and cosmic rays, and radio-activity. A typical number is about 1000 ions per $cm^3$ of atmospheric air. Due to the same causes electrodes will emit some electrons under normal conditions.

Together these sources of charged particles will contribute towards a very small current flow on application of a low voltage, say 1 volt, to the terminal electrodes of a gas-filled tube. At first the current will increase linearly with the applied voltage, thus displaying an ohmic $I/V$ characteristic. These conditions persist as long as the supply of charged particles is not significantly depleted by the sweeping action of the applied field.

The limited stock of charge carriers means that only moderately increased voltages are enough to exhaust available charged particles as rapidly as they are created by the natural agencies mentioned above. The current density then reaches a *saturation* value of the order of $10^{-12}$ A/cm$^2$. This can be increased by artificial irradiation of the gas sample and electrodes, but even then it remains below levels of the order of $10^{-9}$ A/cm$^2$. The relationship between the current and voltage under these conditions is summarized in Fig. 9.3 where the voltage

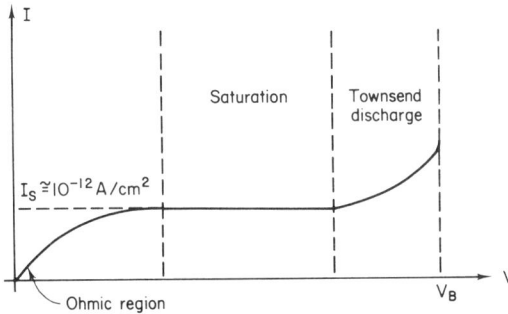

Fig. 9.3. Low current in a gas.

axis extends up to some hundreds of volts. The ohmic region is confined to the lowest values near the origin.

It is emphasized that the condition of saturation corresponds to relatively low applied fields. Such fields cannot impart enough kinetic energy to electrons and ions between collisions to enable them to ionize additional atoms, which in turn could contribute to the transport current. The latter condition can be reached by increasing the voltage sufficiently. The resulting change in the electrical properties of the gas is of a fundemental nature and is discussed in the next section.

## 9.9   The Townsend Discharge

Let the idealized experiment of the preceding section be continued under conditions of a relatively low gas pressure, say 1 mm Hg. On increasing the applied voltage to values of some hundreds of volts the electrode current will be found to grow above the saturation value at a rapidly increasing rate. This condition is called the *Townsend discharge*. It is symptomatic of a state in which the electric field is strong enough to accelerate electrons between inelastic collisions to the point where they gain sufficient energy to ionize gas atoms. The electrons thus liberated become available to enhance the process further, and so on until an *avalanche* condition is set up in the gas, characterized by an exponential growth of the number of charge carriers available for conduction.

The situation lends itself to mathematical analysis. We consider the simplest configuration of parallel plane electrodes of unit area. The sample of gas is confined to the space between them and the effects of field fringing at the edges of the electrodes are assumed to be negligible.

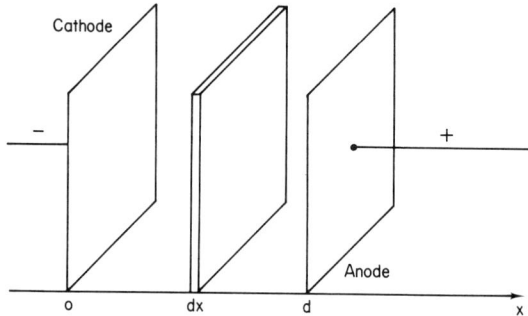

**Fig. 9.4.** Configuration of parallel plane electrodes in the analysis of the Townsend discharge.

Referring to Fig. 9.4 we consider a thin slice of interelectrode space of thickness $dx$, where the x-axis has its origin at the cathode. At first we consider only electrons, which are visualized as moving from cathode to anode under the influence of an applied field. The analysis to follow is carried out in terms of the *number of electrons N crossing unit area of a plane normal to the x-axis per second*. The quantity is related to the local concentration of electrons $n$ and their transport velocity $v$ by the equation

$$N = nv. \tag{9.9.1}$$

The current density is expressed in terms of this quantity as follows

$$J = |\mathbf{J}| = en\mu_n|\mathscr{E}| = env = eN. \tag{9.9.2}$$

The number of electrons crossing the plane at $x$ is increased over the small distance $dx$ by the increment $dN$, which can be written in the form

$$dN = \alpha N dx \tag{9.9.3}$$

where $\alpha$ is the probability of ionizing collisions, or the number of ionizing collisions made by one electron over unit distance (Section 9.2). In the present context $\alpha$ is usually referred to as Townsend's *first ionization coefficient*.

Equation (9.9.3) is readily integrated over the distance from the cathode to the plane at $x$ if $\alpha$ is assumed independent of $x$.

$$\int_{N_0}^{N} \frac{dN}{N} = \int_0^x \alpha\,dx \qquad (9.9.4)$$

where $N_0$ is the number of electrons leaving unit area of cathode per second, also called the *primary electrons* in this context. Carrying out the integration we find

$$N = N_0\,e^{\alpha x}. \qquad (9.9.5)$$

Hence the electron current density at distance $x$ from the cathode is, by eqn (9.9.2)

$$J = eN = eN_0\,e^{\alpha x} = J_0\,e^{\alpha x} \qquad (9.9.6)$$

where $J_0$ is the primary electron current density at the cathode. Thus the electron current grows exponentially with distance, a process usually referred to as an *avalanche*.

To obtain the number of electrons reaching the anode we set $x = d$ in eqn (9.9.5). Hence

$$N = N_0\,e^{\alpha d}. \qquad (9.9.7)$$

The electron current at the anode has the value

$$J = J_0\,e^{\alpha d}. \qquad (9.9.8)$$

Theoretically this expression includes both electrons and negative ions formed by attachment. However the numbers of negative ions are insignificant and it is correct to assume for practical purposes that the current consists purely of electrons, and that it is the total current flowing through the gas.

The conditions at the cathode are somewhat different in that the total current there will be made up of positive ions reaching the cathode, less the primary electrons leaving it. Hence

$$\text{Positive ion current at cathode} = J - J_0 \qquad (9.9.9)$$

where only *magnitudes* of the currents are used.

It must be emphasized that the relations derived above apply to one particular point of the $I/V$ characteristic of Fig. 9.3, corresponding to a voltage somewhat below the value marked $V_B$. A change of voltage will entail a change of the collision probability $\alpha$, caused by a different accelerating field.

The conduction current $J$ derived above depends on the presence of the primary current $J_0$ emitted by the cathode, apart from a sufficiently strong electric field. Although it was convenient to base the above analysis on this assumption, it will be clear from the discussion of Section 9.5 that some electrons are generated in the body of the gas. This can be taken into account in a modified calculation, but although the resulting expression is somewhat different, the dominant term is still the exponential $e^{ax}$. Although photo-ionization requires higher energy than photoemission (say 10 eV as against 4 eV) it acts over a relatively large volume of gas, while photoemission is only operative over the area of electrodes. For this reason it is difficult to state in general terms which source of primary electrons is more important to the maintenance of the Townsend discharge.

A plot of log $J$ against electrode separation $d$ should yield a straight line according to eqn (9.9.8), provided the applied voltage is adjusted to maintain a constant electric field and hence a constant value of $\alpha$. In fact the validity of this result is verified experimentally only for low values of $d$ as suggested in Fig. 9.5.

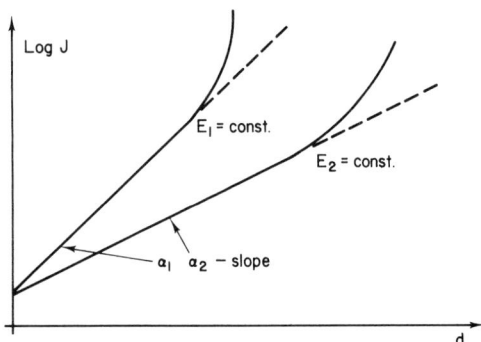

Fig. 9.5. Townsend discharge current as a function of electrode spacing (at constant field).

As the electrode separation is increased the current is found to grow more rapidly than expected on the basis of the ionization mechanism. Clearly additional electrons must be forthcoming from somewhere, and secondary emission from the cathode due to ion impact is the likeliest source (see Section 9.6). However, photoemission caused by radiation from a glowing gas may contribute, as well as other *secondary effects* having their source in the ionized gas itself. It may be noted in passing that attachment and/or recombination would have the opposite effect. In fact, in some gases a decrease of the current below the straight line relationship of eqn (9.9.8) is observed.

The effect of secondary emission is readily evaluated. We start by defining Townsends *second ionization coefficient* which is in fact a coefficient of secondary emission (see Section 6.6).

$$\gamma = \frac{\text{secondary electrons}}{\text{incident ions}} \qquad (\ll 1).$$

We further define the following quantities:

$N_+$ = electrons leaving unit area of cathode per second due to secondary emission.

$N_a$ = electrons arriving per unit area of anode per second from all sources.

Hence

$N_0 + N_+$ = all electrons leaving unit area of cathode per second

$N_+/\gamma$ = ions arriving per unit area of cathode per second.

In the steady state the number of ions arriving at the cathode must equal the difference between the electrons arriving at the anode minus the electrons leaving the cathode. Thus

$$N_+ \, \gamma = N_a - (N_0 + N_+). \tag{9.9.10}$$

The multiplication of electrons due to ionization of the gas still follows the exponential law of eqn (9.9.7) which now assumes the form

$$N_a = (N_0 + N_+)\,e^{\alpha d}. \tag{9.9.11}$$

Eliminating the secondary electrons $N_+$ between the above equations we obtain the following relation

$$N_a = N_0 \, \frac{e^{\alpha d}}{1 - \gamma(e^{\alpha d} - 1)} \tag{9.9.12}$$

which is the generalized version of eqn (9.9.7) including the effect of secondary emission. The corresponding current equation is

$$J_a = J_0 \, \frac{e^{\alpha d}}{1 - \gamma(e^{\alpha d} - 1)}. \tag{9.9.13}$$

Comparison with eqn (9.9.8) shows that the effect of secondary electrons is indeed to increase the current reaching the anode if the term $\gamma(e^{\alpha d} - 1)$ is significant by comparison with unity. Although $\gamma$ may be of the order of $10^{-2}$ the exponential term usually becomes sufficiently great for growing values of electrode separation $d$ to cause the effect indicated in Fig. 9.5.

Apart from accounting for the effects of secondary emission, eqn (9.9.13) suggests a much more far-reaching possibility. This is that a current may flow through the gas without the need to be sustained by the primary electrons from the cathode represented by $J_0$. However, this possibility will be discussed at length in the next section.

## 9.10 The Townsend Criterion of Breakdown and the Self-Sustaining Discharge

The current flow considered in the preceding section is caused by the application of a field to the gas which is strong enough to cause ionization by collision between electrons and neutral gas atoms. However, it must be sustained by the availability of a few primary electrons $N_0$. Without them there can be no current.

As it happens the current flow phenomena described in the preceding two sections do not exhaust all the possibilities of electrical discharges in gases. At least two more distinct processes may take place and, to be sustained, neither of them requires the primary current $J_0$.

To see this we refer back to eqn (9.9.13), and find on inspection that the anode current $J_a$ may retain a finite value in the limit, while $J_0$ and the denominator both tend to zero. Although in practice the primary current $J_0$ is always present, a condition can arise when the denominator tends to zero. This happens when the applied field is strong enough to produce positive ions which in turn cause a sufficient number of secondary electrons to be emitted by the cathode. At some point the latter are enough to render the discharge independent of the primary electrons, or to make it *self-sustaining*. This condition is usually referred to as *breakdown*, and the voltage at which it sets in is the *breakdown* voltage marked $V_B$ in Fig. 9.3. Once the breakdown voltage is reached the discharge conditions become unstable and the current will increase indefinitely unless limited by external circuitry.

The criterion of breakdown is obtained by setting the denominator of eqn (9.9.13) equal to zero.

$$1 - \gamma(e^{\alpha d} - 1) = 0$$

$$\gamma e^{\alpha d} = 1 + \gamma.$$

As pointed out earlier $\gamma \ll 1$, hence the criterion assumes the form

$$\gamma e^{\alpha d} = 1 \tag{9.10.1}$$

which can be given a very appealing physical interpretation. The exponential factor $e^{\alpha d}$ represents the number of positive ions produced as a consequence of one primary electron leaving the cathode (eqn 9.9.5). On reaching the cathode the ions produce secondary electrons numbering $\gamma e^{\alpha d}$. Hence the above criterion means that the breakdown condition is reached when one secondary electron is emitted for each primary electron.

## 9.11 The Ionization Coefficients and Paschen's Law

The discussion of the preceding sections has been conducted in terms of the ionization coefficients $\alpha$ and $\gamma$, quantities whose relevance to a practical experimental situation is not readily grasped. For this reason the breakdown criterion of eqn (9.10.1) is somewhat intangible, despite the neat physical interpretation it can be given. We shall now consider the coefficients rather more closely to see whether they can be related to readily measurable quantities such as gas pressure, electrode separation and the applied field.

As a first step we make the simplifying assumption that the temperature $T$ remains substantially unaffected by changes in the electrical conditions of the gas. This is justified by the fact that extraneous electrical energy is transported to the gas largely through the mechanism of accelerating electrons, which in turn give up very little of their energy to the gas (see Sections 9.2 and 9.3). This leaves the gas pressure $p$ as the principal gas parameter affecting the ionizing collision probability $\alpha$.

By definition the collision probability and hence the number of ionizing collisions per unit length of path is proportional to the gas pressure (see Section 9.2). Hence we can write

$$\alpha \sim p.$$

Secondly the number of ionizing collisions depends functionally on the energy gained by electrons between successive collisions which is

$$E = e\mathscr{E}\lambda \sim e\mathscr{E}/p$$

where $\lambda \sim 1/p$ is the mean free path. Combining the foregoing observations the ionization coefficient $\alpha$ can be expressed in the form

$$\alpha = pF(e\mathscr{E}\lambda) = pF_1(e\mathscr{E}/p)$$

or alternatively

$$\frac{\alpha}{p} = \phi(\mathscr{E}/p). \tag{9.11.1}$$

The last form emphasizes the fact, well borne out by experiment, that the ratio $\alpha/p$ is related to the energy gained by electrons between ionizing collisions.

The second ionization coefficient $\gamma$ is concerned with the impact of ions on the cathode. Hence it will depend on the energy of the ions gained over the last mean free path before impact and will therefore be a function of $\mathscr{E}\lambda$ or $\mathscr{E}/p$. Thus

$$\gamma = \psi(\mathscr{E}/p). \tag{9.11.2}$$

Substituting the above equations into the breakdown criterion (9.10.1) we find

$$\psi(\mathscr{E}/p)\,e^{pd\phi(\mathscr{E}/p)} = 1.$$

Assuming that the field between the electrodes is uniform, we can substitute from $V_B = \mathscr{E}d$ where $V_B$ is the breakdown voltage. Hence

$$\psi\left(\frac{V_B}{pd}\right) e^{\,pd\phi\,(V_B\,/\,pd)} = 1.$$

Treating the product $pd$ as a single variable we find that this expression defines implicitly a functional relation between the variables $V_B$ and $pd$. Explicitly the function can be written in the general form

$$V_B = f(pd). \tag{9.11.3}$$

This is *Paschen's law*. No satisfactory theoretical prediction of the precise form of the function $f(pd)$ has so far been obtained, but there is ample evidence, mainly experimental, that it has a well-defined minimum. For electrode separations of 1 cm or therabouts this occurs at gas pressures of the order of 1 mm Hg. Graphically the function has usually the form shown in Fig. 9.6. The following points are worth noting.

For values of the product $pd$ in excess of $(pd)_{min}$ a higher voltage is required for breakdown for one or both of two reasons. If the pressure is kept constant while the electrode separation $d$ is increased, a higher voltage must be applied to maintain the same accelerating field between collisions. On the other hand an increase in gas pressure $p$ reduces the mean free path and hence the energy gained between collisions. Thus a higher field is required to compensate for that. For values of $pd$ below the minimum of Paschen's curve the opposite considerations apply.

The graph of Fig. 9.6, plotted experimentally for specific gases, is of considerable importance in applications. Two distinct situations usually arise. On the one hand it may be necessary to avoid breakdown in a volume of gas,

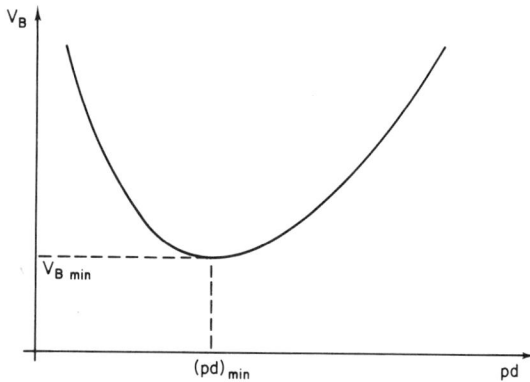

**Fig. 9.6.** Paschen's law.

subjected to a given voltage for one reason or another. Figure 9.6 then allows a prediction of gas pressure which will ensure immunity against breakdown. On the other hand the objective may be to secure breakdown at the lowest possible voltage. The gas conditions are then adjusted for operation at the minimum of the breakdown curve.

It is desirable to note at this juncture a method of reducing the breakdown voltage which is of importance in applications of gas discharges. This is achieved by arranging for a copious supply of electrons by thermionic emission from an externally heated cathode. The numbers of electrons obtainable by this method are far in excess of secondary electrons and give rise to a self-sustaining discharge at breakdown voltages of the order of 30 volts. These possibilities will be further discussed in a later section in the context of applications.

## 9.12  The Glow Discharge

As explained in Section 9.10 the electrical breakdown of the gas ushers in a new conduction regime in which the flow of current is independent of the availability of primary electrons. The self-sustaining characteristics of this regime entail a degree of instability in the discharge which causes the current density to rise by several orders of magnitude without the necessity to increase the applied voltage at all. The current may rise to values of the order of 1 mA/cm$^2$ and can increase further to assume catastrophic proportions, unless it is controlled by external circuitry. The catastrophic condition corresponds to yet another conduction mechanism, called the arc discharge, to be discussed in the next section. For the present we shall be concerned with the *glow discharge* which sets in immediately after breakdown. The name is justified by the fact that the conducting gases emit visible radiation which forms the basis of applications of this phenomenon for lighting purposes. However, as there are many other applications of the glow discharge it is desirable to study some of its relevant properties.

We begin with Fig. 9.7 which depicts the full $I/V$ characteristic of gases. This covers the complete range of conduction phenomena in a gas beginning with the ohmic region, through saturation and the Townsend discharge to the glow discharge and the arc. Such a characteristic is best obtained experimentally on a volume of gas at a pressure of the order of 1 mm Hg, confined within a glass tube as suggested in Fig. 9.8. Under these conditions the gas is near the minimum of Paschen's curve of Fig. 9.6 and the breakdown voltage is a few hundred volts. It is emphasized that the curve of Fig. 9.7 cannot be obtained using just one set of experimental equipment. This is because to measure the vast range of current values instruments of differing sensitivities are required.

The glow discharge itself covers several orders of magnitude of the current. Its full extent depends on the method of measurement, since increasing or decreasing the current will be accompanied by different voltage drops across the tube near the transitions to the arc at the one end and to the Townsend discharge at the other. Two segments can be distinguished in the glow discharge region. The *normal glow* is characterized by a constant voltage drop which is somewhat less than the breakdown voltage. In this regime the current density at the surface of the cathode remains constant. The variation of the total current is effected by utilizing an increasing or decreasing portion of the cathode surface.

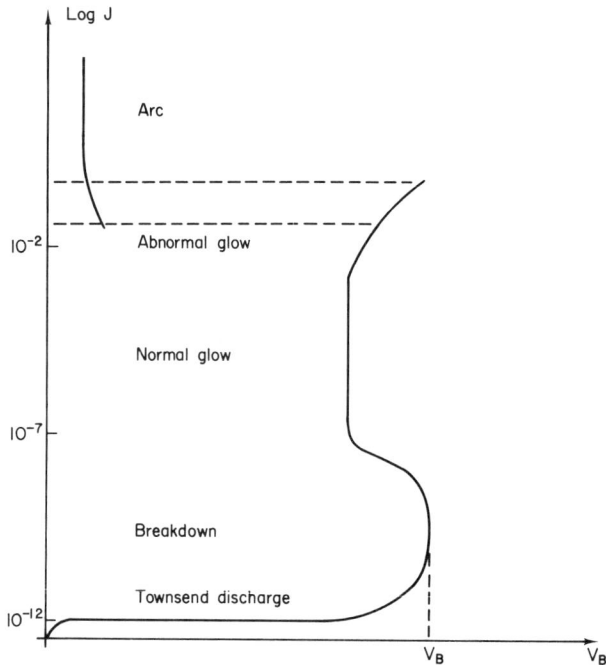

**Fig. 9.7.** Complete $I/V$ characteristic of a gas discharge.

Otherwise the condition of the gas in the discharge remains substantially unchanged. When the full area of the cathode is used up by the constant density discharge current, any further increase in the current requires a somewhat increased applied voltage. This leads to the *abnormal glow*, so called because pronounced changes in the condition of the conducting gas sample can be observed. These are best described by reference to Fig. 9.8 which depicts schematically a gas discharge tube.

The column of glowing gas subdivides into several zones of differing luminosities, each characterized by the predominance of one or other process of ionization, recombination, excitation or emission. In the present context it is unnecessary to study these in detail and for practical purposes it is usually sufficient to note two major regions, as distinguished by the potential distribution of Fig. 9.8(b). *The positive column* next to the anode is highly luminous, and sometimes exhibits *striations*, successive zones of high and low luminosity. The potential drop over the extent of the positive column is small, that is the region is exposed to only very weak electric fields. The net space

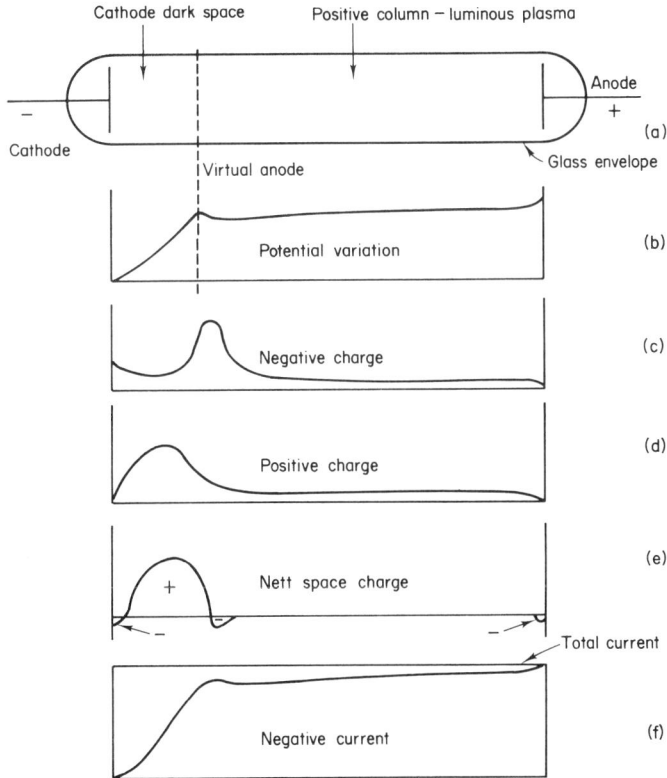

Fig. 9.8. Electrical properties of a glow discharge.

charge is zero and the current is carried largely by electrons which respond much more readily to the weak field. In the immediate proximity of the anode a small negative space charge is built up by the positive terminal which attracts electrons and repels positive ions.

The other major zone of the discharge takes up almost the entire potential drop, and although it consists of several distinct layers, some darker than others, it is usually referred to as the *cathode dark space* or *cathode fall*. Ionization of the gas takes place largely in this region which contains a net positive charge, since electrons are rapidly swept away by the strong electric field present.

The situation depicted in Fig. 9.8 corresponds to the normal glow. If the anode is moved closer to the cathode under these conditions, it is found that the only effect is to shorten the positive column. The slight potential maximum at the end of the cathode dark space retains its magnitude and position and is therefore referred to as the *virtual anode*. As pointed out above, increasing the discharge current does not affect gas conditions until the whole cathode area has been taken up by this current of constant density. To increase the current above that necessitates a higher voltage which characterizes the abnormal glow. However, the potential gradient in the cathode dark space still retains its value. This is because it has assumed an optimum corresponding to the minimum value, $(pd)_{min}$, of Paschen's law, Fig. 9.6. At a given pressure $p$ the virtual anode is at a distance $d$ from the cathode consistent with the condition $pd = (pd)_{min}$, assuming that the pre-breakdown condition corresponded to $pd > (pd)_{min}$. As the applied voltage is inreased, the virtual anode recedes towards the actual anode, thus shortening the positive column. When the position of the virtual anode passes that of the actual anode, the discharge is extinguished if a current limitation is built into the equipment, otherwise it makes a transition to the arc regime.

The conditions in the discharge tube are also affected by gas pressure, and again the effect is best understood in terms of the position of the virtual anode. Since this represents the optimum discharge condition, an increase in pressure is equivalent to starting with a higher pre-breakdown value of $pd$, necessitating an even lower value of $d_{(virtual)}$. Thus the cathode fall length varies roughly inversely with pressure. If the pre-breakdown condition corresponds to a value of $pd$ below the Paschen minimum, the resulting discharge is in the abnormal glow region from the start. Increasing the pressure will have the effect of bringing it within the normal glow range and again decreasing the cathode dark space in length.

### 9.13 The Arc

The transition from the abnormal glow to the arc discharge is shown discontinuous in Fig. 9.7, but in many cases it is possible to trace out a continuous change from the high voltage glow to the low voltage and high current arc. The discharge current may again increase by several orders of magnitude, while the potential drop assumes values of the order of tens of volts.

The electrical structure of the conducting column of gas resembles glow discharge conditions. There is a positive column and a cathode fall, but this time the latter is very short, giving rise to intense electric fields adjacent to the cathode surface. The high field values give rise to two possible explanations of the arc discharge: (i) thermionic emission and (ii) high field emission.

The increasing bombardment of the cathode by positive ions in the abnormal glow region is bound to raise the cathode temperature which may well reach sufficiently high values to cause significant emission of thermionic electrons. Once this process sets in the stream of electrons leaving the cathode will be far greater than anything associated with secondary emission and their presence must be expected to affect the conduction processes in the gas. In any case there is now a considerable body of experimental data consistent with the presence of thermionic emission in gas discharges terminating in electrodes made of high melting point refractory metals such as tungsten or tantalum.

The position assumes a somewhat different complexion when the cathode is made of a low boiling point metal such as copper, or even consists of a pool of liquid metal such as mercury. The boiling point of copper is less than 2600 °C, a temperature at which thermionic emission is negligible: and yet an arc can be set up in a tube having copper electrodes without evaporating the latter. The same argument applies to mercury arc rectifiers which contain a pool of liquid metal in a large glass bulb. When in operation the temperature of the envelope and the mercury pool does not rise appreciably above ambient conditions. Discharges of this nature are usually referred to as *cold cathode* arcs. The discharge is assumed to be sustained by an intense flow of electrons from the cathode, produced by field emission.

A characteristic common to all arc discharges is the concentration of the current in relatively small areas of the cathode. The current density at these cathode spots is consequently very high, but it still differs as between thermionic and cold cathode arcs. The former have the lower density of $10^3$ A/cm$^2$ while the latter have densities of the order of $10^6$ A/cm$^2$. The two types of arc also differ in that the transition from the glow discharge can be made continuously in the thermionic arc, and the cathode spot remains stationary. The transition to the cold cathode arc is abrupt and the cathode spot wanders around unless special measures are taken to keep it fixed, such as arranging for a metallic pin to protrude above the surface of a mercury pool.

In addition to the above types of self-sustaining arc it is possible to maintain an arc discharge with the help of an externally heated thermionic cathode. Several devices of considerable practical importance based on this arrangement will be mentioned in a subsequent section dealing with applications.

## 9.14 The Effect of High Pressures

The discussion of the preceding sections was tacitly limited to relatively low gas pressures, say 1 mm Hg. This was done because phenomena at low pressures are more readily analysed and hence their incidence can be reliably predicted and incorporated in useful applications. Against this discharges at higher pressures, say 1 atm or more, are frequently undesirable and even destructive, e.g. lightning and sparking phenomena associated with high voltage switchgear.

In the first place increasing pressures make it necessary to bring the electrodes closer together to effect breakdown of the gas at the minimum of Paschen's law. The limit of this procedure is reached by the time pressures of a few cm Hg are reached and at atmospheric conditions breakdown can only be secured by increased voltages. However, the breakdown voltage increases roughly linearly with pressure for constant electrode separation, thus satisfying Paschen's law. At still higher pressures the trend is not followed and Paschen's law is therefore not valid. Moreover, there is an increasing tendency for the discharge to pass directly into the arc regime.

The presence of non-uniform fields at high pressure gives rise to various types of *corona*. This is a self-sustaining discharge which carries a significant current ($\cong$ 1 mA) and is thus of importance in high voltage installations by virtue of the power loss it causes.

### 9.15 A.C. Discharges

The effect of a.c. fields on gases depends very much on the relative duration of the period of the applied voltage and the time of transit of charged particles, particularly slow ions, between the electrodes. If the transit time is only a small fraction of the a.c. period the conditions can be said to be quasistatic in the sence that the d.c. processes described in preceding sections are followed through for a sufficient length of time to establish themselves at each point of the a.c. cycle. In these circumstances a.c. currents will flow in the form of electrons reaching one and then the other electrode while positive ions do the opposite. Thus at frequencies below some 100 Hz the application of a sufficiently high voltage will induce the gas to pass through all the stages up to a glow or arc discharge. Off course, the current waveshape will not in any way resemble a sinusoidal applied voltage but will be dictated by the $I/V$ characteristic of the discharge. A point of practical importance in the operation of fluorescent lighting and a.c. arcs is to ensure, by a reactive outside circuit, that peak voltage is applied between electrodes immediately after the current has reached zero.

At frequencies in the KHz range there is not enough time for the gas to become deionized at the minima of the current, and it becomes effectively an ohmic resistance passing a sinusoidal current in response to a sinusoidal voltage.

At sufficiently high frequencies the electrons will not have enough time to reach any one electrode but will oscillate within the volume of gas in response to the applied field. If the latter is intense enough, breakdown will be effected even in the absence of electrodes. This effect is utilized in some gas lasers which can be pumped by this method. In the absence of electrodes a state of equilibrium is reached at some stage in which the generation of charge carriers is exactly balanced by their recombination, as no current flow in and out of the tube is possible. If the gas is connected to an outside circuit, a current flow is possible by virtue of diffusion of charge carriers, primarily electrons, from regions of high space charge density towards the electrodes.

## 9.16 Some Applications of Gas Discharges

It will be clear from Section 9.8 that below breakdown a gas is a good *insulator*. This fact is utilized in the design of high voltage equipment intended for operation under atmospheric conditions when the breakdown voltage is about 33 KV/cm. In situations where it is necessary to have the highest possible breakdown voltage, special measures can be taken to operate under conditions far removed from the minimum of the Paschen curve of Fig. 9.6. One possibility is to operate under vacuum conditions. This involves rather elaborate and costly equipment and is of necessity limited to the operation of small and specialist equipment. On the other hand it is possible to operate at increased pressure. This method is used in microwave waveguides carrying high power radar signals. To prevent breakdown by the intense electric fields present in the centre of a rectangular waveguide, the latter is filled with a selected gas, usually argon, and pressurized to say 3 atm. Although joints in such a system have to be made gas tight this imposes less stringent conditions than vacuum operation, since a slight leak will not contaminate the insulated space.

The application of a glow discharge in *fluorescent lighting* imposes opposite requirements. The gas conditions must be adjusted to secure minimum breakdown voltage so that the devices can be operated directly from the a.c. mains supply. Even so, it is not possible to secure breakdown without an auxiliary reactive circuit, designed to provide an adequate instantaneous voltage in the right phase relationship to the electrode current. As mentioned in the preceding section this means the peak voltage must be applied to electrodes immediately after a current zero.

As mentioned in Section 9.6 bombardment by positive ions can dislodge cathode atoms which propagate by diffusion through the gas and deposit themselves as thin films on neighbouring objects. This process of *sputtering* is now widely used in a great variety of forms for both research and commercial purposes. The simplest system, sometimes called diode sputtering, uses a d.c. glow discharge with a pure inert gas in order to minimize contamination. The gas and voltage conditions are adjusted to obtain maximum impact energy of ions and a high current density at the cathode. This usually means operating on the abnormal glow region at pressures of a fraction of 1 mm Hg.

The constant voltage drop characteristic of a normal glow is utilized for purposes of *voltage stabilization*. The basic voltage stabilizing circuit is shown in Fig. 9.9. Variations of the source voltage are accompanied by changes in the resistor voltage drop which in turn are due to current changes. The latter are taken up by the stabilizer tube without affecting the voltage across the load. The operation of this circuit is limited to the range of currents over which voltage variations remain small enough for a particular application. This is usually insufficient for direct use in power supplies. Instead a tube is used as a *voltage reference* in a feedback circuit designed to perform the required control function.

The existence of gas-filled rectifier tubes does not mean that the gas itself has any rectifying properties. The function of the gas is to provide a low resistance, high current density path under conditions of an arc discharge, while the rectifying function is fulfilled by the properties and configuration of electrodes and particularly the cathode. It was pointed out on p. 153 that an arc discharge can be sustained by an externally heated thermionic cathode. The presence of

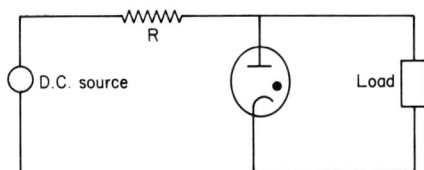

**Fig. 9.9.** Simple voltage-stabilizing circuit.

the electron emitting cathode prevents a reversal of the current when an a.c. voltage is applied to the tube, just as it does in the vacuum diode of Section 6.2. The positive ions serve to neutralize the space charge of emitted electrons outside the cathode, rather than to carry current. With the electron space charge removed there is a low resistance path between the cathode and anode, characterized by a voltage drop of about 20 volts which is largely independent of the current. Thus the tube can handle high currents as compared with a vacuum valve without incurring comparable losses. The gas filling is usually mercury at its saturation pressure, which is some 5 mm Hg at 40 °C, the sort of ambient temperature at which gas-filled diodes can be expected to operate. There are two main limitations on the operation of gas-filled thermionic tubes. Firstly the current carrying capacity is limited by the emissivity of cathodes. Secondly cathode coatings disintegrate due to gas action, much faster than in vacuum valves.

These limitations are surmounted by the use of a pool of mercury as a cold cathode. The latter cannot be influenced adversely by mercury vapour, hence its life is indefinite. Also, as pointed out on p. 153 the current densities obtainable from field emission are orders of magnitude higher than from thermionic emission. Against these gains must be set some technical difficulties. In the absence of the sustaining action of thermionic electrons provision must be made for initiating the discharge with the help of an auxiliary electrode situated close to the mercury pool. Secondly, measures must be taken to prevent a reversal of the current flow on the negative half cycle of the applied voltage. This is achieved by operating the valve in a multiphase circuit, with each phase connected to a separate anode. Thus there is always a positive voltage on some anode and the arc connects the cathode to it in a jump.

The above outline neglects numerous ramifications of the applications mentioned and omits altogether many others. Readers interested in any of them are referred to specialist texts and papers for further details.

*Chapter 10*

# Properties of Dielectrics

## 10.1 Introduction

Properties of insulators were touched upon in Chapter 1 in the context of the discussion of the energy bands in solids. It was then emphasized that insulators were characterized by a very wide forbidden gap, far too wide to permit significant activation of electrons from the valence band. Consequently no conduction of electricity can take place in insulators through the medium of free electrons. Since the discussion of subsequent chapters was largely concerned with the conductive properties of solids and gases, no further mention of insulators was relevant. However, electrical engineering cannot exist without materials which can effectively prevent the passage of currents through certain regions, such as the space between the conductors of a cable or the plates of a capacitor. For this reason we now turn our attention to insulators.

Insulators are not only used to confine the passage of electric currents as required. They perform a more positive function in capacitors by enhancing the capacitance through their *dielectric properties* which will form the main subject matter of the present chapter.

Although there are virtually no free conduction electrons which can give rise to dissipation through the mechanism of ohmic resistance, some loss is always present when an electric field is applied to an insulator or dielectric. In most conditions this is sufficiently small not to matter, but occasionally a situation arises when the *dielectric loss* is significant and must be taken into account. In the sections to follow we shall consider how this loss arises and when it assumes serious proportions.

Another property of insulators which is important in applications is their *breakdown strength*. This is the value of the electric field above which the insulator ceases to insulate and gives way to electric conduction currents, usually with catastrophic results. We shall touch upon this problem at the close of the chapter and attempt to relate it to breakdown phenomena discussed in earlier chapters.

We shall start by considering the dielectric properties of gases. Although gases are of less practical importance as dielectrics than solids or liquids, they can be used to introduce the basic concepts by relatively simple methods which do not require a recourse to quantum mechanics. With the help of these concepts a qualitative account of the properties of solid dielectrics can be given which meets the requirements of many practical situations.

Before starting on the problem of how the properties of dielectrics arise from atomic or molecular structure, we summarize briefly the electromagnetic relations applicable to dielectrics as continuous media. Dielectric materials are described in terms of three vector quantities: the *electric field* $\mathbf{E}_0$, the *electric flux density* or *electric displacement* $\mathbf{D}$, and the *polarization* $\mathbf{P}$. The vectors are related by the equation

$$\mathbf{D} = \epsilon_0 \mathbf{E}_0 + \mathbf{P} \tag{10.1.1}$$

where $\epsilon_0$ is the permittivity of free space. Our preoccupation in this chapter will be with the polarization $\mathbf{P}$ which is defined as the electric dipole moment per unit volume of the material. It is the sum of the electric dipole moments of individual atoms or molecules. The electric field associated with each dipole modifies the externally applied field and the flux density. As we shall see the polarization of a substance is linearly related to the applied field except in a few rare situations. Hence we can write

$$\mathbf{P} = \chi \, \epsilon_0 \mathbf{E}_0 \tag{10.1.2}$$

where $\chi$ is a constant called the *electric susceptibility* or *polarizability* of the material. Combining the above equations one obtains

$$\mathbf{D} = \epsilon_0 (1 + \chi) E_0 = \epsilon \epsilon_0 E_0 \tag{10.1.3}$$

where $\epsilon = 1 + \chi$ is the *relative permittivity* or *dielectric constant* of the material.

In many cases a dielectric is made up of two or more distinct types of molecules or atoms which are polarized in different ways by an applied field. It is then desirable to treat each species separately but it is necessary to know how to include their combined effect in a single dielectric constant. Assuming that the polarization has two components $\mathbf{P}_i$ and $\mathbf{P}_p$, we can write $\mathbf{P} = \mathbf{P}_i + \mathbf{P}_p$. The constitutive relation (10.1.1) then assumes the form

$$\mathbf{D} = \epsilon_0 E_0 + \mathbf{P} = \epsilon_0 E_0 + \mathbf{P}_i + \mathbf{P}_p$$

$$= \epsilon_0 E_0 (1 + \chi_i + \chi_p)$$

$$= \epsilon_0 \mathbf{E}_0 (1 + \Sigma \chi_i).$$

Thus the susceptibilities are added and the dielectric constant has the form

$$\epsilon = 1 + \Sigma \chi_i. \tag{10.1.4}$$

If the vectors appearing in eqn (10.1.1) are parallel, the susceptibility $\chi$ is a single scalar constant and the dielectric is said to be *isotropic*. However, there are materials in which this does not apply. The relations between the field vectors are still linear, but have the form of matrix equations with $\chi$ being a square matrix of order 3 x 3 and thus consisting of 9 constants. Dielectric materials of this type are said to be *anisotropic*. Readers are urged to compare this summary with the analogous discussion of magnetic materials in Section. 20.1.

Perhaps the most familiar use of dielectrics is as insulation between capacitor plates. It is well known from electrostatics that the capacitance of a given configuration is enhanced by a factor equal to the relative permittivity, if air is replaced by some other insulating medium. Thus the capacitance $C_0$ (with air as insulation) is increased to $C = \epsilon C_0$.

The application of an a.c. voltage to the capacitor will cause the flow of a current which is expressed by the relation

$$I = j\omega CV = j\omega \epsilon C_0 \, V. \tag{10.1.5}$$

If $\epsilon$ is a real constant, the current and voltage are in quadrature, and there is no dissipation of energy in the capacitor. In practice there is always an in-phase component of current, however small, and some power loss does occur. This effect is most conveniently included in eqn (10.1.5) by writing $\epsilon$ as a complex number

$$\epsilon = \epsilon' - j\epsilon'' \tag{10.1.6}$$

where $\epsilon''$ relates the in-phase component of the current to the applied voltage and is thus a measure of the power loss. $\epsilon'$ is the "real" part of the dielectric constant, that is the actual factor which enhances the capacitance. The reactive performance of a capacitor is usually defined in terms of the loss angle $\delta$ defined by

$$\tan \delta = \frac{I \text{ loss}}{I \text{ reactive}} = \frac{\epsilon''}{\epsilon'}. \tag{10.1.7}$$

At higher a.c. frequencies, in fact as soon as the microwave range is reached, the a.c. fields applied to a dielectric sample of macroscopic dimensions must be visualized as waves. It is then sometimes convenient to describe the medium in terms of the refractive index $(n - jk)^2 = \epsilon' - j\epsilon''$ where $n$ is the *refractive index* proper and $k$ is the *absorption coefficient*.

## 10.2  Polar and Non-Polar Atoms and Molecules

A neutral atom contains a balance of positive and negative charge, the former concentrated on the nucleus and the latter distributed over the surrounding space. Moreover, as will be discussed in Part II the positioning of electrons which constitute the negative charge is symmetrical about the nucleus, with the result that there is no electric dipole moment. This observation applies even to the hydrogen atom. Although this has only one electron, the negative charge distribution is still symmetrical about the nucleus. The absence of a dipole moment on isolated atoms should be contrasted with the fact that many atoms have a resultant magnetic moment as will be shown in later chapters.

The application of an electric field has a distorting effect on the electronic charge distribution, tending to push it in one direction while attracting the nucleus in the opposite direction. The initial symmetry is upset and a dipole moment appears. Summed over the atoms contained in a unit volume the dipoles yield a polarization **P**.

The foregoing remarks also apply to molecules held together by covalent forces which involve the sharing of one or more electrons between the constituent atoms. In such cases the charge distribution retains the symmetry which is inconsistent with an electric dipole moment. However, polarization appears on application of a field. Best examples of molecules in this category are provided by homonuclear pairs such as $N_2$, $O_2$ and other gas molecules. Dielectrics made up of atoms and molecules which do not naturally possess an electric dipole moment are said to be *non-polar*.

There are molecules which possess a natural dipole moment in the absence of any field, due to a lack of symmetry of their negative charge distributions. The best examples are alkali halides, e.g. KCl. These molecules are held together by electrostatic forces between ions, e.g. $K^+$ and $Cl^-$. An effective transfer of one electron has taken place from one atom to another resulting in a dipole moment which can be expected to approach the product of the electronic charge and the distance between the nuclei. Such an estimate is in fact quite close to measured dipole moments of alkali halides but is found to be much too great in the case of other ionic compounds, e.g. hydrogen halides such as HCl or HBr. This is because the transferred electronic charge distributes itself around individual nuclei in a more complicated fashion than envisaged by the above simple model. Quantum mechanical calculations of the electron distributions in such molecules indicate a smaller dipole moment. Leaving aside individual variations the fact is that there are many diatomic molecules which have a natural dipole moment and are therefore said to be *polar*.

A d.c. electric field applied to polar molecules will tend to align them parallel to itself. Although this tendency is counteracted by thermal agitation the net result is the appearance of a polarization **P** whose magnitude is proportional to the field, as anticipated in eqn (10.1.2).

Of greater practical importance than the static conditions considered so far is the application of a.c. fields to atoms and molecules. Although this problem will constitute a large part of the subject matter of the remaining parts of this book, it is necessary to approach it in the context of dielectrics in this chapter. As long as the frequency of the applied field is low, and the a.c. mains frequency is certainly in this category, the static polarization follows the instantaneous value of the field in a linear fashion. However, in some conditions the time variation of the applied field approaches the internal orbiting frequency of electrons in atoms and molecules and a phenomenon of forced resonance manifests itself. The resonance is characterized by sharply increased loss ($\epsilon''$ of eqn (10.1.5)) and rapid variation of the dielectric constant with frequency ($\epsilon'$ of eqn (10.1.5)). Details of this behaviour will be discussed in subsequent sections when it will be seen that the net effect is a decrease in the dielectric constant of insulating materials with increasing frequency. The scale of frequencies over which this happens is very great, extending from audio frequencies in some cases to optical and ultraviolet wavelengths.

### 10.3 Static Dielectric Constant of Gases

Let us first consider non-polar molecules. These are polarized by an applied field which displaces the electronic charge in relation to the nucleus. As explained in the preceding section there is ample experimental evidence to justify the assumption that the induced dipole moment $\mathbf{p}_i$ is proportional to the electric field. Hence

$$\mathbf{p}_i = \alpha \mathbf{E} \tag{10.3.1}$$

where $\alpha$ is the *molecular polarizability*. $\mathbf{E}$ is the *local field*. This differs from the externally applied field, to be denoted by $\mathbf{E}_0$, due to the effect of the field generated by the neighbouring dipoles themselves. In gases under normal pressures the contribution of the dipoles to the field is negligible. Hence we can write

$$\mathbf{p}_i = \alpha \mathbf{E}_0. \tag{10.3.2}$$

If there are $n_0$ identical molecules per unit volume of gas the induced polarization is

$$\mathbf{P}_i = n_0 \alpha \mathbf{E}_0. \tag{10.3.3}$$

From eqns (10.1.1) and (10.1.3) we can also write

$$\mathbf{P}_i = (\epsilon_i - 1) \epsilon_0 \mathbf{E}_0. \tag{10.3.4}$$

Combining the above equations we find

$$\chi_i = \epsilon_i - 1 = \frac{n_0 \alpha}{\epsilon_0}. \tag{10.3.5}$$

The molecular polarizability $\alpha$ and dielectric constant $\epsilon_i$ can be related through the usual gas constants on multiplication of eqn (10.3.5) by $M/\rho$ where $M$ is the molecular weight and $\rho$ is the density of the gas. Noting also that $Mn_0/\rho = N$ is Avogadro's number we can write

$$\chi_i \frac{M}{\rho} = (\epsilon_i - 1) \frac{M}{\rho} = \frac{N\alpha}{\epsilon_0}. \tag{10.3.6}$$

The above equation gives the relative permittivity or dielectric constant of non-polar gases at normal pressures. The values of $\epsilon_i$ barely exceed unity; the susceptibility $\epsilon_i - 1 = \chi_i$ is of the order of $10^{-3}$ or less. Equation (10.3.5) also gives the contribution to the dielectric constant of polar gases due to induced dipole moments, since polar molecules are subject to the same charge distorting effects when a field is applied.

However, polar gases also have a component of polarization due to the permanent dipole moments of the molecules. As the next step we propose to estimate this contribution to the dielectric constant.

As a first step it is essential to realize that the application of a field $E_0$ does not align all the dipoles parallel to $E_0$. The complete ordering of all dipoles is prevented by thermal agitation. To obtain an insight into these relationships we consider the energy of an electric dipole in a field, which is given by the expression

$$W = -\mathbf{p} \cdot \mathbf{E}. \tag{10.3.7}$$

According to quantum mechanical principles the dipole can assume only two orientations relative to the applied field: parallel or anti-parallel. Consequently its energy has two values $W = \pm pE$, which give two energy levels for the gas molecules. The population of molecules is distributed over the two levels according to the Maxwell-Boltzmann distribution which has the same exponential form as the approximate Fermi-Dirac distribution of Section 3.2. Hence we can write

$$n_1 = A\,e^{pE/kT}; \quad n_2 = A\,e^{-pE/kT} \tag{10.3.8}$$

where $n_1$ and $n_2$ are the populations of the lower and upper level respectively and $A$ is a constant of proportionality. It should be noted that in the present chapter energy is denoted by $W$ to distinguish it from the magnitude of the electric field $E$.

To evaluate the contribution of the permanent dipoles to the polarization of the gas we must know the numbers $n_1$ and $n_2$ of molecules which are aligned and anti-aligned with the field. The difference $n_1 - n_2$ gives the net density of dipoles aligned with the field. It is these dipoles which contribute to the polarization of the gas. The constant $A$ of eqn (10.3.8) is eliminated by forming the ratio $n_2/n_1 = e^{2pE/kT}$. Recalling that $n_1 + n_2 = n_0$, where $n_0$ is the number density of molecules, we have two equations from which to determine the difference $n_1 - n_2$ in terms of $n_0$. After some manipulation the following expression is found

$$n_1 - n_2 = n_0 \frac{e^{pE/kT} - e^{-pE/kT}}{e^{pE/kT} + e^{-pE/kT}}$$

$$n_1 - n_2 \simeq n_0 \frac{pE}{kT}. \tag{10.3.9}$$

The final result is obtained on expansion of the exponentials, and retention of the linear terms only on the assumption that $pE \ll kT$. Hence the polarization of the gas due to polar molecules is

$$\mathbf{P}_p = (n_1 - n_2)\mathbf{p} = \frac{n_0\, pE}{kT}\mathbf{p}. \tag{10.3.10}$$

Assuming that the field vectors are parallel and that the local field $\mathbf{E}$ does not differ significantly from the applied field $\mathbf{E}_0$, we obtain the following relation for the polar contribution to the susceptibility (by eqn (10.1.2))

$$\chi_p = \frac{\mathbf{P}_p}{\epsilon_0\,\mathbf{E}_0} = \frac{n_0\, p^2}{\epsilon_0\, kT}. \tag{10.3.11}$$

The dielectric constant of a polar gas is made up of the contributions $\chi_i$ and $\chi_p$ according to eqn (10.1.4).

The foregoing is a simplified argument which neglects some effects, and yields a result differing somewhat from a formula obtained by a more thorough derivation. In fact the only difference is a factor of $\frac{1}{3}$ and the correct expression is

$$\chi_p = \frac{n_0\, p^2}{3\epsilon_0\, kT}. \tag{10.3.12}$$

As regards the assumption that $pE \ll kT$ this is amply justified, as the reader may verify by assuming a dipole moment consisting of two charges $\pm e$ separated by a distance of atomic dimensions, say 1Å, subject to an electric field readily realizable in the laboratory.

The polar contribution to the static susceptibility of a gas is substantial and may exceed the induced susceptibility by an order of magnitude or more. Consequently the dielectric constant of polar gases is significantly higher. There is ample experimental evidence that the local field $\mathbf{E}$ differs very little from the applied field $\mathbf{E}_0$ in all gases under normal pressures. However, at increased pressures of the order of $10^2$ atm, the distinction becomes significant and is reflected in values of dielectric constant which diverge from the predictions of eqns (10.3.5) and (10.3.12).

The local or internal field $E$ can be estimated, in the case of non-polar gases, by a method devised by Lorentz, and leads to the Clausius-Mossotti relation.

$$\frac{\epsilon_i - 1}{\epsilon_i + 2} = \frac{n_0\,\alpha}{3\epsilon_0}. \tag{10.3.13}$$

It can be seen that eqn (10.3.5) is obtained on setting $\epsilon_i + 2 = 3$. The Lorentz method is based on electrostatics as are the more elaborate methods used to estimate the local field in polar dielectrics. This problem arises largely in connection with solid and liquid dielectrics and will be discussed in a later section.

A comment regarding terminology may be desirable at this point. The polarizability of atoms and non-polar molecules is frequently referred to as *electronic polarizability* because it is based on the distortion of the electronic charge distributed around the positive charge of the nucleus. The polarizability of polar molecules, which is due to the alignment of permanent dipoles in relation to a field, is said to be *orientational* or *dipolar*. In solids and liquids there is also *ionic polarizability* due to dipoles formed by oppositely charged ions.

### 10.4 A.C. Dielectric Properties of Gases

The application of an a.c. field of a sufficiently low frequency will produce the effects discussed in the preceding section except that the polarization $P$ will vary with time in unison with the field. However, as the frequency is increased resonance phenomena supervene. These are due to interactions between the applied field and the internal motions of electrons within atoms and molecules at their natural frequencies.

The non-polar case is the easiest to understand in classical terms. As a first step we shall set up the equation of motion and consider its solution in this case, and then extend the conclusions by a qualitative argument to the dipolar case. As soon as an electronic charge is displaced from its equilibrium position a restoring force appears which is proportional to the displacement. The situation is entirely analogous to an elastic oscillator, hence the restoring force can be written in the form $-m\omega_c^2 r$ where $m$ is the electronic mass, $\omega_c$ the natural resonant frequency of the oscillator and $r$ the displacement vector of the electron from its equilibrium position. The electron motion will be subject to energy losses or damping caused by collisions with other atoms and possibly by radiation. Such losses can be expressed as a retarding force proportional to the electron velocity in the form $-m\gamma(dr/dt)$ where $\gamma$ is constant. The equation of motion includes the inertia term $m(d^2r/dt^2)$ and the force of the a.c. field $E(t) = Ee^{j\omega t}$. Thus

$$m\left(\frac{d^2 r}{dt^2} + \gamma\frac{dr}{dt} + \omega_c^2 r\right) = -eEe^{j\omega t}. \tag{10.4.1}$$

As we are only interested in the steady-state motion, leaving the transient term until a later stage, we can assume a solution of the form $r(t) = re^{j\omega t}$. Substitution in eqn (10.4.1) then yields the relation

$$\mathbf{r} = -\frac{e\mathbf{E}}{m\{(\omega_c^2 - \omega^2) + j\gamma\omega\}}. \tag{10.4.2}$$

This instantaneous deflection of the electron is associated with a dipole moment $\mathbf{p} = -e\mathbf{r}$ which in turn contributes towards a polarization $\mathbf{P} = n_0\mathbf{p}$, if there are $n_0$ molecules per unit volume. Hence we find

$$\mathbf{P} = \frac{n_0 e^2 \mathbf{E}}{m} \cdot \frac{1}{(\omega_c^2 - \omega^2) + j\gamma\omega}. \tag{10.4.3}$$

As in the preceding section we neglect the difference between the internal and external fields and obtain the following expression for the a.c. dielectric constant (by eqns (10.1.2) and (10.1.3))

$$\epsilon = 1 + \chi = 1 + \frac{n_0 e^2}{\epsilon_0 m} \cdot \frac{1}{(\omega_c^2 - \omega^2) + j\gamma\omega}. \tag{10.4.4}$$

To put this result in the complex form of eqn (10.1.6) the fractional expression is rationalized under the simplifying assumptions usually made in connection with resonant circuits. These apply when the resonance phenomenon covers a small *bandwidth* or has a narrow *line*. Then $\omega \cong \omega_c$ and we can write

$$(\omega_c^2 - \omega^2) = (\omega_c + \omega)(\omega_c - \omega) \cong 2\omega(\omega_c - \omega).$$

Using the symbol $\Delta\omega$ to denote the damping factor $\gamma/2$ the following expression is found for the complex dielectric constant or permittivity of the gas

$$\epsilon' - j\epsilon'' = 1 + \frac{n_0 e^2}{2m\epsilon_0 \omega} \cdot \frac{\omega_c - \omega}{(\omega_c - \omega)^2 + (\Delta\omega)^2} - j\frac{n_0 e^2}{2m\epsilon_0 \omega} \cdot \frac{\Delta\omega}{(\omega_c - \omega)^2 + (\Delta\omega)^2}.$$

$$\tag{10.4.5}$$

This expression makes it clear that in the neighbourhood of resonance dielectric properties vary rapidly with frequency. In particular the loss part $\epsilon''$ builds up to the sharp peak shown in Fig. 10.1. The accompanying absorption of a.c. energy affects crucially the usefulness of the dielectric as an insulator. The dielectric constant proper, $\epsilon'$, also undergoes rapid changes near resonance. Equation (10.4.5) and the curve of Fig. 10.1 are only applicable near resonance by virtue of the simplifying assumptions made at the outset. A more detailed analysis shows that the dielectric constant tends to a lower limiting value above resonance than below. Bearing in mind that the electron orbits of atoms and molecules have many resonant frequencies, we can conclude that there is a *gradual net decrease in the value of the dielectric constant with increasing frequency*, superimposed on the sharp variations at resonance. At this point it should be noted that the resonant frequencies of electronic motions in atoms and molecules are concentrated in the visible and ultraviolet range of the electromagnetic spectrum. The curve representing $\epsilon'$ in Fig. 10.1 shows that the

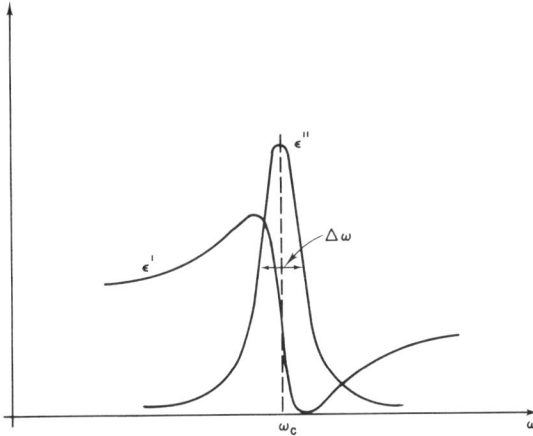

Fig. 10.1. Relative permittivity near resonance (arbitrary scales).

dielectric constant and hence the refractive index of an insulating material varies greatly over a relatively narrow range of wavelengths. Hence waves having wavelengths within this band are refracted by the material at varying angles, a phenomenon referred to as *dispersion*.

The half-power linewidth $\Delta\omega$ of a resonance is only a small percentage fraction of the resonant frequency $\omega_c$. A convenient measure of the sharpness of resonance is provided by the $Q$-factor, which is defined by the ratio $Q = \omega_c/\Delta\omega$. Typical values are of the order of $10^3$.

The foregoing discussion has been framed to apply to non-polar molecules. Polar molecules differ from this only in the nature of the force acting on them. Although an applied electric field still has a distorting effect on the electronic charge distribution, there will also be a couple acting on the dipole moment. The consequent rotation of the molecule will be superimposed on the natural *rotational motions of free molecules* and at the right frequency a resonance phenomenon will take place. Such resonances usually occur at lower frequencies than the electronic phenomena discussed above. In fact they can be observed at far infrared and microwave wavelengths.

A third source of resonance phenomena in dielectrics are *internal vibrations* of molecules. These are due to movements of the constituent nuclei in relation to each other. Such movements may entail variations in the instantaneous value of the molecular dipole moment and will be subject to forced resonance in the presence of an a.c. field having the right frequency. The resonant frequencies of molecular vibrations are situated in the infrared and microwave bands of the spectrum.

## 10.5 Static Dielectric Properties of Solids and Liquids

In the condensed phase the atoms and molecules are tightly packed and the polarization which appears on application of a field is very much higher than in gases. Consequently the local or internal field differs substantially from the externally applied field and must not be neglected in any estimate of the dielectric constant. The evaluation of the internal field under the varying conditions encountered in solids and liquids constitutes a problem in electrostatics and will not be attempted in the present context. Instead we shall be content with a summary of the principal physical effects and results, formulated against the background of preceding sections.

The distinction between non-polar and polar materials still applies but with some elaborations. At one extreme are the inert gases in their condensed phases, which necessarily means at low temperatures. In this class of substances the electron configurations of individual atoms remain virtually undisturbed on condensation. Hence the effect of an applied electric field is the same as in isolated gas atoms and the dielectric constant is approximated quite well by the Clausius-Mossotti formula, eqn (10.3.12) which includes the effect of the internal field. The actual value of the dielectric constant is of course higher due to the higher number density $n_0$ of atoms in a condensed phase.

The counterpart of polar gases are polar liquids which consist of, or contain, molecules having permanent dipoles. The presence of the permanent dipoles, in addition to induced dipoles, introduces extra complications in the evaluation of the local field. The method of Lorentz, which applies to induced dipoles, can no longer be used and a more elaborate calculation has been proposed by Onsager. This provides a reasonable approximation to the static dielectric constant of water, found to be about 100, higher than the measured value of 80. The effect of closely packed permanent dipoles cannot be fully accounted for by a classical electrostatic calculation because short-range quantum mechanical forces are also present.

Ionic crystals display effects which have no counterpart in polar gases or liquids. Such crystals can be visualized as consisting of two *interlocking sublattices*, one having negative ions on the lattice sites and the other positive ones. The application of an electric field displaces one lattice in relation to the other. The result is a substantial polarization and a high value of the dielectric constant. An effort to estimate the latter is subject to the same difficulties which apply to liquids.

## 10.6 A.C. Permittivity of Liquid and Solid Dielectrics—Relaxation Processes

Of the three types of resonant phenomena listed in Section 10.4 the rotational motion of dipoles is excluded in solid dielectrics by virtue of the binding forces present. However, molecular vibrations and electronic resonances are present and occur at frequencies not differing greatly from those in a gas. Consequently the dielectric constant of solids is largely independent of the frequency until the lowest resonances are approached. These are due to molecular vibrations in the upper microwave and infrared bands. As explained in Section 10.4 they cause a net reduction in the dielectric constant or real part of the complex permittivity $\epsilon'$. At the same time they cause resonant losses in the insulator, associated with a peak in the imaginary part of the permittivity $\epsilon''$. The effect of the electronic resonances is to continue the trend of diminishing dielectric constant (or refractive index) over the visible and ultraviolet range of the spectrum, where these resonances are situated.

Permanent dipoles in liquids have some freedom to reorient themselves in response to an applied a.c. field and one would expect to detect some resonances due to the rotational motion of polar molecules. However, the interaction between the dipoles and the surrounding liquid is so strong that any resonance phenomena are observed at much lower frequencies than expected for isolated molecules. Since the effect of the liquid body is so crucial let us consider the interaction rather more closely.

The torque associated with an electric field applied to a dipole rotates a polar molecule, but the rotational motion tends to be frustrated by the random influence of neighbouring molecules. This random bombardment is due to the Brownian motion of liquid molecules and is energetic enough to delay the alignment of the polar molecule and the field. It is quite correct to visualize the dipole as attempting to rotate in response to the applied field against the viscous resistance of the surrounding liquid. A degree of alignment is eventually achieved, but only after a relatively long time interval has elapsed. This time interval, required by the polar molecules to relax into a minimum energy position, is usually denoted by $\tau$ and is called the *relaxation time*. Its value is characteristic of a liquid under given conditions of temperature and pressure.

The above argument concerns a liquid to which a d.c. electric field was applied suddenly, following a step function of time. Let us now consider the effect of an a.c. field. If the frequency is too high, i.e. $\omega \gg 1/\tau$, the viscous resistance will prevent the dipole from following the field and there will be no contribution to the a.c. polarization from the polar molecules. Hence the dielectric constant, or real part of the permittivity $\epsilon'$, will have the value $\epsilon_i$, due to induced polarization only, as indicated in Fig. 10.2. At sufficiently low frequencies, $\omega \ll 1/\tau$, the dipoles have ample time to follow the a.c. field and the dielectric constant will assume the static value $\epsilon_s$ which contains contributions from both the induced polarization $\mathbf{P}_i$ and the dipolar polarization $\mathbf{P}_p$ (see Section 10.3).

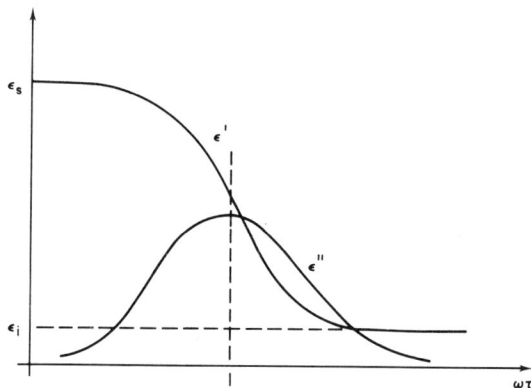

Fig. 10.2. Effect of relaxation on the a.c. permittivity of a polar liquid.

At frequencies approximately equal to the reciprocal of the relaxation time, $\omega \cong 1/\tau$, a change takes place which gradually eliminates the dipolar contribution to the polarization, and the dielectric constant drops from $\epsilon_s$ to $\epsilon_i$. At these frequencies the dipoles are able to follow the a.c. field, but with a slight time lag. This means that a component of polarization in quadrature with the field appears, giving rise to losses and a significant value of the imaginary part of the permittivity $\epsilon''$. Indeed the behaviour of the liquid at these frequencies has the characteristics of resonance, as indicated in Fig. 10.2. However, it is emphasized that the resonance phenomenon is really associated with the Brownian motion of the liquid and has little to do with the internal structure of the polar molecules or their natural rotational motion. Whereas the rotational frequencies of isolated molecules are in the upper microwave and infrared bands, the relaxation frequencies of polar liquids containing these molecules are in the radio frequency range. Indeed in the case of some highly viscous organic liquids, having very long relaxation times, the relaxation frequencies are found in the audio range.

The foregoing physical arguments can be put on a quantitative basis through the equation of (rotational) motion of the polarization $\mathbf{P}_p$. We consider only the polar contribution to the polarization since at the low frequencies in question the induced polarization has the static value of eqn (10.3.4) and its contribution to the final result will be taken into account at a later stage. The motion of the polarization is so heavily damped that the inertial term is assumed to be negligibly small. Hence the second derivative, $d\mathbf{P}^2/dt^2$, will not appear in the equation of motion. The viscous resistance to the motion is proportional to the first derivative and will be represented by the term $\tau d\mathbf{P}/dt$. The tendency of the Brownian motion to restore equilibrium conditions is clearly proportional to the polarization and will be represented by $\mathbf{P}_p$. The force of the applied field applies an a.c. torque to the polarization and will be represented by the term $(n_0 p^2/3kT)\ \mathbf{E}\ e^{j\omega t}$, where $\mathbf{E}$ is the local field and eqn (10.3.10) has been applied. Hence we can write

$$\tau \frac{d\mathbf{P}_p}{dt} + \mathbf{P}_p = \frac{n_0 p^2}{3kT} \mathbf{E} e^{j\omega t}. \tag{10.6.1}$$

As we are interested in the steady state we can assume that the polarization will vary with time as $\mathbf{P}_p\ e^{j\omega t}$. Substitution in the above equation yields the result

$$\frac{\mathbf{P}_p}{\epsilon_0 \mathbf{E}} = \chi_p = \frac{n_0 p^2}{3\epsilon_0 kT} \frac{1}{1 + j\omega\tau} \tag{10.6.2}$$

where $\chi_p$ is the a.c. susceptibility due to polar molecules only. By eqn (10.1.4) the overall a.c. susceptibility is $\epsilon = 1 + \chi_i + \chi_p$ where $\chi_i$ is the induced susceptibility assumed to have the d.c. value under present conditions. Noting also that by the results of Section 10.3 we can write $n_0 p^2/3\epsilon_0 kT = \epsilon_s - \epsilon_i$, where $\epsilon_s$ is the overall static dielectric constant, eqn (10.6.2) can be put in the form

$$\frac{\epsilon - \epsilon_i}{\epsilon_s - \epsilon_i} = \frac{1}{1 + j\omega\tau}. \tag{10.6.3}$$

Separating real and imaginary parts we find the following expressions for $\epsilon = \epsilon' - j\epsilon''$.

$$\epsilon' = \epsilon_i + \frac{\epsilon_s - \epsilon_i}{1 + \omega^2 \tau^2} ; \quad \epsilon'' = \frac{(\epsilon_s - \epsilon_i)\omega\tau}{1 + \omega^2 \tau^2}. \tag{10.6.4}$$

These results are plotted in Fig. 10.2 and confirm the conclusions of the qualitative argument.

As a final comment it may be desirable to point out that some viscous or supercooled liquids may have more than one relaxation time, giving rise to the behaviour depicted in Fig. 10.2 at two distinct points on the frequency axis.

## 10.7 Breakdown in Dielectrics

We have now learned about some dielectric properties of insulators and in particular about their behaviour near resonance. We noted the sharp increase in losses which is observed under these conditions and we found that this has nothing to do with conduction currents based on free electrons. In this section we take up the latter possibility to find that it is frequently destructive and sets a limit to the insulating properties of dielectrics.

Dealing first with gases we recall that the question of breakdown has been considered at length in Chapter 9 within the framework of conduction in gases. Since breakdown of a gas entails a high degree of ionization, it can be said to cause a change in its structure and properties. However, the change is usually reversible and the gas returns to its insulating properties upon removal of the applied voltage. All the same, if the gas is used as an insulator, breakdown may cause destructive changes in the surrounding equipment by bombardment and heating, particularly if the arc condition is reached.

Of considerable practical importance in the context of insulation is breakdown in solid dielectrics. Perhaps the first observation to make is that breakdown is not caused by a single phenomenon but is based on a number of distinct processes. In what follows it is proposed to describe some of them in terms of the concepts developed in earlier chapters.

As explained in Chapter 1 the band structure of a pure crystalline dielectric is characterized by a very wide forbidden gap, say, 6 eV. An estimate based on the methods of Chapter 3 will disclose that the concentration of conduction electrons is negligibly small and hence the application of an electric field produces a current flow which is barely detectable by the most sensitive instruments available. Moreover, in a regular lattice the electrons accelerated by the field have a very low probability of colliding with the lattice and causing impact ionization (Section 5.8) or appreciable heating of the dielectric. For these reasons a pure crystalline insulator will have a very high breakdown strength if actually used. In point of fact practically all dielectrics in use are neither pure nor mono-crystalline and these facts influence decisively their properties.

The effects of impurities and crystal defects on the properties of dielectrics are best understood in terms of the energy band structure, as in the case of semiconductors. The principal feature is the presence of discrete electron states within the forbidden gap. Most of these are in the nature of donor states situated relatively close to the conduction band, but by contrast with semiconductors these states do not occupy a well-defined donor level (see Chapter 2). In that case the donor impurities were carefully selected and introduced under controlled conditions into a single crystal specimen in the course of its growth. Instead the extra electron states in glassy or plastic insulators are largely due to crystal defects which take a variety of forms, and the resulting distortions of crystalline fields give rise to varying energy levels. However, the net result is similar to semiconductors in that there is a supply of electrons which are much easier to excite into the conduction band than electrons from the valence band.

The free conduction electrons are subject to three types of collision processes, which serve as a medium through which energy can be interchanged and thermodynamic equilibrium can be established within the specimen.

(i) Collisions between free electrons in the conduction band.

(ii) Collisions between conduction electrons and the vibrating lattice (phonons).

(iii) Collisions between conduction electrons and lattice defects or impurities (traps).

A pure crystal is by definition one in which the last collision mechanism is negligible. The relative probability of the first two mechanisms depends on temperature. At low temperatures collisions with the lattice predominate because the density of conduction electrons is so low that a meeting of two free electrons is extremely unlikely. As the temperature increases the density of conduction electrons increases more rapidly than the intensity of lattice vibrations and a point is reached when electron-electron collisions predominate. The division between the two temperature ranges is sometimes termed the *critical temperature* $T_c$. In impure and amorphous dielectrics the density of traps and conduction electrons is very much higher and the third collision mechanism becomes dominant at normal temperatures.

Let us now consider the effect of applying a field to a pure, crystalline dielectric at a temperature below $T_c$. The dominant collisions are with the lattice and the net effect is at first a transfer of energy from the free, accelerating electrons to the lattice which becomes warmer. As the electric field is increased the kinetic energy of some electrons becomes sufficient to ionize some lattice atoms by impact. The free electrons thus generated become available for further ionization, the regenerative process leading to an *avalanche* and *breakdown*, resembling avalanche breakdown in semiconductors and gases. The above process is usually referred to as *low-temperature breakdown*.

In impure dielectrics the dominant process are collisions between conduction electrons and trapped electrons. The latter are relatively easy to excite to the conduction band where they join the free electrons, enhance the conductivity $\sigma$ and hence heating. The latter causes more electrons to be activated from the traps, and ultimately from the valence band, and breakdown follows. This process is usually referred to as *high-temperature breakdown*.

A distinct type of breakdown, called *thermal breakdown*, is caused by a reduction of the energy gap with increasing temperature. Resistive heating and a.c. dielectric loss may cause a substantial rise in the sample temperature when its thermal conductivity is insufficient to conduct away the generated heat. The reduced gap leads to an enhanced activation of electrons into the conduction band which in turn increases the heating until breakdown follows.

Most practical dielectrics have a degree of porosity which means that small cavities filled with a gas are interspersed throughout the body of the dielectric.

Such cavities are a cause of *discharge breakdown*, because a gas discharge is usually induced in the cavities long before any other breakdown condition is reached by the material. This is due to two factors: (a) The breakdown voltage of the gas is lower than that of the surrounding solid. (b) An electrostatic calculation shows that the electric field is higher inside a cavity of lower relative permittivity than the surrounding medium. A detailed analysis shows that a cavity discharge is in the nature of an intermittent spark which tends to erode the opposite walls of the cavity and thus to elongate it until a hollow channel is formed throughout the thickness of the material. In organic materials the heat generated may carbonize the material and cause a short circuit.

Dielectric breakdown in liquids is usually induced by *field emission* of electrons from the cathode. The emitted electrons are rapidly accelerated by the high field present and the energy they gain between collisions with liquid molecules is greater than the energy loss during collisions. As a result a state is reached in which liquid molecules are ionized on impact and an avalanche is precipitated.

# Part II

*Chapter 11*

# Schroedinger's Equation

## 11.1 Introduction of Schroedinger's Equation

The fundamental laws of classical mechanics are Newtons Laws, summarized in the equation

$$m \frac{d^2 \mathbf{R}}{dt^2} = \mathbf{F}. \tag{11.1.1}$$

These laws predict accurately the dynamical behaviour of large mass particles and rigid bodies, including the movements of planets and other celestial bodies as well as man made machines. In fact Newton's laws were formulated as the outcome of a long history of astronomical observation.

With the discovery of atomic and nuclear physics it was found that Newton's laws do not describe accurately the dynamics of small particles, which constitute atoms and molecules. It became necessary to formulate a new theoretical basis to deal with microscopic particles. As a result quantum mechanics emerged as a more general theory than classical mechanics. This newly established discipline is capable of describing and predicting the dynamical states of atomic particles, but it does more than this. It contains classical mechanics as a special case within itself.

Just as classical mechanics is based on Newton's laws, quantum mechanics can be developed from Schroedinger's equation. We introduce this equation as an axiomatic statement, without attempting to prove or derive it in any way. In this connection the reader is reminded that he was never given a proof or derivation of Newton's equations of motion. Instead their physical meaning and application to the solution of dynamical problems was explained.

To begin with we shall only consider the motion of a single particle of mass $m$ in a potential field $V = V(x, y, z, t)$. Schroedinger's equation in this case has the form

$$-\frac{\hbar^2}{2m} \nabla^2 \psi + V\psi = i\hbar \frac{\partial \psi}{\partial t} \tag{11.1.2}$$

where

$$\nabla^2 = \frac{\partial^2}{\partial x^2} + \frac{\partial^2}{\partial y^2} + \frac{\partial^2}{\partial z^2} = \text{Laplacian operator}$$

$h = $ Planck's constant

$$\hbar = \frac{h}{2\pi}.$$

The above relation is a partial differential equation which includes the mass of the particle and the potential function among its coefficients. On solving eqn (11.1.2) we obtain as a rule many functions $\psi = \psi(x,y,z,t)$, in general complex, which somehow describe the motion of the particle.

Here we come up against a fundamental difference between the classical equations of motions and Schroedinger's equation. Whereas solutions of the former are functions which define the position of the particle at any instant of time, solutions of the latter have no such direct physical interpretation. Before meaningful physical results are reached, solutions of Schroedinger's equation must be further processed.

The nearest expression which can be given a physical meaning is the squared modulus, $|\psi|^2 = \psi^*\psi$, where the asterisk denotes the complex conjugate. This new function is interpreted as a *position probability distribution*. The expression $|\psi|^2\, dx\, dy\, dz$ is the *probability that the particle under consideration can be found in the volume element dx dy dz centred on the point $(x, y, z)$*. Since the function contains the time $t$ as an independent variable, the probability distribution $|\psi|^2$ must, in general, be expected to be a function of time also. However, since the particle must be found somewhere *within the region of its motion*, the integral of the probability distribution over the volume of that region must be constant.

$$\int_V |\psi(x, y, z, t)|^2\, dx\, dy\, dz = \text{const.} \qquad (11.1.3)$$

This relation imposes a major restriction on the form of the solution $\psi(x, y, z, t)$ of Schroedinger's equation and will be applied repeatedly to the specific examples discussed in subsequent chapters.

The foregoing is only the first step of processing solutions of Schroedinger's equation to obtain dynamical information. In the chapters to follow many more mathematical steps will be taken and their physical significance explained, and this will constitute the bulk of the principles of quantum mechanics. However, before we pass on to this work, we consider some aspects of solving Schroedinger's equation in the following section.

## 11.2 Elimination of the Time Variable

Schroedinger's equation resembles in form partial differential equations en-countered in connection with electromagnetic waves and other types of wave motion in continuous media. There are differences, however, and the most important of them is the presence of the potential term $V\psi$, particularly when $V$ is a function of all the independent variables of the problem. In such cases it is difficult to obtain a general solution of Schroedinger's equation in closed analytical form. In fact, general solutions of this type are available only for a few simple cases. Various approximate methods exist, which can be used to obtain numerical solutions in other cases. However, since the object of this treatment is to present the fundamentals of the theory and to relate them to physical phenomena, we shall limit ourseleves almost entirely to a thorough discussion of a few typical cases. Fortunately enough these are sufficient to provide a general guide to more complicated problems.

There is one important simplification of eqn (11.1.2) which can be introduced in general terms at this stage. The time variable can be eliminated by the procedure of separation of variables, provided the potential $V$ is assumed to be independent of time. As this will be the case in all problems to be dealt with in the present treatment, we assume once and for all that $V$ is of the form

$$V = V(x, y, z). \tag{11.2.1}$$

Following the method of separation of variables we further assume that the solutions of Schroedinger's equation can be written in the form

$$\psi(x, y, z, t) = u(x, y, z) T(t) \tag{11.2.2}$$

where $\psi$ is a product of a function $u$ of the space variables only and of a function $T$ of time only. Equation (11.1.2) can now be rewritten as follows

$$T\left(-\frac{\hbar^2}{2m}\nabla^2 u(x, y, z) + V(x, y, z) u(x, y, z)\right) = i\hbar \frac{dT}{dt} u$$

$$\frac{1}{u}\left(-\frac{\hbar^2}{2m}\nabla^2 u + Vu\right) = i\hbar \frac{1}{T}\frac{dT}{dt} = E = \text{const.} \tag{11.2.3}$$

We are now left with two seperate differential equations, one including the time and the other space variables only.

$$i\hbar \frac{dT}{dt} = ET \tag{11.2.4}$$

$$\left(-\frac{\hbar^2}{2m}\nabla^2 + V\right) u = Eu. \tag{11.2.5}$$

Equation (11.2.4) is of a simple type and its solution is readily written down

$$T(t) = C e^{-i(E/\hbar)t} \qquad (11.2.6)$$

where $C$ is a constant of integration.

It now remains to solve eqn (11.2.5), and it is this equation that will be the subject of the following chapters. We note for future reference that all solutions of (11.2.5) are to be multiplied by (11.2.6) to obtain the full solution $\psi$ of Schroedinger's eqn (11.1.2). Thus

$$\psi(x, y, z, t) = C u(x, y, z) e^{-i(E/\hbar)t}. \qquad (11.2.7)$$

The value of the separation constant $E$ will be determined from the boundary conditions to be imposed on eqn (11.2.5). Its physical significance will form a recurrent theme in all our subsequent work.

At this point a few general remarks can be made regarding the probability distributions introduced at the end of the preceding section. Writing the probability distribution in terms of the partial solution (11.2.7) we find

$$|\psi|^2 = \psi^* \psi = |u(x, y, z)|^2 = u^* u. \qquad (11.2.8)$$

Thus in this case the time-dependent exponential factor drops out, and the probability distribution is a function of the space variables only. Hence the condition (11.1.3) assumes the simpler form

$$\int_V |\psi(x, y, z, t)|^2 \, dx \, dy \, dz = \int_V |u(x, y, z)|^2 \, dx \, dy \, dz. \qquad (11.2.9)$$

$$= \text{const.}$$

It is in the last form that the above condition will be applied to the examples discussed in the following chapters.

*Chapter 12*

# Basic Concepts of Quantum Mechanics and the Linear Harmonic Oscillator

In this chapter Schroedinger's equation will be applied to the problem of the linear oscillator. The example will be used as a vehicle whereby some of the fundamental concepts of quantum mechanics will be introduced. This work will then be continued in Chapter 13, on the more complicated example of the hydrogen atom. Both these problems illustrate solutions of Schroedinger's equation in closed analytical form and provide suitable illustrations of the basic principles.

### 12.1 The Equation of Motion of the Linear Oscillator and its Solutions: Position Probability Distributions

We consider a particle of mass $m$ constrained to move in one dimension, say along the $x$-axis. The particle is hemmed in between two springs, each having

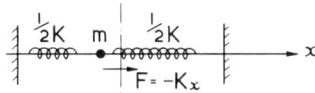

Fig. 12.1.

spring constant $\frac{1}{2}K$, as shown in Fig. 12.1. If the particle is displaced from the position of equilibrium by a distance $x$ the springs will exert a restoring force given by

$$F = -Kx. \tag{12.1.1}$$

By integration of (12.1.1) we obtain the corresponding elastic potential in which the particle is effectively immersed.

$$V(x) = \tfrac{1}{2}Kx^2. \tag{12.1.2}$$

After the particle has been displaced from its equilibrium position and then released, it will continue moving according to the classical equation of motion

$$m \frac{d^2 x}{dt^2} = -Kx \qquad (12.1.3)$$

if it is sufficiently heavy, or according to Schroedinger's equation if its mass is of an atomic order of magnitude.

To write down Schroedinger's equation of motion we use the potential function (12.1.2) and note that it is independent of time. Hence the time-independent equation (11.2.5) is directly applicable with the further simplification that only one space variable $x$ is involved. The partial differential equation therefore reduces to the ordinary differential equation

$$-\frac{\hbar^2}{2m} \frac{d^2 u}{dx^2} + \tfrac{1}{2}Kx^2 u = Eu. \qquad (12.1.4)$$

Before an attempt is made to solve the above equation, boundary conditions must be imposed. Since the solutions are to be processed into probability distributions according to the principle explained in the preceding chapter we utilize eqn (11.2.9). When adapted to the present case the latter assumes the form

$$\int_{-\infty}^{\infty} |u(x)|^2 \, dx = \text{const.} \qquad (12.1.5)$$

The integration extends over the length of the $x$-axis, since the particle must be somewhere there. Subject to this boundary condition the equation of motion (12.1.4) can be solved by the method of expansion in power series. Although the analysis is not difficult, it is rather lengthy, and since it does not affect the physical argument it will be omitted here. Interested readers are referred to other textbooks on the subject.

The main features of the solutions of eqn (12.1.4) are as follows. Solutions satisfying the condition (12.1.5) can only be obtained if the undetermined constant of separation $E$ is given discrete values according to the equation

$$E_n = (n + \tfrac{1}{2})\hbar \sqrt{\frac{K}{m}} = (n + \tfrac{1}{2})\hbar\omega_c \qquad (12.1.6)$$

where $n$ assumes all positive integral values 0, 1, 2, 3 ... and $\omega_c = \sqrt{K/m}$ is the frequency of a classical oscillator having mass $m$ and elastic constant $K$.

Corresponding to each value $E_n$ a solution of eqn (12.1.4) can be constructed and is denoted by $u_n(x)$. The solutions have the form of finite polynomials multiplied by exponential factors given by the expression

$$u_n(x) = N_n H_n(\alpha x) e^{-(1/2)\alpha^2 x^2} \qquad (12.1.7)$$

where $N_n$ is a constant of integration still to be determined, $H_n(\alpha x)$ is a Hermite polynomial of order $n$ and

$$\alpha^2 = \frac{\sqrt{mK}}{\hbar} \qquad (12.1.8)$$

is a subsidiary constant. Some Hermite polynomials of low order are listed in Table 12.1.

**Table 12.1**

Eigenfunctions of the linear oscillator

| $n$ | 0 | 1 | 2 | 3 |
|---|---|---|---|---|
| $H_n(\xi)$ | 1 | $2\xi$ | $4\xi^2 - 2$ | $8\xi^3 - 12\xi$ |

$$H_n(\xi) = (-1)^n e^{\xi^2} \frac{d^n}{d\xi^n} e^{-\xi^2} \; ;$$

$$N_n = \sqrt{\frac{\alpha}{2_n n! \sqrt{\pi}}} \; ; \quad \alpha x = \xi \; ;$$

At this point it is necessary to introduce some special terms which are continually used in quantum theory. The values of the separation constant $E_n$ are called *eigenvalues* of the differential eqn (12.1.4). Its solutions $u_n(x)$ are called *eigenfunctions*. Each eigenfunction is said to belong to the corresponding eigenvalue. The suffix $n$, which labels the eigenvalues and eigenfunctions, will be referred to as a *quantum number*. Each eigenvalue and the corresponding eigenfunction satisfy the equation

$$\left( -\frac{\hbar^2}{2m} \frac{d^2}{dx^2} + \tfrac{1}{2}Kx^2 \right) u_n(x) = E_n u_n(x) \qquad (12.1.4a)$$

which is called an *eigenvalue equation*. It is identical in form with Schroedinger's equation (12.1.4), except that a specific eigenfunction and eigenvalue have been substituted for the unknown function $u$ and constant of separation $E$. Equations of the form (12.1.4a) are of great importance in quantum theory. The important fact to note about (12.1.4a) is that it shows how a fairly complicated differential operator affects certain functions. *Its operation of the function* $u_n(x)$ *is equivalent to multiplication by a constant,* the eigenvalue $E_n$.

We now take the first steps towards the physical interpretation of the solutions of Schroedinger's equation. Each eigenfunction $u_n(x)$ identifies a distinct *dynamical state* of motion, called an *eigenstate* of the oscillating particle. The eigenfunctions are mathematically analogous to modes of vibration of a string, or modes of propagation of electromagnetic waves in a waveguide. Physically, however, they have no further significance.

To obtain physically meaningful results we apply the postulate stated in the first chapter. This means that the *square of the modulus of an eigenfunction gives the position probability density* of the particle in the corresponding state of motion. Thus the expression

$$|u_n(x)|^2 \, dx \tag{12.1.9}$$

is assumed to be the probability that the particle is found in the interval $dx$ at a distance $x$ from the origin. The foregoing postulate is justified by the fact that results derived from it agree with experiment.

Since the squares of eigenfunctions are to be interpreted as probability distributions, their integrals over the entire range of the independent variable should be equal to unity, as the particle must be somewhere. It is this fact that made it necessary to impose the boundary condition (12.1.5) on solutions of Schroedinger's equation. Moreover, with the help of this condition we can determine the constants of integration appearing in eqn (12.1.7). Thus

$$\int_{-\infty}^{\infty} |u_n(x)|^2 \, dx = N_n^2 \int_{-\infty}^{\infty} H_n^2(\alpha x) e^{-\alpha^2 x^2} \, dx = 1. \tag{12.1.10}$$

From the properties of Hermite polynomials it is possible to evaluate the integral in (12.1.10) and hence to obtain a general expression for $N_n$.

$$N_n = \sqrt{\frac{\alpha}{\sqrt{\pi} 2^n n!}}. \tag{12.1.11}$$

The reader should verify the validity of eqn (12.1.11) for some of the low-order wave functions listed in Table 12.1.

Once the values of the constants of integration given by (12.1.11) are substituted in eqn (12.1.7) the eigenfunctions are said to be *normalized*, which means that they satisfy eqn (12.1.10).

At this point we note a most important property of the eigenfunctions: their *orthogonality*. By this we mean that the normalized eigenfunctions satisfy the following integral relation.

$$\int_{-\infty}^{\infty} u_n^*(x)\, u_m(x)\, dx = \delta_{nm} \tag{12.1.12}$$

where $\delta_{nm}$ is the Kronecker $\delta$- symbol having the property

$$\delta_{nm} = \begin{matrix} 1 \text{ for } n = m \\ 0 \text{ for } n \neq m \end{matrix}$$

When $n = m$ eqn (12.1.12) reduces to the normalizing condition (12.1.10). Although the eigenfunctions of the harmonic oscillator are real, and their complex conjugates equal themselves, this is not always so with solutions of Schroedinger's equation. The possibility of complex eigenfunctions is taken into account in the definition of orthogonality, eqn (12.1.12). The orthogonality relation for eigenstates is fundamental in quantum theory and will be constantly applied in the work to follow.

Having written down solutions of Schroedinger's equation for the linear oscillator and having laid down the first postulate regarding their physical interpretation let us now consider them in graphical and tabular form. Table 11.1 lists some of the lowest order Hermite polynomials together with some additional information.

Some eigenfunctions are displayed graphically in Fig. 12.2. Their most noteworthy feature is their wavelike shape, which becomes more apparent the higher the quantum number. As they are reminiscent of standing waves on a

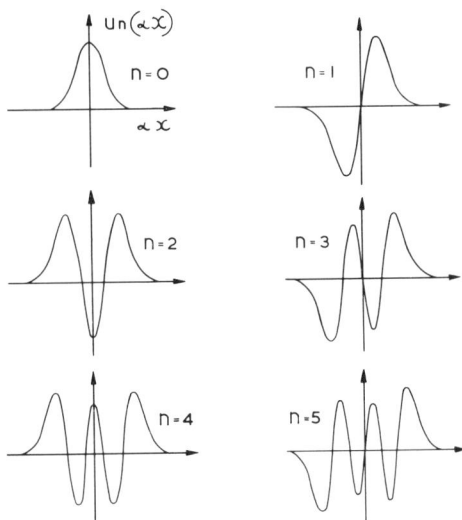

Fig. 12.2. Some eigenfunctions of the linear harmonic oscillator.

string, or in a resonant section of a waveguide, the eigenfunctions are frequently called *wave functions*. In subsequent chapters we shall see that reasons for the use of this term go beyond the superficial resemblance to standing waves. It should be noted that the wave functions are either even or odd functions. This fact is frequently referred to in quantum theory by the terms *symmetric* or *antisymmetric* wave functions, or functions having *even* or *odd parity*.

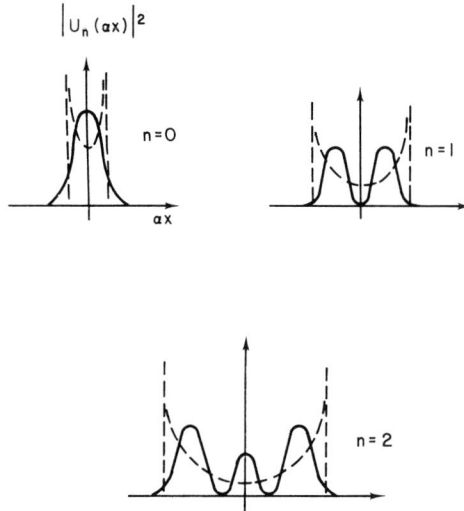

**Fig. 12.3.** The position probability distributions of some states of the harmonic oscillator.

The position probability distributions of some eigenstates of the harmonic oscillator are shown graphically in Fig. 12.3. It is clear from these graphs that the oscillating particle is unlikely to be found outside a certain range of the deflection $x$, marked by vertical dotted lines. These dotted lines mark the maximum deflection from equilibrium of a classical oscillator having the same energy as the corresponding quantum mechanical oscillator. The precise relationship between the classical and quantal motion will be discussed in the following section. In the meantime we note that within the limits of oscillation of the particle its position probability varies between zero and a number of maxima. The higher the number of a given eigenstate the greater the number of maxima. Again it is useful to note at this point that the probability distributions are reminiscent of the distribution of stored energy along a standing wave pattern or interference pattern.

## 12.2 Dynamical Variables as Operators and their Average Values

So far we have obtained only one item of physical information regarding the motion of the linear oscillator on quantum principles: we have plotted the probability of finding it at given points along the line of its motion. As a rule we shall want to know the position, momentum, energy and perhaps other, electrical characteristics of motion of particles. In this section we explain some of the methods whereby such information can be evaluated using the eigen functions of a dynamical system.

From statistics we recall that given the probability distribution of a random variable we can evaluate the average values or expectations of any function of the random variable. Thus, given the position probability density $|u_n(x)|^2$ of the $n$-th state of the linear oscillator its average position $\langle x \rangle_n$ can be calculated from the expression

$$\langle x \rangle_n = \int_{-\infty}^{\infty} x |u_n(x)|^2 \, dx = 0. \tag{12.2.1}$$

Hence the average position of the particle in any eigenstate is zero. This agrees entirely with our knowledge of linear harmonic oscillations about the origin on classical principles.

It may be noted in passing that no calculation is required to arrive at the result (12.2.1). We only have to note that the integrand is an odd function, being the product of an even function $|u_n(x)|^2$ and $x$, and that the interval of integration is symmetric about the origin.

By the same principle we can evaluate the average value of the elastic potential energy of the particle which is a function of $x$ only: $-V(x) = \frac{1}{2}Kx^2$.

$$\langle V \rangle_n = \int_{-\infty}^{\infty} \frac{1}{2}Kx^2 |u_n(x)|^2 \, dx = \frac{1}{2}(n + \frac{1}{2})\hbar\omega_c = \frac{1}{2}E_n. \tag{12.2.2}$$

This integral is more difficult to evaluate than (12.2.1) and we shall be content to accept the result as stated. The reader is urged, however, to check it for some low quantum numbers using the wave functions listed in Table 12.1.

The above result leads directly to some of the most important facts of quantum mechanics. We note that the average value of the potential energy equals half the eigenvalue of the state under consideration. Furthermore, we can verify by a straightforward calculation that the average value of the potential energy of a classical oscillator equals the average value of its kinetic energy. It is tempting to jump to the conclusion that the same applies to the quantum mechanical oscillator, that the sum of its potential and kinetic energies equals the eigenvalue of the corresponding eigenstate. However, before this result can be established convincingly, new methods and terminology must be introduced.

First of all, quantities descriptive of the dynamical state of a particle such as position, momentum or energy are usually called *dynamical variables*. Their average values with respect to given wave functions are calculated by the methods exemplified above in eqns (12.2.1) and (12.2.2).

Secondly, in quantum theory it is customary, and as we shall see necessary. to write integrals like (12.2.2) rather differently. The probability density is written as the product $u_n^*(x)u_n(x)$, and the function whose average value is required is placed in the centre of the product. Thus

$$\langle V \rangle_n = \int_{-\infty}^{\infty} u_n^*(x)\, V(x)\, u_n(x)\, dx. \tag{12.2.3}$$

This makes no difference when averages of functions of space variables are computed, but it is crucial to the evaluation of average values of other functions, which we now introduce.

Dynamical variables, such as momentum or kinetic energy, are not functions of position but of velocity. Hence it is impossible to evaluate their averages with respect to probability densities such as $|u_n(x)|^2$. To get over this difficulty one of the major postulates of quantum theory is put forward. *It is assumed that for purposes of evaluating average values of dynamical variables components of momentum can be replaced by differential operators with respect to the corresponding space variables as follows:*

$$p_x = -i\hbar\, \frac{\partial}{\partial x}, \quad p_y = -i\hbar\, \frac{\partial}{\partial y}, \quad p_z = -i\hbar\, \frac{\partial}{\partial z},$$

or

$$\mathbf{p} = -i\hbar\left( \hat{i}\, \frac{\partial}{\partial x} + \hat{j}\, \frac{\partial}{\partial y} + \hat{k}\, \frac{\partial}{\partial z} \right) = -i\hbar\ \text{grad}. \tag{12.2.4}$$

These operators are then placed between the complex conjugate of a wave function and the wave function proper like the potential in eqn (12.2.3). After the indicated differential operation has been effected, the integral can be evaluated. The ultimate justification for this procedure lies in the fact that it leads to results which agree with experiment.

As an example let us compute the average momentum of the linear oscillator in the $n$-th eigenstate. Since this is a one-dimensional problem the partial differential operators of (12.2.4) reduce to a single ordinary differential operator.

$$p = -i\hbar\, \frac{d}{dx}.$$

Hence the average value of the momentum is

$$\langle p \rangle_n = \int_{-\infty}^{\infty} u_n^*(x)\, p u_n(x)\, dx$$

$$= -i\hbar \int_{-\infty}^{\infty} u_n^*(x)\, \frac{d}{dx}\, u_n(x)\, dx = 0. \tag{12.2.5}$$

The foregoing result will be seen to hold when it is recalled that differentiation changes the parity of a function. Thus $u_n(x)$ is changed from, say, an even into an odd function by the operator $d/dx$. After multiplication by $u_n^*(x)$, which remains even, we are left with an integrand of odd parity. Integration over an interval which is symmetric about the origin then reduces to zero. The reader is urged to satisfy himself by direct integration that the result (12.2.5) holds for eigen states of low quantum number $n$.

Again we note that the above result agrees with what we expect of a classical linear oscillator. Since momentum is a vector quantity it should average out to zero for harmonic motion symmetric about the origin.

The method of calculating average values of momenta, exemplified by eqn (12.2.5), is easily extended to other dynamical variables which can be expressed as functions of momenta. In every case the appropriate differential operator is substituted for a component of momentum, and the dynamical variable is allowed to operate on the eigenfunction in question, before an integral of the form (12.2.5) is evaluated. In general, given a dynamical variable which is a function of space coordinates and momenta we write

$$F(x, y, z, p_x, p_y, p_z) = F\left(x, y, z, -i\hbar\,\frac{\partial}{\partial x}, -i\hbar\,\frac{\partial}{\partial y}, -i\hbar\,\frac{\partial}{\partial z}\right).$$

Its average value for a given eigenstate $u_n$ is

$$\langle F \rangle_n = \int_V u_n^*\, F u_n\, d\tau \tag{12.2.6}$$

where $d\tau$ is an element of volume and $V$ is the region over which the integration is to be carried out.

The above procedure uses a *dynamical variable as an operator* and provides a reason why in quantum theory it is customary to call all dynamical variables operators. As our treatment unfolds we shall see further reasons for this practice.

An example of an operator or dynamical variable which is a function of a coordinate and a momentum is provided by the total energy of the linear oscillator. We write

$$H = \frac{p^2}{2m} + \tfrac{1}{2}Kx^2$$

$$= -\frac{\hbar^2}{2m}\frac{d^2}{dx^2} + \tfrac{1}{2}Kx^2 \qquad (12.2.7)$$

where $H$ is the energy operator or *Hamiltonian operator*. In this case it is the sum of the kinetic and potential energies of the oscillating particle. In more complicated problems, which will be encountered in later chapters, the Hamiltonian operator, or briefly *the Hamiltonian, will include all contributions to the energy of a dynamical system, including electrical and magnetic energies.*

Let us now find the average value of the Hamiltonian operator, or energy, for the $n$-th eigenstate of the harmonic oscillator.

$$\langle H \rangle_n = \int_{-\infty}^{\infty} u_n^*(x)\, H u_n(x)\, dx$$

$$= \int_{-\infty}^{\infty} u_n^*(x)\left( -\frac{\hbar^2}{2m}\frac{d^2}{dx^2} + \tfrac{1}{2}Kx^2 \right) u_n(x)\, dx.$$

The value of this integral can be found without any actual computation, only by applying results already established. As the method is of fundamental importance in quantum theory, every step of the argument to be given should be carefully followed.

To begin with we note that the factor $Hu_n(x)$ is identical with the left-hand side of the eigenvalue equation eqn (12.1.4a) p. 183. Hence we can substitute

$$Hu_n(x) = E_n u_n(x) \qquad (12.2.8)$$

in the integral above.

$$\int_{-\infty}^{\infty} u_n^*(x)\, H u_n(x)\, dx = \int_{-\infty}^{\infty} u_n^*(x)\, E_n u_n(x)\, dx = E_n.$$

The constant $E_n$ is taken outside the integral, and the latter reduces to unity by virtue of the normalization condition eqn (12.1.10). Hence we obtain

$$\langle H \rangle_n = E_n \qquad (12.2.9)$$

*The average value of the energy of the oscillator in the n-th eigenstate is thus equal to the corresponding eigenvalue*, as anticipated above.

The eigenvalues $E_n$ are called the *energy levels* of a dynamical system, in this case of the oscillator. One of the most important properties of the harmonic oscillator, as described by quantum theory, is the fact that it can assume only discrete energy levels. On classical theory the energy of the oscillating particle can have any one of a continuum of values, as imparted to it by the initial deflection against the elastic force of the spring.

Referring back to eqn (12.1.6) for the eigenvalues of the linear oscillator, we see that its energy levels are spaced evenly on an ascending scale at intervals of $\hbar\omega_c$. The lowest energy level is not zero, however, as would be expected from classical considerations. The value $E_0 = \frac{1}{2}\hbar\omega_c$ is called the *zero point energy* and is a phenomenon peculiar to quantum mechanics. When the oscillator is in the $n$-th eigenstate it is sometimes said to have *n quanta* of energy. Since the energy has its origin in the elastic force of the spring, and the oscillations are therefore of acoustic nature, the quanta of energy are frequently called *phonons*.

It may be remarked at this juncture that quite complicated oscillatory systems may be described by equations identical in form with Schroedinger's equation of the single harmonic oscillator. For example the normal modes of vibrations of crystal lattices obey the same equations.

Let us now return to Fig. 12.3. in which a comparison between the properties of the oscillator on the classical and quantum theory was suggested. To compare the limits of linear deflection on the two theories one plots the position probability density for a given quantal eigenstate. Then the amplitude of the classical oscillator is found, after an amount of energy equal to the energy level of the given eigen state has been assigned to it. This is done by choosing the mass of the particle and the elastic constant of the spring so that $\sqrt{K/m} = \omega_c$. Given $K$ and $m$, Newton's equation of motion can be solved. Once the position $x$ of the particle has been determined as a function of time, its position probability distribution is easily worked out. The results are marked in Fig. 12.3. by dotted lines.

On the classical theory the particle is most likely to be found near the extremes of its deflection, since it is moving slowly there. This is in marked disagreement with the lowest quantum state where the particle is most likely to be in the centre of its movement. However, as the quantum number $n$ increases, the probability distributions on the two theories approach each other. For high values of $n$ the agreement is quite good, provided the rapid oscillations of the quantal probability are ignored.

## 12.3  Operators, Matrices and their Physical Interpretation

In the preceding section we have combined the eigenfunctions of Schroedinger's equation and operators representing dynamical variables to obtain average values of the latter. The average values took the form of integrals such as eqn (12.2.6).

This equation represents one of the most important stages through which solutions of Schroedinger's equations are processed, to arrive at results having physical significance.

In this section we take another step in this processing procedure by introducing integrals of a more general form than (12.2.6). We include under the integral sign two different eigenfunctions as follows:

$$F_{nm} = \int_{-\infty}^{\infty} u_n^* F u_m \, dx. \tag{12.3.1}$$

The symbol $F_{nm}$ is merely an abbreviation for the complete integral. We note that by inserting all possible eigenfunctions into the integral we obtain a two-dimensional array of values $F_{nm}$ which can be formed into a matrix.

$$[F_{nm}] = \begin{bmatrix} F_{11} & F_{12} & \cdots \\ F_{21} & F_{22} & \cdots \\ \cdots\cdots\cdots\cdots \end{bmatrix} \tag{12.3.2}$$

The *matrix elements* of an operator as defined above have various physical interpretations, which will be given in stages as the subject develops. To begin with we observe that the diagonal elements of the matrix (12.3.2) are the average values of the operator $F$ with respect to the eigenstates in question.

As an example let us write down the matrix of the position $x$ of the harmonic oscillator. The diagonal elements were all found to be zero in eqn (12.2.1). It remains now to evaluate the off-diagonal elements, which we denote by the symbol $x_{nm}$.

$$x_{nm} = \int_{-\infty}^{\infty} u^*(x)\, x u_m(x)\, dx = \begin{cases} \dfrac{1}{\alpha} \sqrt{\dfrac{n+1}{2}}, & m = n+1 \\[2ex] \dfrac{1}{\alpha} \sqrt{\dfrac{n}{2}}, & m = n-1. \\[2ex] 0 & , \text{ otherwise} \end{cases}$$

$$\tag{12.3.3}$$

This integral can be evaluated in general from the properties of the eigenfunctions, but for our purpose it is not imperative to trace the calculation in full. One or two values of $x_{nm}$ should, however, be computed for low quantum numbers, using Table 12.1.

The above results, written out in full, are as follows:

$$[x_{nm}] = \frac{1}{\alpha} \begin{bmatrix} 0 & \frac{1}{\sqrt{2}} & 0 & 0 & \cdots \\ \frac{1}{\sqrt{2}} & 0 & 1 & 0 & \cdots \\ 0 & 1 & 0 & \sqrt{\frac{3}{2}} & \cdots \\ 0 & 0 & \sqrt{\frac{3}{2}} & 0 & \cdots \\ & & \cdots & & \end{bmatrix} \qquad (12.3.4)$$

It is clear from the foregoing that only very few of the matrix elements of the position operator do not vanish.

The physical significance of the off-diagonal matrix elements is connected with the possibility of the oscillator changing from one state to another under the influence of some extraneous agency, or spontaneously. This subject will be discussed more thoroughly at a later stage.

As another example let us write down the matrix of the Hamiltonian or energy operator of the oscillator.

$$H_{nm} = \int_{-\infty}^{\infty} u_n^* H u_m \, dx = E_m \delta_{nm}. \qquad (12.3.5)$$

Written out in full, the matrix has the following form:

$$[H_{nm}] = \hbar \omega_c \begin{bmatrix} \frac{1}{2} & 0 & 0 & \cdots \\ 0 & 1\frac{1}{2} & 0 & \cdots \\ 0 & 0 & 2\frac{1}{2} & \cdots \\ & \cdots & & \end{bmatrix}$$

The diagonal form of this matrix follows from eqn (12.2.8) and the argument preceding it. It clearly results from the fact that the eigenvalue equation of the Hamiltonian $H$ was utilized in evaluating the integral (12.3.5). Furthermore the diagonal elements themselves are equal to the eigenvalues of the operator which were shown on p. 190 to be the average values of the operator.

Although the foregoing result has been obtained on the special case of the Hamiltonian of the linear oscillator, it exemplifies a general principle which applies to any operator and its eigenstates. By this we mean that given an operator $A$, we can set up an equation of the form

$$Av = \alpha v \tag{12.3.6}$$

where $\alpha$ is assumed to be constant. If a set of functions $v_n$ can be found in conjunction with constants $\alpha_n$ which satisfy the eigenvalue relation

$$Av_n = \alpha_n v_n \tag{12.3.7}$$

then the matrix of $A$ with respect to the $v_n$ will be diagonal. This follows by the procedure applied above to the Hamiltonian operator, i.e.

$$A_{nm} = \int v_n^* A v_m \, dx = \alpha_m \delta_{nm}. \tag{12.3.8}$$

The existence of the eigenvalue relation (12.3.7) makes the evaluation of the matrix elements a simple problem, and ensures that the matrix is diagonal.

Let us now turn to the physical interpretation which is to be attached to eigenvalue relations and the resulting diagonal matrices of operators.

The measurement or observation of a dynamical variable of a quantum mechanical system in a given state is only accomplished by causing the system to change into another state. Whatever measuring procedure is adopted its effect is identifiable with the action of the corresponding operator on the given state. Now, if the state is an eigenstate of the dynamical variable to be measured, the result of the operation is to produce the corresponding eigenvalue every time the measurement is carried out. Hence the average value of the dynamical variable under observation will be its eigenvalue, also obtained by calculation using the method of the preceding section. On this basis the appearance of *a diagonal matrix of an operator is taken as an indication that observations of the operator will inevitably yield the diagonal elements,* the latter being of course the eigenvalues.

What significance is to be attached to the diagonal elements of a non-diagonal matrix? According to the postulates laid down in the preceding section they must still represent average values of the corresponding dynamical variable. However, since they are no longer the eigenvalues, they are not expected to be the result of *every* measurement. In fact they are true *statistical averages,* obtained after many observations in which each individual measurement yields a different value.

The above interpretation of matrices is well illustrated by the energy and position operators of the oscillator. The zero diagonal elements of the matrix (12.3.4) must be seen as averages over a number of measurements, while the eigenvalues appearing in the diagonal matrix (12.3.5) will be the certain result of every individual measurement of energy of the oscillator. It is emphasized again, however, that this state of affairs only applies to energy eigenstates of the oscillator, as derived in the opening sections of this chapter.

### 12.4  Dirac's Notation and Summary of Chapter 12

Before closing this chapter we introduce a special type of notation devised by Dirac. The notation is highly adapted to the handling of quantum mechanical states, dynamical variables and probabilities, and greatly helps to understand and remember relationships between these concepts.

Although initially the notation will be introduced as a substitute for the functional form of states, as obtained by solving Schroedinger's equation, it is, in fact, more general. It is applicable to states which are not readily expressible in the form of wave functions, such as spin states.

States of a dynamical system are denoted by pointed brackets thus $|\,\rangle$. A label identifying the state in question is inserted into the bracket, e.g. $|\,\psi_i\,\rangle$. Frequently the context makes it possible to discard the symbol $\psi_i$ and merely retain its subscript or quantum number: $|\,i\,\rangle$. The complex conjugate of the state $\psi_i$ is denoted by a bracket pointing in the opposite direction: $\langle\,\psi_i|$ or simply $\langle\,i|$.

The *integral* of the product of a wave function by its complex conjugate, $\int\psi_i^*\psi_j\,dx$, is written as follows:

$$\langle\psi_i|\psi_j\rangle = \langle i|j\rangle. \qquad (12.4.1)$$

This expression is in general a complex number; its complex conjugate is obtained by reversing the sequence of factors in the bracket expression (12.4.1)

$$\langle i|j\rangle^* = \langle j|i\rangle. \qquad (12.4.2)$$

If $i = j$ in (12.4.2) we have

$$\langle i|i\rangle^* = \langle i|i\rangle$$

which is real and is another way of writing the integral $\int|\psi_i|^2\,dx$. We assume that this expression is normalized to unity, hence

$$\langle i|i\rangle = 1.$$

The orthogonality and normalization conditions or *orthonormality con-ditions* of a set of states are expressed as follows:

$$\langle\psi_i|\psi_j\rangle = \langle i|j\rangle = \delta_{ij}. \qquad (12.4.3)$$

The application of an operator F to a state $\psi_j$ is written as follows

$$F|\psi_j\rangle = F|j\rangle. \qquad (12.4.4)$$

As a rule the effect of an operator on a state is to change it into another state. However, if the state happens to be an eigenstate of the operator, with eigenvalue $f_j$, we write

$$F|j\rangle = f_j|j\rangle. \qquad (12.4.5)$$

In the special case of the Hamiltonian operator $H$ and its eigenstates $|u_n\rangle = |n\rangle$ the eigenvalue relation assumes the form

$$H|n\rangle = E_n|n\rangle \qquad (12.4.6)$$

which is another way of writing eqn (12.1.4a). Matrix elements of an operator, which are integrals of the form $\int \psi_i^* F \psi_j \, dx$, are denoted by the symbols

$$\langle \psi_i|F|\psi_j\rangle = \langle i|F|j\rangle. \qquad (12.4.7)$$

If an operator happens to be just a constant number, say $k$, it can be taken outside the bracket symbol as follows:

$$\langle i|k|j\rangle = k\langle i|j\rangle \qquad (12.4.8)$$

which is equivalent to taking a constant outside an integral. In the case of eigenstates of an operator the eigenvalues can also be taken outside the bracket expression.

$$\langle \psi_i|F|\psi_j\rangle = \langle i|F|j\rangle = \langle i|f_j|j\rangle = f_j\langle i|j\rangle = f_j\delta_{ij}. \qquad (12.4.9)$$

Sums of operators when operating on states, or sums of states, obey the distributive laws.

$$(A + B)|i\rangle = A|i\rangle + B|i\rangle$$

$$A(|i\rangle + |j\rangle) = A|i\rangle + A|j\rangle \qquad (12.4.10)$$

where $A, B$ are any operators and $|i\rangle$, $|j\rangle$ are any states.

Products of operators obey the associative law.

$$ABC|i\rangle = A(BC)|i\rangle = (AB)C|i\rangle. \qquad (12.4.11)$$

The above rules will be easily appreciated if it is remembered all the time that the operators are in general linear differential operators and the states are functions to be differentiated. In this connection it is emphasized that products of operators do not in general obey the commutative law, i.e.

$$AB|i\rangle \neq BA|i\rangle \qquad (12.4.12)$$

or more briefly $AB \neq BA$. There are, of course special cases of operators which do happen to commute; they will be the subject of discussion in a later chapter.

Dirac's notation will be used from time to time in situations where it is particularly advantageous. Its scope will be extended as the need arises.

*Chapter 13*

# The Hydrogen Atom

## 13.1 The Equation of Motion of an Electron in a Central Field of Force and its Solution

The hydrogen atom consists of a nucleus having a positive charge and an electron having a negative charge of equal magnitude. Since the nucleus is very much heavier than the electron the motion of the latter is quite accurately represented on the assumption that the nucleus is at rest. The electron performs orbital motions in which the electrical attractive force between it and the nucleus is balanced by the centrifugal force due to the electron's kinetic energy.

To write down Schroedinger's equation of motion of the electron in this central field of force we must first obtain an expression for its electrical potential energy. We start with a somewhat more general case than the hydrogen atom, in that we assume that the nucleus has a positive charge equal to $Ze$, where

$e$ = numerical value of the electron charge,
$Z$ = atomic number of nucleus (= number of positive charges on nucleus).

Hence

$$V(x, y, z) = V(r) = -\frac{Ze^2}{4\pi\epsilon_0 r}. \qquad (13.1.1.)$$

The potential energy $V$ of the electron is here given in m.k.s. units, and is expressed in polar coordinates. This is done because the potential has spherical symmetry and the problem is therefore easier tackled in spherical coordinates. Before the potential (13.1.1) is substituted into Schroedinger's equation the latter must first be transformed into spherical coordinates, which is done with

197

the help of standard expressions for the Laplacian operator appearing in eqn (11.2.5) of Chapter 11. The formula is

$$\nabla^2 = \frac{\partial^2}{\partial x^2} + \frac{\partial^2}{\partial y^2} + \frac{\partial^2}{\partial z^2}$$

$$= \frac{1}{r^2}\frac{\partial}{\partial r}\left(r^2\frac{\partial}{\partial r}\right) + \frac{1}{r^2\sin\theta}\frac{\partial}{\partial\theta}\left(\sin\theta\frac{\partial}{\partial\theta}\right) +$$

$$+ \frac{1}{r^2\sin^2\theta}\frac{\partial^2}{\partial\phi^2}. \tag{13.1.2}$$

The Cartesian and spherical coordinate systems are summarized for convenient reference in Fig. 13.1.

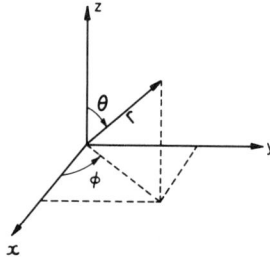

Fig. 13.1.

Schroedinger's equation for the central force problem has the form

$$\left(-\frac{\hbar^2}{2m_e}\nabla^2 - \frac{Ze^2}{4\pi\epsilon_0 r}\right)u(r,\theta,\phi) = Eu(r,\theta,\phi) \tag{13.1.3}$$

where $m_e$ = mass of electron
  $\epsilon_0$ = permittivity of free space.

To solve this partial differential equation we proceed to separate variables in two stages. At first the radial part of the wave function $u(r\,\theta,\,\varphi)$ is separated, and then the angular variables $\theta$ and $\varphi$ are separated. We write

$$u(r,\theta,\phi) = R(r)\,Y(\theta,\phi) = R(r)\Theta(\theta)\Phi(\phi). \tag{13.1.4}$$

Substituting the second expression of eqns (13.1.4) into (13.1.3) and performing the indicated differentiations we obtain

$$-\frac{\hbar^2}{2m_e}\left\{\frac{Y}{r^2}\frac{\partial}{\partial r}\left(r^2\frac{\partial}{\partial r}\right)R+\frac{R}{r^2\sin\theta}\frac{\partial}{\partial\theta}\left(\sin\theta\frac{\partial}{\partial\theta}\right)Y+\right.$$

$$\left.+\frac{R}{r^2\sin^2\theta}\frac{\partial^2 Y}{\partial\phi^2}\right\}-\frac{Ze^2}{4\pi\epsilon_0 r}RY=ERY. \tag{13.1.5}$$

To separate the radial variable we divide (13.1.5) by $u = RY$ and by $-\hbar^2/2m_er^2$ and rearrange the equation

$$\frac{1}{R}\frac{d}{dr}\left(r^2\frac{d}{dr}\right)R+\frac{2m_er^2}{\hbar^2}\left(\frac{Ze^2}{4\pi\epsilon_0 r}+E\right)=$$

$$=-\left\{\frac{1}{Y\sin\theta}\frac{\partial}{\partial\theta}\left(\sin\theta\frac{\partial}{\partial\theta}\right)Y+\frac{1}{Y\sin^2\theta}\frac{\partial^2 Y}{\partial\phi^2}\right\}. \tag{13.1.6}$$

Setting eqn (13.1.6) equal to a constant of separation $\lambda$ we obtain an ordinary differential equation in $r$, which we leave aside for a while to continue the process of separation on the angular variables. In the equation

$$\frac{1}{\sin\theta}\frac{\partial}{\partial\theta}\left(\sin\theta\frac{\partial}{\partial\theta}\right)Y+\frac{1}{\sin^2\theta}\frac{\partial^2 Y}{\partial\phi^2}+\lambda Y=0 \tag{13.1.7}$$

we substitute $Y = \Theta\Phi$ from eqn (13.1.4), divide throughout by $Y$ and multiply by $\sin^2\theta$ to obtain two ordinary differential equations in $\theta$ and $\varphi$.

$$\frac{\sin\theta}{\Theta}\frac{d}{d\theta}\left(\sin\theta\frac{d\Theta}{d\theta}\right)+\lambda\sin^2\theta=-\frac{1}{\Phi}\frac{d^2\Phi}{d\phi^2}=m^2. \tag{13.1.8}$$

The constant of separation has been set equal to $m^2$.

Let us now summarize the three ordinary differential equations which remain to be solved. They are obtained from (13.1.6) and (13.1.8) after some additional manipulation.

$$\frac{1}{r^2}\frac{d}{dr}\left(r^2\frac{dR}{dr}\right)+\left\{\frac{2m_e}{\hbar^2}\left(\frac{Ze^2}{4\pi\epsilon_0 r}+E\right)-\frac{1}{r^2}\lambda\right\}R=0$$

$$\frac{1}{\sin\theta}\frac{d}{d\theta}\left(\sin\theta\frac{d\Theta}{d\theta}\right)+\left(\lambda-\frac{1}{\sin^2\theta}m^2\right)\Theta=0$$

$$\frac{d^2\Phi}{d\phi^2}+m^2\Phi=0. \tag{13.1.9}$$

The above equations are to be solved subject to the same conditions as were imposed on the solutions of Schroedinger's equation of the linear oscillator. The integral of the squared modulus of the solutions over the whole range of the space variables must be finite. In terms of spherical coordinates this is

$$
\int_V |u(r, \theta, \phi)|^2 \, d\tau = \int_0^\infty \int_0^\pi \int_0^{2\pi} |u|^2 \, r^2 \sin \theta \, dr \, d\theta \, d\phi
$$

$$
= \int_0^\infty \int_0^\pi \int_0^{2\pi} |R|^2 \, |\Theta|^2 \, |\Phi|^2 \, r^2 \sin \theta \, dr \, d\theta \, d\phi
$$

$$
= \int_0^\infty |R|^2 \, r^2 \, dr \int_0^\pi |\Theta|^2 \sin \theta \, d\theta \int_0^{2\pi} |\Phi|^2 \, d\phi
$$

$$
= \text{finite}. \tag{13.1.10}
$$

From the last integral expression in (13.1.10) it follows that the solutions of each separate equation (13.1.9) are subject to the same conditions as the wave functions of the linear oscillator (see p. 182). Since the ranges of integration of the variables $\theta$ and $\varphi$ are finite, eqn (13.1.10) means that the functions $\Theta(\theta)$ and $\Phi(\varphi)$ must not have any infinities within these ranges.

One more point should be noted at this stage. Since the function $|u|^2$ is to be interpreted as a probability density, the integral (13.1.10) will be normalized to unity, as is customary in quantum mechanics (see p. 184). From the factored form of (13.1.10) it then follows that each of the functions $R$, $\Theta$, and $\Phi$ must be separately normalized. The above observations will be further elaborated as the solutions of eqn (13.1.9) are written down.

To obtain solutions of eqn (13.1.9) it is best to start with the last one, since it is of very simple form. Solutions which are finite within the range $\varphi = 0$ to $2\pi$ of the independent variable can be given exponential form.

$$
\Phi_m(\phi) = A \, e^{im\phi}.
$$

Moreover, the solutions must satisfy *cyclic boundary conditions*, which means that they must have the same value at $\varphi = 0$ and $\varphi = 2\pi$. This is in effect a requirement of continuity, as it means that there must be no sudden change in the value of the wave function as we move anticlockwise around the $z$-axis and pass through the plane $\varphi = 0$ (see Fig. 13.1). The cyclic boundary condition is satisfied by giving $m$ positive or negative integral values.

$$
m = \ldots -2, -1, 0, 1, 2, \ldots.
$$

Finally, the function $\Phi(\varphi)$ must be normalized.

$$\int_0^{2\pi} |\Phi|^2 \, d\phi = 1. \tag{13.1.11}$$

It is easily verified by direct integration that this condition constrains the constant of integration $A$ to the value $1/\sqrt{2\pi}$. Hence the function $\Phi$ has the form

$$\Phi_m(\phi) = \frac{1}{\sqrt{2\pi}} e^{im\phi}$$

$$m = \ldots -2, -1, 0, 1, 2, \ldots$$

(positive or negative integer)

$$\tag{13.1.12}$$

We note that solutions of the last of eqns (13.1.9) have features in common with the eigenfunctions of the linear oscillator: there is an infinite sequence of wave functions, each associated with a discrete value of the constant of separation. The eigenfunctions are now literally wave functions as they have the form of complex exponentials, while the eigenvalues are equally simple in that they are integral numbers.

Next we consider the second of eqns (13.1.9) remembering that the constant of separation $m$ has already been assigned specific values. We note that this is Legendre's differential equation whose solutions are Legendre functions or polynomials. To obtain functions which have no infinities for values of $\theta$ within the range 0 to $\pi$ it is found necessary on analysis to restrict the constant of separation $\lambda$ to the sequence

$$\lambda = l(l+1), \quad l = 0, 1, 2, 3 \quad \text{(positive integer)}. \tag{13.1.13}$$

Moreover, it is found that the eigenvalues of eqns (13.1.12) must be numerically less than or equal to $l$, otherwise the Legendre functions vanish.

$$|m| \leqslant l. \tag{13.1.14}$$

Bearing in mind these restrictions, the solutions of the second of eqns (13.1.9) can be written in the form

$$\Theta_{lm}(\theta) = N_{lm} P_l^m(\cos\theta)$$

where $P_l^m (\cos \theta)$ is the Legendre polynomial belonging to the eigenvalues $m$ and $l$, and $N_{lm}$ is a constant of integration to be fixed by normalization. In textbooks dealing with the properties of Legendre functions it is shown that the $N_{lm}$ must be given the values

$$N_{lm} = \sqrt{\frac{(2l+1)}{2} \cdot \frac{(l-|m|)!}{(l+|m|)!}} \tag{13.1.15}$$

hence the solutions of the second of eqns (13.1.9) have the form

$$\Theta_{lm}(\theta) = \left[\frac{2l+1}{2}\frac{(l-|m|)!}{(l+|m|)!}\right]^{1/2} P_l^m(\cos\theta). \qquad (13.1.16)$$

Some Legendre polynomials are listed in Table 13.1 for convenient reference.

Finally we consider the solution of the first of eqns (13.1.9), for the radial part of the hydrogen atom wave functions. Like the second equation, the first cannot be solved in terms of elementary functions and series expansions must be used. On analysis it is found that only terminating series or polynomials can satisfy the boundary conditions (13.1.10), and that they are only obtainable for discrete values of the separation constant $E$, given by the expression

$$E_n = -\left(\frac{Ze^2}{4\pi\epsilon_0}\right)^2\frac{m_e}{2\hbar^2(n'+l+1)^2} = -\left(\frac{Ze^2}{4\pi\epsilon_0}\right)^2\frac{m_e}{2\hbar^2}\frac{1}{n^2} \qquad (13.1.17)$$

where $n = n' + l + 1$.

The number $n'$ can assume all positive integral values including zero, hence $n$ assumes all positive integral values starting with unity

$$n = 1, 2, 3, \ldots. \qquad (13.1.18)$$

It should be noted from (13.1.17) that the quantum number $n$ is always greater than the quantum number $l$ by at least unity. From our study of Schroedinger's equation of motion of the linear oscillator we know that the $E_n$ are eigenvalues of energy. Hence (13.1.17) gives the possible energy levels of the electron in the hydrogen atom, or briefly energy levels of the hydrogen atom.

The radial wave functions of the hydrogen atom, as obtained on solving eqn (13.1.9), depend on both the quantum numbers $n$ and $l$, and have the following form after normalization.

$$R_{nl}(r) = -\left\{\left(\frac{2Z}{na_0}\right)^3\frac{(n-l-1)!}{2n[(n+l)!]^3}\right\}^{1/2} e^{-\frac{1}{2}\rho}\rho^l L_{n+l}^{2l+1}(\rho)$$

$$a_0 = \frac{4\pi\epsilon_0\hbar^2}{m_e e^2}; \ \rho = \frac{2Z}{na_0}r. \qquad (13.1.19)$$

The symbol $L_{n+l}^{2l+1}(\rho)$ in eqn (13.1.19) represents a *Laguerre polynomial* in the variable $\rho$ given by the expression

$$L_{n+l}^{2l+1}(\rho) = \sum_{k=0}^{n-l-1}(-1)^{k+1}\frac{[(n+l)!]^2}{(n-l-1-k)!(2l+1+k)!k!}\rho^k.$$

$$(13.1.20)$$

Some of the low-order radial wave functions are listed in Table 13.1.

**Table 13.1**
Some Legendre functions of low order

| $l$ | $m$ | $P_{lm}(\cos\theta)$ |
|---|---|---|
| 0 | 0 | 1 |
| 1 | 0 | $\cos\theta$ |
| 1 | 1 | $\sin\theta$ |
| 2 | 0 | $\frac{1}{2}(3\cos^2\theta - 1)$ |
| 2 | 1 | $3\sin\theta\cos\theta$ |
| 2 | 2 | $3\sin^2\theta$ |
| 3 | 0 | $\frac{1}{2}(5\cos^2\theta - 3\cos\theta)$ |
| 3 | 1 | $\frac{3}{2}\sin\theta(5\cos^2\theta - 1)$ |

Some radial wave functions of low order

| $n$ | $l$ | $R_{nl}(r)$ |
|---|---|---|
| 1 | 0 | $(Z/a_0)^{3/2},\ 2\,e^{-\rho/2}$ |
| 2 | 0 | $\dfrac{(Z/a_0)^{3/2}}{2\sqrt{2}}\,(2-\rho)\,e^{-\rho/2}$ |
| 2 | 1 | $\dfrac{(Z/a_0)^{3/2}}{2\sqrt{6}}\,\rho\,e^{-\rho/2}$ |
| 3 | 0 | $\dfrac{(Z/a_0)^{3/2}}{9\sqrt{3}}\,(\rho^2 - 6\rho + 6)\,e^{-\rho/2}$ |
| 3 | 1 | $\dfrac{(Z/a_0)^{3/2}}{9\sqrt{6}}\,(4-\rho)\,\rho\,e^{-\rho/2}$ |

$$\rho = \frac{2Z}{na_0}\,r.$$

We now collect together the solutions (13.1.12), (13.1.16), and (13.1.19) for the three individual space variables, according to eqn (13.1.4), and obtain the energy eigenstates of the hydrogen atom.

$$u_{nlm} = R_{nl}(r)\,\Theta_{lm}(\theta)\,\Phi_m(\phi) = R_{nl}(r)\,Y_{lm}(\theta,\phi). \qquad (13.1.21)$$

Each eigenfunction is associated with three quantum numbers $n$, $l$, $m$ related to each other by the inequality

$$n > l \geqslant |m|. \qquad (13.1.22)$$

Before passing on to a more detailed consideration of the *hydrogen atom wave functions* it should be noted that they form *orthogonal sets of eigenfunctions*. This applies to the factor functions $R_{nl}$, $\Theta_{lm}$, $\Phi_m$ separately, and to the complete eigenstates of energy $u_{nlm}$.

The orthogonality of the set $\Phi_m$ is easily proved by direct integration of the exponentials.

$$\int_0^{2\pi} \Phi_n^*,\,\Phi_m\,d\phi = \frac{1}{2\pi}\int_0^{2\pi} e^{-in\phi}\,e^{im\phi}\,d\phi = \delta_{n'm}. \qquad (13.1.23)$$

The corresponding proof for the set $\Theta_{lm}$ is more difficult, not only because it requires a knowledge of the properties of Legendre functions, but because two indices $l$ and $m$ are involved. In texts dealing with Legendre functions it is shown that the orthogonality conditions take on the form

$$\int_0^\pi \Phi_{l'\,m''}^*\,\Theta_{lm}\sin\theta\,d\theta = \delta_{l'\,l}\,\delta_{m'\,m}. \qquad (13.1.24)$$

Hence integrals of the above form will vanish, unless both pairs of corresponding indices are equal. Similar orthogonality conditions apply to the raradial functions.

$$\int_0^\infty R_{n'\,l'}^*\,R_{nl}\,r^2\,dr = \delta_{n'\,n}\,\delta_{l'\,l}. \qquad (13.1.25)$$

Once orthogonality conditions for sets of functions of a single variable have been established it is easy to prove that their products are also orthogonal. For the angular part of the eigenstates we can write

$$\int_0^\pi\int_0^{2\pi} Y_{l'\,m'}^*\,Y_{lm}\,\sin\theta\,d\theta\,d\phi =$$

$$= \int_0^\pi \Theta_{l'\,m'}^*\,\Theta_{lm}\,\sin\theta\,d\theta\int_0^{2\pi}\Phi_m^*,\,\Phi_m\,d\phi = \delta_{l'\,l}\,\delta_{m'\,m}. \qquad (13.1.26)$$

Continuing this argument to include the radial functions we obtain the orthogonality relation for the energy eigenstates of the hydrogen atom.

$$\int_0^\infty \int_0^\pi \int_0^{2\pi} u^*_{n'\,l'\,m'}\, u_{n\,l\,m}\, r^2 \sin\theta\, dr\, d\theta\, d\phi = \delta_{n'\,m}\,\delta_{l'\,l}\,\delta_{m'\,m}. \qquad (13.1.27)$$

## 13.2  Discussion and Interpretation of the Hydrogen Atom Wave Functions

The energy eigenfunctions of the hydrogen atom obtained in the preceding section are considerably more complicated than those of the linear oscillator. This is only to be expected as three independent space variables are involved and three ordinary differential equations had to be solved. Although the interpretation of the wave functions as sources of position probability distributions follows the same principle as with the linear oscillator, extra complications arise because of the three-dimensional nature of the problem.

Since the square of the modulus of the normalized wave function is the position probability density the expression

$$|u_{n\,l\,m}(r, \theta, \phi)|^2\, r^2 \sin\theta\, dr\, d\theta\, d\phi \qquad (13.2.1)$$

gives the probability of finding the electron in the volume element $r^2 dr \sin\theta$ $d\theta\, d\varphi = d\tau$ at the point $(r,\ \theta,\ \varphi)$. The three-dimensional distribution (13.2.1) is not easily visualized as it is not possible to plot it graphically. For this reason we go back to the factored form of $u_{nlm}$ and consider the distributions with respect to each space variable separately. We integrate the expression (13.2.1) with respect to the angular variables but leave $r$ as a parameter.

$$\int_0^\pi \int_0^{2\pi} |u_{nlm}|^2\, r^2\, dr\, \sin\theta\, d\theta\, d\phi = |R_{nl}(r)|^2\, r^2\, dr \int_0^\pi |\Theta(\theta)|^2 \sin\theta\, d\theta \int_0^{2\pi} |\Phi(\phi)|^2\, d\phi.$$

By the normalization conditions the two integrals with respect to the angular variables integrate to unity leaving the expression

$$|R_{nl}(r)|^2\, r^2\, dr \qquad (13.2.2)$$

as the probability of finding the electron in the radial range $dr$ at a distance $r$ from the nucleus. To put it geometrically eqn (13.2.2) gives the probability of finding the electron in a spherical shell of thickness $dr$ and radius $r$, as shown in Fig. 13.2. The function $|R_{nl}(r)|^2 r^2$ is the *radial probability density*. In Fig. 13.3 we plot some of the radial distributions for the hydrogen atom, that is for an atom of atomic number $Z = 1$. The maxima of these curves are at distances from the nucleus at which the electron is most likely to be found.

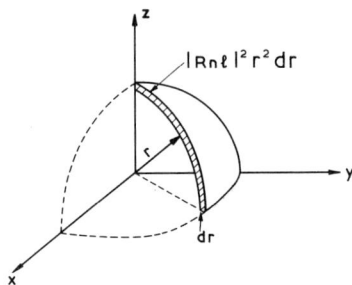

Fig. 13.2.

At this point it will be instructive to calculate the average radial distance of the electron as this is an indication of the physical size of the atom. Following the methods of finding average values of dynamical variables explained in Section 12.2 we have to evaluate an integral of the form

$$\langle r \rangle_{nlm} = \int_0^\infty \int_0^\pi \int_0^{2\pi} u_{nlm}^* r u_{nlm} r^2 \, dr \sin\theta \, d\theta \, d\phi. \qquad (13.2.3)$$

Since the dynamical variable in question, $r$, does not depend on the angular variables, integrals with respect to the latter reduce to unity by virtue of normalization. There remains an integral involving only the radial part of the wave function and this can be evaluated from a detailed knowledge of Laguerre

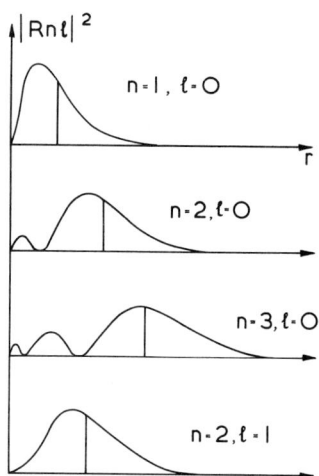

Fig. 13.3.

polynomials. Here we merely state the general result, leaving it to the reader to verify it for some of the lower states.

$$\langle r \rangle_{nlm} = \frac{n^2 a_0}{Z} \left\{ 1 + \tfrac{1}{2} \left[ 1 - \frac{l(l+1)}{n^2} \right] \right\}. \tag{13.2.4}$$

In the *ground state*, $n = 1, l = 0$, the *average* radius of the hydrogen atom is

$$\langle r \rangle_{100} = \tfrac{3}{2} a_0 = 0 \cdot 81 \text{ Å} ; \quad (Z = 1)$$

by eqn (13.1.19). For excited states of the atom, that is states with higher energy and hence quantum number $n$, the radius is greater. The average radius is marked by a vertical line for each distribution plotted in Fig. 13.3.

It should be noted that the average value of the radius is not the same as the median value which corresponds to the maximum of the radial probability distribution curve. The latter is easily found by differentiation of the probability function. After some calculation the value $a_0 = 0 \cdot 53$ Å is arrived at for the ground state.

As can be seen from Fig. 13.3 all radial probability distributions tend to zero with increasing values of the radius. This vanishing property is ensured by the exponential factor present in eqn (13.1.19). The latter defines the outermost extremities of the atom and limits the range at which individual atoms can directly interfere with each other. This fact will be utilized in later chapters in connection with ferromagnetism.

As the next step it would be natural to consider the angular probability distributions of the electron moving in a central field of force. These are associated with the angular part of the wave functions, $Y_{lm} (\theta, \varphi)$. It is to be expected that for states for which $Y_{lm}$ does not reduce to a constant the electron will be more likely to be found in certain preferred directions in space, and hence the probability distributions will not have spherical symmetry. However, it is preferable to delay consideration of these questions until angular momentum is discussed.

## 13.3  Energy Levels of the Hydrogen Atom–Degeneracy

As explained in Section 12.3, the eigenvalues of Schroedinger's equation, or the Hamiltonian operator, are the energy levels of the dynamical system under consideration. The energy levels of an electron moving in a central field of force, or of the hydrogen atom, are stated in eqn (13.1.17). These levels depend only on the quantum number $n$, called the total or *principal quantum number*.

It should be noted that the energy has a negative sign. This has its origin in the conventions of electrostatics, where it is assumed that the potential energy of two charges, an infinite distance apart, is zero. As two charges of opposite sign are brought closer together their energy decreases, hence it must have a negative sign.

Because of the negative sign the energy level having the greatest *numerical* value is the lowest. This is quite convenient as it corresponds to the lowest possible value of the quantum number $n = 1$. As $n$ increases the energy levels form a sequence tending to zero. The energy level scheme is shown graphically in Fig. 13.4.

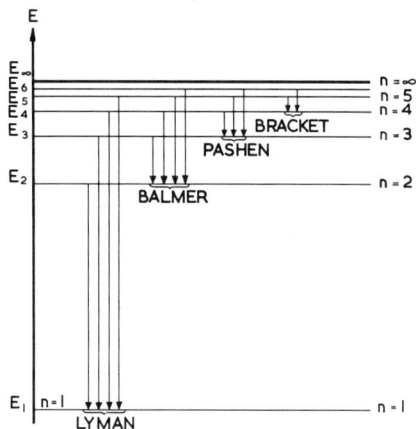

Fig. 13.4. Basic energy level scheme of the hydrogen atom.

The energy levels of atomic hydrogen were first discovered by spectroscopic methods. Various workers observed series of spectral lines which correspond to transitions between groups of energy levels as shown in Fig. 13.4.

The first successful attempt to provide a theory of the energy level scheme of hydrogen was made by Bohr. He derived eqn (13.1.17) for the energy levels many years before Schroedinger proposed his equation of motion.

We shall learn later that the energy level scheme derived in this chapter is only an approximation, although quite a good one. A number of smaller effects, not included in the theory presented so far, but disclosed by spectroscopic observations, will have to be accounted for. These problems will form the subject matter of following chapters in which quantum theory will be further developed.

Before closing the present chapter we note an important theoretical point, which constitutes a difference between the energy levels and eigenstates of the linear oscillator and the hydrogen atom. Whereas each energy eigenvalue of the oscillator was associated with only one eigenfunction this is not the case with the hydrogen atom. To each energy eigenvalue, as characterized by the quantum number $n$, there belongs in general a number of eigenfunctions. With each value of $n$ a number of eigenfunctions $u_{nlm}$ are associated, corresponding to the possible values of the quantum numbers $l$ and $m$. This feature of eigenstates is called *degeneracy*. Each group of eigenfunctions, belonging to one and the same eigenvalue, are said to be degenerate. Sometimes this statement is made in reverse: it is said that an eigenvalue is degenerate, and it is meant that several eigenstates of the corresponding dynamical variable belong to it.

It is not difficult to take a count of the degenerate states belonging to a given energy eigenvalue of the hydrogen atom. Corresponding to a specific value $n$ of the principal quantum number, we can have $n$ different values of $l$, including 0 and $n - 1$. Within each value of $l$ there are $2l + 1$ possible values of $m$, ranging from $-l$ to $+l$. Hence the total number of eigenfunctions belonging to $n$ is:

$$\sum_{l=0}^{n-1} (2l + 1) = 2 \sum_{l=0}^{n-1} l + n = n^2. \tag{13.3.1}$$

In connection with the concept of degeneracy it is worth while writing down the eigenvalue equation of the Hamiltonian operator of the hydrogen atom

$$Hu_{nlm} = E_n u_{nlm}. \tag{13.3.2}$$

It is clear that the above equation holds for all eigenfunctions belonging to quantum numbers $l$ and $m$, subject to the restriction (13.1.22).

The significance of degeneracy will become apparent, at least in part, in subsequent chapters.

## 13.4 Dirac's Notation for States Characterized by More Than One Quantum Number

At the close of the preceding chapter Dirac's notation was introduced on the example of the eigenstates of the linear oscillator. There the position was simple in that each state belonged to just one quantum number which was used to label the pointed bracket representing that state.

States characterized by more than one quantum number are represented by a succession of brackets written side by side, each labelled by one eigenvalue, or else by a single bracket containing several eigenvalues. Thus the state $u_{nlm}$ is written

$$|n\rangle|l\rangle|m\rangle = |nlm\rangle. \tag{13.4.1}$$

Its complex conjugate, $u_n^*lm$, is denoted by brackets pointing in the opposite direction

$$\langle n|\langle l|\langle m| = \langle nlm|. \tag{13.4.2}$$

An integral of the form (13.1.27), expressing the orthonormality of the states $u_{nlm}$, is written in the form

$$\langle n'\, l'\, m'|nlm\rangle = \delta_{n'\, m}\, \delta_{l'\, l}\, \delta_{m'\, m}. \tag{13.4.3}$$

The average value of a dynamical variable, say the radius $r$, has the form (see eqn (13.2.3))

$$\langle r\rangle_{nlm} = \langle nlm|r|nlm\rangle \tag{13.4.4}$$

while the eigenvalue relation (13.3.2) is written

$$H|nlm\rangle = E_n|nlm\rangle. \tag{13.4.5}$$

It should be noted that operators operate only on states which include the corresponding independent variables. Thus the operation of $r$ on one of the eigenstates of the hydrogen atom can be satisfactorily represented by the symbol $r|nl\rangle$, since only the quantum numbers $n$ and $l$ enter into the radial state $R_{nl}(r)$. The Hamiltonian operates on all the variables in the state $u_{nlm}$, hence none of the symbols in eqn (13.4.5) is redundant. These observations are obvious once Dirac's abbreviated symbols are compared with the corresponding functional relations developed in the course of this chapter.

When it comes to off-diagonal matrix elements of operators with respect to states characterized by several quantum numbers, the position becomes rather more complicated in the general case. If, however, all except one of the quantum numbers are kept fixed, we still get a two-dimensional array of numbers. Thus the symbols

$$\langle n'\, l_1\, m_1|F|nl_2\, m_2\rangle = \langle n'|F|n\rangle = F_{n'\, n} \tag{13.4.6}$$

represent the matrix of the operator $F$ with respect to the states $|n\rangle$, while the quantum numbers $l$ and $m$ are kept fixed to the label shown, provided always that $F$ operates on the state $|n\rangle$ only. It is possible to reduce most problems to the form (13.4.6). In any specific case the context will make the position clear.

In the chapters to follow Dirac's notation will be used with increasing frequency.

*Chapter 14*

# Angular Momentum and Magnetic Moment

## 14.1 Orbital Angular Momentum Operators–Their Eigenstates and Eigenvalues

In this chapter we shall consider a dynamical variable of the greatest importance in quantum mechanics: angular momentum. Related to angular momentum is the magnetic moment on an atomic scale which is the ultimate source of the magnetic properties of matter. The operators of angular momentum are of a very general nature, but in this chapter we shall consider them only in connection with the hydrogen atom, this being an example susceptible of reasonably simple theoretical description.

Any particle moving along a curved path has *angular momentum*, or moment of momentum, about the centre of curvature of its path. Since the electron in a hydrogen atom moves in some sort of orbit it must have angular momentum. The latter is a dynamical variable of the hydrogen atom which we shall now formulate as an operator.

The angular momentum or moment of momentum of a particle of mass $m$, moving with velocity $\mathbf{v}$ is

$$\mathbf{L} = m\mathbf{r} \times \mathbf{v} \times = \mathbf{r} \times \mathbf{p} \tag{14.1.1}$$

where $\mathbf{p}$ is the linear momentum of the particle and $\mathbf{r}$ its position vector (see Fig. 14.1).

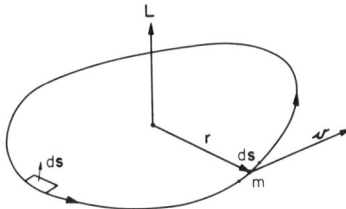

Fig. 14.1.

211

Since we are treating the angular momentum **L** as a dynamical variable or operator, whose matrix elements we intend to evaluate, the components of momentum **p** must be replaced by differential operators according to eqn (12.2.4). Taking Cartesian components of eqn (14.1.1) we write

$$L_x = yp_z - zp_y = -i\hbar\left(y\frac{\partial}{\partial z} - z\frac{\partial}{\partial y}\right)$$

$$L_y = zp_x - xp_z = -i\hbar\left(z\frac{\partial}{\partial x} - x\frac{\partial}{\partial z}\right)$$

$$L_z = xp_y - yp_x = -i\hbar\left(x\frac{\partial}{\partial y} - y\frac{\partial}{\partial x}\right). \qquad (14.1.2)$$

In addition to the components of momentum we shall require the square of its magnitude.

$$|\mathbf{L}|^2 = L^2 = L_x{}^2 + L_y{}^2 + L_z{}^2. \qquad (14.1.3)$$

As is usual squares of differential operators are interpreted as second-order differential operators.

As the eigenstates of the hydrogen atom have been formulated in spherical coordinates the operators (14.1.2) and (14.1.3) cannot be applied to them straight away, but must be transformed into spherical coordinates first.

$$L_x = i\hbar\left(\sin\phi\frac{\partial}{\partial\theta} + \cot\theta\cos\phi\frac{\partial}{\partial\phi}\right)$$

$$L_y = i\hbar\left(-\cos\phi\frac{\partial}{\partial\theta} + \cot\theta\sin\phi\frac{\partial}{\partial\phi}\right)$$

$$L_z = -i\hbar\frac{\partial}{\partial\phi}. \qquad (14.1.4)$$

After some manipulation the square of the magnitude of angular momentum can be put in the form

$$L^2 = -\hbar^2\left\{\frac{1}{\sin\theta}\left(\frac{\partial}{\partial\theta}\sin\theta\frac{\partial}{\partial\theta}\right) + \frac{1}{\sin^2\theta}\frac{\partial^2}{\partial\phi^2}\right\}. \qquad (14.1.5)$$

We are now ready to compute the average values of angular-momentum operators for the electron orbits of hydrogen. Starting with the squared magnitude (14.1.5) we can write, using Dirac's notation:

$$\langle nlm|L^2|nlm\rangle = \int_0^\infty \int_0^\pi \int_0^{2\pi} u_{nlm}^* L^2 u_{nlm} r^2 \sin\theta \, dr \, d\theta \, d\phi$$

$$= \int_0^\pi \int_0^{2\pi} Y_{lm}^* L^2 Y_{lm} \sin\theta \, d\theta \, d\phi. \qquad (14.1.6)$$

The last step can be taken because $L^2$ operates on the angular part of the wave functions only, and the integral over the radial part reduces to unity by normalization.

At this point the operator $L^2$ should be compared with Schroedinger's equation of the hydrogen atom, eqns (13.1.2) and (13.1.3) of Chapter 13. We note that $L^2$ is identical with the angular part of the Laplacian operator appearing in those equations, except for the factor $-\hbar^2$. Hence $L^2$ has the same differential form as the differential eqn (13.1.7), which is rewritten here as an eigenvalue relation.

$$-\left\{ \frac{1}{\sin\theta} \frac{\partial}{\partial\theta} \left( \sin\theta \frac{\partial}{\partial\theta} \right) + \frac{1}{\sin^2\theta} \frac{\partial^2}{\partial\phi^2} \right\} Y_{lm} = l(l+1) \, Y_{lm}$$

$$L^2 \, Y_{lm} = \hbar^2 \, l(l+1) \, Y_{lm}$$

$$L^2|lm\rangle = \hbar^2 \, l(l+1)|lm\rangle. \qquad (14.1.7)$$

Substitution of eqn (14.1.7) into the integral (14.1.6) yields the following result:

$$\langle lm|L^2|lm\rangle = \hbar^2 \, l(l+1), \quad l = 0, 1, 2, \ldots. \qquad (14.1.8)$$

This is the average value of the squared magnitude of angular momentum of an electron orbit in the state having quantum number $l$.

For this reason $l$ *is frequently called the angular momentum quantum number.* The expression (14.1.8) is briefly referred to as the angular momentum of the electron orbit, although strictly speaking this term means the square root of (14.1.8).

$$|\mathbf{L}| = \hbar \sqrt{l(l+1)}.$$

This quantity is of little use in quantum theory.

From eqn (14.1.7) and the orthogonality properties of the functions $Y_{lm}$ it follows that the matrix of the operator $L^2$ is diagonal. This we write in Dirac's notation as follows:

$$\langle l' m'|L^2|lm\rangle = \langle l' m'|\hbar^2 \, l(l+1)|lm\rangle$$

$$= \hbar^2 \, l(l+1)\delta_{l' \, l} \delta_{m' \, m}. \qquad (14.1.9)$$

Hence by the principles laid down on p. 194 the *functions $Y_{lm}$ are eigenstates of the operator $L^2$ with eigenvalues $\hbar^2 l(l + 1)$.* As the operator $L^2$ leaves the radial functions $R_{nl}$ unaffected, we conclude that the *eigenstates $u_{nlm}$ of energy are simultaneously eigenstates of the squared magnitude of angular momentum.*

As the next step let us evaluate the average values of the $z$-component of angular momentum.

$$\langle nlm|L_z|nlm\rangle = \int\limits_0^\infty \int\limits_0^\pi \int\limits_0^{2\pi} u_{nlm}^* L_z u_{nlm}\, r^2 \sin\theta\, dr\, d\theta\, d\phi$$

$$= \int\limits_0^{2\pi} \Phi_m^* L_z \Phi_m\, d\phi$$

$$= \langle m|L_z|m\rangle. \tag{14.1.10}$$

The last line follows from the fact that the operator $L_z$ acts on the coordinate $\varphi$ only, and the integrals over the remaining variables reduce to unity by virtue of normalization.

Recalling now the form of the function $\Phi_m$ from eqn (13.1.12) we find on differentiation with respect to $\varphi$

$$L_z \Phi_m = \hbar m \Phi_m$$

$$L_z|m\rangle = \hbar m|m\rangle \tag{14.1.11}$$

whence

$$\langle m|L_z|m\rangle = \hbar m; \quad m = -2, -1, 0, 1, 2. \tag{14.1.12}$$

According to eqn (14.1.12) $\hbar m$ is the average value of the $z$-component of angular momentum of an electron orbit in the state having quantum number $m$.

Furthermore, we note from eqn (14.1.11), which has the form of an eigenvalue equation, that the matrix of the operator $L_z$ is diagonal, and the *functions $\Phi_m$ are eigenstates of the $z$-component of angular momentum* belonging to the eigenvalues $\hbar m$.

$$\langle m'|L_z|m\rangle = \langle m'|\hbar m|m\rangle = \hbar m\, \delta_{m'm}. \tag{14.1.13}$$

By the same argument as applies to the operator $L^2$ we arrive at the conclusion that the *eigenstates $u_{nlm}$ of energy are simultaneously eigenstates of the $z$-component of angular momentum.*

Having evaluated the average values of some angular momentum operators for electron orbits in hydrogen let us now return to the position probability distributions and consider their angular dependence (see p. 207). We shall find that these two subjects are closely related.

The angular probability distributions are given by the functions $| Y_{lm} |^2 = | \Theta_{lm} |^2 | \Phi_m |^2$. From the form of the function $\Phi_m(\varphi)$ it is clear that the distributions depend on the angle $\theta$ only, hence we shall consider the functions $| \Theta_{lm}(\theta) |^2$ only. The expression $(1/2\pi) | \Theta_{lm}(\theta) |^2 \sin \theta \, d\theta \, d\varphi = (1/2\pi) | \Theta_{lm}(\pi) |^2$ $d\sigma$ is equal to the probability of finding the electron in the element $d\sigma$ of solid angle in the direction $\theta$ relative to the $z$-axis.

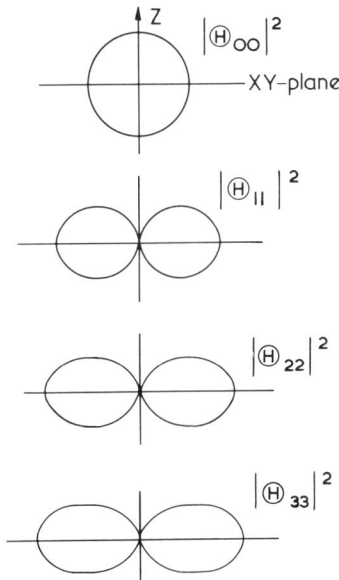

Fig. 14.2.

In Fig. 14.2 we plot a sequence of the functions $| \Theta_{lm} |^2$ in polar form. In the state $l = 0$, $m = 0$, the electron is equally likely to be found in any direction, and this accords with the fact that the angular momentum vanishes. As the value of the quantum numbers $l$ and $m$ increases the distributions lose their spherical symmetry and become increasingly concentrated about the $XY$-plane and extend farther away from the origin. We may think that in these states the electron orbits are more clearly defined and located in a plane normal to the $z$-axis. Again this accords with the fact that the magnitude of angular momentum and its $z$-component increase for increasing values of the quantum numbers $l$ and $m$.

Figure 14.3 shows a slightly different sequence of the probability distributions $| \Theta_{lm} |^2$. $l$ is kept fixed at $l = 3$ and $m$ varies over the allowed range from 0 to $\pm 3$. These graphs show how the $z$-component of angular momentum of the electron increases as the probability density is concentrated in a plane

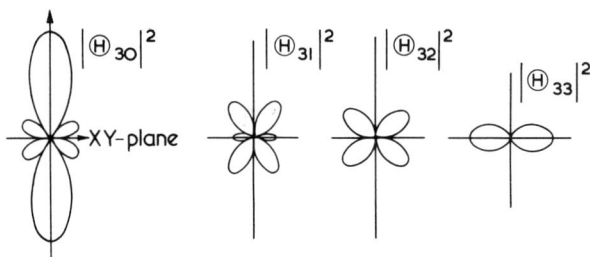

Fig. 14.3.

perpendicular to the $z$-axis while the magnitude of the momentum remains constant.

The results of this section may be summarized as follows. *The angular momentum* of electron orbits *is quantized* just as their energy is quantized. The magnitude of the angular momentum is related to the quantum number $l$ through eqn. (14.1.8), but it is customary to call $l$ *the* angular momentum.

Apart from being quantized in magnitude, the *angular momentum is quantized in direction*, as revealed by the quantum number $m$ of its $z$-component. Thus an orbit with an angular momentum given by $l = 2$ may have $z$-components given by any one of the quantum numbers $m = -2, -1, 0, +1, +2$ (see p. 201). The relationship between the angular momentum vector and its $z$-components is usually represented by the diagram of Fig. 14.4.

## 14.2  Orbital Magnetic Moment

Directly related to the angular momentum is a dynamical variable of the greatest practical importance—magnetic moment. In this section we show that the magnetic moment operators differ from angular momentum operators by a

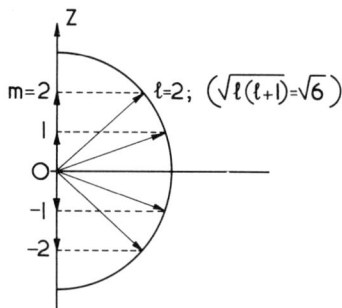

Fig. 14.4.

constant factor only. Once this is established all the properties of angular momentum can be extended to magnetic moments.

Since the electron orbiting in the hydrogen atom constitutes a current loop, it has a magnetic moment. The moment can be expressed in either of the forms[1]

$$\mathbf{m} = \int id\mathbf{S}$$
$$= -\int \tfrac{1}{2} i(d\mathbf{s} \times \mathbf{r}) \tag{14.2.1}$$

where $d\mathbf{S}$ is an element of area of the loop, $d\mathbf{s}$ is an element of the path, and $\mathbf{r}$ is the position vector of the electron (see Fig. 14.1). The current element $id\mathbf{s}$ can be replaced by a charge element through the equation $id\mathbf{s} = (dq/dt)\, d\mathbf{s} = dq\mathbf{v}$ where $\mathbf{v}$ is the velocity of the moving charge. Hence eqn (14.2.1) assumes the form

$$\mathbf{m} = -\tfrac{1}{2} \int dq(\mathbf{v} \times \mathbf{r}). \tag{14.2.2}$$

By eqn (14.1.1) the vector product under the integral sign is proportional to the angular momentum of the electron. As this is constant in the absence of external forces, it can be taken outside the integral sign.

$$\mathbf{m} = \frac{1}{2m_e} \mathbf{L} \int dq$$
$$= -\frac{e}{2m_e} \mathbf{L}$$
$$= -\gamma \mathbf{L}. \tag{14.2.3}$$

where $\int dq$ is the total circulating charge, in our case the charge of the orbiting electron $-e$, and $m_e$ is the mass of the electron.

From eqn (14.2.3) we see that the magnetic moment of an electron orbit is proportional to its angular momentum. Since the constant of proportionality is negative the two vectors are antiparallel. As the coefficient in eqn (14.2.3) is of great practical importance a separate name has been coined for it. It is called the *gyromagnetic ratio* of the electron orbit.

$$\gamma = \frac{e}{2m_e}. \tag{14.2.4}$$

Now it only remains to work out the average values of the magnetic moment operator (14.2.3) with respect to the states of the hydrogen atom. As this is equivalent to finding the average values of angular momentum we can draw on the results of the preceding section.

[1] The symbol $\mathbf{m}$ should not be confused with the quantum number $m$.

Thus the squared magnitude of the magnetic moment is:

$$\langle lm|\mathbf{m}^2|\,lm\rangle = \left(\frac{e\hbar}{2m_e}\right)^2 l(l+1) \tag{14.2.5}$$

by eqn (14.1.8). The $z$-component of the magnetic moment is, by eqn (14.1.12)

$$\langle m|m_z|m\rangle = -\frac{e\hbar}{2m_e}\,m. \tag{14.2.6}$$

The numerical value of the coefficient of eqn (14.2.6) constitutes a natural atomic unit of magnetic moment. It is called the *Bohr magneton* and is usually denoted by the symbol $\beta$.

$$\beta = \frac{e\hbar}{2m_e}. \tag{14.2.7}$$

The reader will no doubt have noticed that we have not discussed the $x$ or $y$ components of angular momentum or magnetic moment. The reasons for this omission will be discussed in subsequent sections. In the meantime we note the physical interpretation given to the results obtained so far.

It is one of the fundamental assumptions of quantum theory, amply justified by experiment, that *the z-components of magnetic moments* discussed above *will align themselves parallel to an applied d.c. magnetic field.* From this it follows that an atomic magnetic moment can only assume a discrete set of orientations relative to an applied field. In other words the relative direction of atomic magnets and external fields is quantized.

## 14.3  Commuting Operators and Simultaneous Eigenstates

The study of the hydrogen atom, through Schroedinger's equation of motion in the preceding sections, has now provided material which can be used as a basis for the introduction of further theoretical concepts and methods, and their interpretation.

In Section 14.1 we have noted that the eigenstates of the hydrogen atom are simultaneously eigenstates of the Hamiltonian, the squared magnitude of angular momentum and the $z$-component of angular momentum. The observation was based on the fact that the matrices of these operators were diagonal (see p. 194), or alternatively that they satisfied eigenvalue equations. Let us collect the latter together for convenience.

$$H|nlm\rangle = E_n|nlm\rangle$$
$$L^2|nlm\rangle = \hbar^2\, l(l+1)|nlm\rangle$$
$$L_z|nlm\rangle = \hbar m|nlm\rangle. \tag{14.3.1}$$

The reader will readily satisfy himself, by working through a simple special case, that the $x$- and $y$-components of angular momentum do not form diagonal matrices with respect to the eigenstates of the hydrogen atom, and therefore do not satisfy eigenvalue equations including these eigenstates.

What is it then that gives the operators of eqn (14.3.1) the special characteristic of being able to share the same eigenstates? The answer lies in the fact that *they commute*. By this we mean that they satisfy the following relations:

$$[H, L^2] = HL^2 - L^2 H = 0$$
$$[L^2, L_z] = L^2 L_z - L_z L^2 = 0$$
$$[L_z, H] = L_z H - HL_z = 0. \qquad (14.3.2)$$

The square brackets are abbreviated symbols usually called *commutators*.

Equation (14.3.2) may be verified by substituting the appropriate differential expressions for the operators and letting them act on an arbitrary function $f(r, \theta, \varphi)$. It is then found, after a certain amount of manipulation that the effect of, say, the operator $HL_z$ is the same as the effect of $L_z H$.

Quite apart from the foregoing examples it is possible to prove in general that *operators which have simultaneous eigenstates commute*. Thus two operators $A$ and $B$ which share a set of eigenstates $| n \rangle$, satisfy the following eigenvalue relations:

$$A|n\rangle = \alpha_n |n\rangle, \quad B|n\rangle = \beta_n |n\rangle$$

where $\alpha_n$ and $\beta_n$ are the eigenvalues. To show that $A$ and $B$ commute we proceed as follows:

$$AB|n\rangle = A\beta_n |n\rangle = \beta_n A|n\rangle = \beta_n \alpha_n |n\rangle$$
$$BA|n\rangle = B\alpha_n |n\rangle = \alpha_n B|n\rangle = \alpha_n \beta_n |n\rangle.$$

Taking the difference of the above equations we find (see eqn (12.4.10))

$$AB|n\rangle - BA|n\rangle = (AB - BA)|n\rangle = 0|n\rangle = 0$$

which proves the statement. The converse of this result, to prove that commutability of operators is a sufficient condition for them to share a set of eigenstates, is established as follows. Given that $AB = BA$ and assuming $| n \rangle$ to be eigenstates of $A$ we can write (see eqn (12.4.11))

$$A(B|n\rangle) = BA|n\rangle = B\alpha_n |n\rangle = \alpha_n(B|n\rangle).$$

The bracketed expression $(B| n \rangle)$ is clearly an eigenstate of $A$ with eigenvalue $\alpha_n$ just like the state $| n \rangle$. Hence one must be able to write

$$B|n\rangle = b|n\rangle$$

where $b$ is a constant. But this expression is an eigenvalue equation relating the operator $B$ and state $| n \rangle$ (see eqn (12.4.5)). Hence the latter are also eigenstates of $B$. The foregoing proof is subject to difficulties in cases of degeneracy, where there are more states with the same eigenvalue. However, it is then possible to form linear combinations of the degenerate states which are eigenstates of $B$.

The physical interpretation of commuting dynamical variables and simultaneous eigenstates is a straightforward extension of the principles laid down in Section 12.3.

The fact that eigenvalue relations are satisfied by two or more operators when applied simultaneously to a shared eigenstate means that the corresponding dynamical variables are simultaneously precisely measurable. This implies that the measurement is certain to yield the corresponding eigenvalues. At the same time the matrices of commuting operators with respect to their common eigenstates are diagonal, having eigenvalues for their diagonal elements.

A question naturally arises regarding sets of operators which do not commute. Examples of such cases are readily found. The reader will soon satisfy himself that any two components of angular momentum do not commute, e.g.

$$[L_z, L_x] = L_z L_x - L_x L_z = i\hbar L_y \neq 0.$$

Also the position and momentum of the linear oscillator do not commute.

$$[x, p_x] = [x p_x - p_x x] = i\hbar \neq 0.$$

Such sets of operators have no eigenstates in common. Hence the result of operating with them on a given state is not to produce their eigenvalues simultaneously. From this it follows that a simultaneous measurement of the corresponding dynamical variables will not yield precisely predictable values but rather a scatter of results. Only the averages of many such measurements will equal the diagonal elements of the matrices of the given operators. Such diagonal elements have been worked out for the position and momentum of the oscillator in Section 12.2.

In cases of this type it may be necessary to measure one of the variables with greater accuracy or predictability than the other. Efforts can then be made to find the dynamical system in an eigenstate of that variable. For example the linear oscillator is established in energy eigenstates to measure its energy levels. If this is done the other variable, say momentum, will be subject to increasing scatter in measurement. This phenomenon is a manifestation of the *principle of uncertainty*. This states that a *simultaneous measurement of two incompatible dynamical variables invariably yields a spread of values*. This spread of results can be estimated, but the derivation of the relevant formula will be given at a later stage.

## 14.4 Electron Spin

In addition to orbital angular momentum electrons possess another component of angular momentum called spin. This is to be considered as somehow residing in the electron itself, quite independently of whatever motion the electron might perform.

The source of electron spin has now been shown to be relativistic, it is, therefore, impossible to derive it from Schroedinger's equation as introduced in the present treatment. As it is beyond our scope to go into the relativistic theory of the electron, we shall introduce it into our scheme of states and operators by an ad hoc procedure, which is nevertheless characteristic of some methods to be used in subsequent chapters. We shall use Dirac's notation throughout.

As a dynamical variable the spin angular momentum will be denoted by operators analogous to orbital angular momentum and postulated to have the same properties. The symbol used for the spin vector is usually $S$, while its squared magnitude and $z$-component are denoted by $S^2$ and $S_z$. The operators $S^2$ and $S_z$ are assumed to commute just like the orbital angular momentum operators $L^2$ and $L_z$.

$$S^2 S_z - S_z S^2 = 0. \tag{14.4.1}$$

By virtue of this relation the spin operators will have simultaneous eigenstates in complete analogy with orbital angular momentum. Since we have no wave functions representing eigenstates of the spin operators, we must use Dirac's abstract symbols from the start. Thus the symbol $|sm_s\rangle$ will denote an eigenstate of the operator $S^2$ belonging to the eigenvalue $\hbar^2(s + 1)$, and simultaneously an eigenstate of the operator $S_z$ belonging to the eigenvalue $\hbar m_s$. By analogy with eqn (14.1.7) and (14.1.11) we assume that the spin operators and states satisfy the following eigenvalue equations

$$S^2|sm_s\rangle = \hbar^2\, s(s+1)|sm_s\rangle$$
$$S_z|sm_s\rangle = \hbar m_s|sm_s\rangle. \tag{14.4.2}$$

At this point it is necessary to state some facts regarding the electron spin quantum numbers $s$ and $m_s$. Whereas the orbital quantum numbers could assume any integral values subject to the limitations stated on p. 201 this is not so as regards spin. *The quantum number $s$*, associated with the magnitude of the spin angular momentum, *can assume only the value* $\frac{1}{2}$. The number $m_s$, associated with the $z$-component of the spin, is again subject to the limitation $|m_s| \leqslant s$ and it can only assume values *differing by unity*. Hence its only possible values are

$$m_s = -\tfrac{1}{2}, \tfrac{1}{2}. \tag{14.4.3}$$

The only one eigenvalue of the operator $S^2$ is $\frac{3}{4}\hbar^2$, and the only two eigenvalues of $S_z$ are $\pm \frac{1}{2}\hbar$. The relation between spin quantum numbers is shown graphically in Fig. 14.5. To simplify mathematical expressions it is customary to drop the label $s$ from the eigenstates, because it is understood that it has only one possible value.

In addition to being in eigenstates of energy and orbital angular momentum, the electron in the hydrogen atom is simultaneously in one of the two possible spin states. It now remains to incorporate this fact into the mathematical expressions obtained in the preceding sections. To do this we treat a spin state as

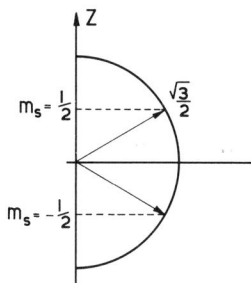

Fig. 14.5.

another eigenfunction which can be appended as an additional factor to the states $|nlm\rangle$. By virtue of the rule laid down on p. 209 we write

$$|nlm\rangle|m_s\rangle = |nlmm_s\rangle. \tag{14.4.4}$$

It is now assumed that the spin operator $S_z$ operates on this state independently, just like the operator $L_z$ did.

$$S_z|nlmm_s\rangle = \hbar m_s|nlmm_s\rangle. \tag{14.4.5}$$

With this result in hand the set of eigenvalue equations (14.3.1) is extended to four.

$$H|nlmm_s\rangle = E_n|nlmm_s\rangle$$
$$L^2|nlmm_s\rangle = \hbar^2\, l(l+1)|nlmm_s\rangle$$
$$L_z|nlmm_s\rangle = \hbar m|nlmm_s\rangle$$
$$S_z|nlmm_s = \hbar m_s|nlmm_s\rangle. \tag{14.4.6}$$

Let us now summarize the position regarding the states of the hydrogen atom. *To specify a state completely it is necessary to give four quantum numbers. Each state is a simultaneous eigenstate of four operators, whose eigenvalues are related to quantum numbers.* The first three operators were shown to form a commuting set. To ensure that the states are simultaneous eigenstates of all four operators, *the spin operator $S_z$ is assumed to commute with the remaining three.*

As the electron is a charged particle, it might be expected that its spin angular momentum would be associated with a magnetic moment. This is indeed so, and the magnetic moment is again proportional to the spin angular momentum, as was the case with orbital moments. However, the constant of proportionality is different. It is, in fact, twice as great as the gyromagnetic ratio defined by eqn (14.2.4). The $z$-component of the spin magnetic moment is therefore given by the equation

$$\langle m_s | m_z | m_s \rangle = -\frac{e\hbar}{m_e} m_s \qquad (14.4.7)$$

which should be compared with eqn (14.2.6).

The spin magnetic moment provides the main contribution to the magnetic properties of matter. Indeed, a ferromagnetic substance contains many tightly packed spins, which align themselves spontaneously to form magnetic domains. Under the influence of a relatively small applied magnetic field the domains get aligned as well, resulting in macroscopic magnetization which can be detected and measured by conventional magnetic methods.

## Chapter 15

# Magnetic Energy

## 15.1 Normal Zeeman Effect

In the preceding chapters we studied the states of a hydrogen atom by considering the motion of a single electron in a central electric field of force, provided by the positive charge of the nucleus, which was assumed to be fixed. Under these conditions the Hamiltonian of the moving electron contained only two terms which accounted for the kinetic and electrical potential energy of the electron (see p. 198). Subsequently we learned that the electron possesses a magnetic moment due to its orbital motion and spin. By virtue of this moment the hydrogen atom is a minute magnet which will have magnetic potential energy when placed in a magnetic field. To study the dynamical states of a hydrogen atom in the presence of the magnetic field an extra term must be included in the potential energy which appears in Schroedinger's equation. It is our immediate object to formulate this extra term and to see what effect it has on the states of motion. *We consider only the orbital moment of the electron neglecting spin for the time being.* The latter introduces complications which we are not yet sufficiently equipped to tackle.

The potential energy $W$ of a magnetic moment $\mathbf{m}$ immersed in a d.c. magnetic field $\mathbf{B}$ is given by the expression

$$W = -\mathbf{m} \cdot \mathbf{B}. \tag{15.1.1}$$

The Hamiltonian of the orbiting electron therefore has the form

$$\mathscr{H} = -\frac{\hbar^2}{2m_e} \nabla^2 + V + W$$

$$= -\frac{\hbar^2}{2m_e} \nabla^2 - \frac{Ze^2}{4\pi\epsilon_0 r} - \mathbf{m} \cdot \mathbf{B} \tag{15.1.2}$$

where a curly $\mathscr{H}$ has been used to denote the Hamiltonian to avoid confusion with the magnetic field vector $\mathbf{H}$.

Before attempting to solve Schroedinger's equation of the form (15.1.2) let us consider the magnetic term more carefully. We assume that the d.c. magnetic field is uniform and is parallel to the $z$-axis. Hence it has only the component $B_z$, and the scalar product (15.1.1) has the form

$$W = -B_z m_z \qquad (15.1.3)$$

where the constant $B_z$ is deliberately written first. We now recall that for an electron orbit the magnetic moment is proportional to the angular momentum by eqn (14.2.3). Equation (15.1.3) assumes the form

$$W = \gamma B_z L_z \qquad (15.1.4)$$

where $L_z$ is the $z$-component of the angular momentum operator (see eqn 14.1.4).

In the preceding chapter we found that energy eigenstates of the hydrogen atom are simultaneously eigenstates of the $z$-component of angular momentum. Hence they are also *eigenstates of the magnetic energy* (15.1.4), and of the extended Hamiltonian (15.1.2). The new energy eigenvalues are readily found.

$$\mathcal{H}|nlm\rangle = \left( -\frac{\hbar^2}{2m_e}\nabla^2 - \frac{Ze^2}{4\pi\epsilon_0 r} + \gamma B_z L_z \right)|nlm\rangle$$

$$= \left( -\frac{\hbar^2}{2m_e}\nabla^2 - \frac{Ze^2}{4\pi\epsilon_0 r} \right)|nlm\rangle + \gamma B_z L_z|nlm\rangle$$

$$= E_n|nlm\rangle + \gamma B_z \hbar m|nlm\rangle$$

$$= (E_n + \gamma\hbar m B_z)|nlm\rangle$$

$$= (E_n + \beta m B_z)|nlm\rangle \qquad (15.1.5)$$

where $E_n$ is given by eqn (13.1.17).

From eqn (15.1.5) we see that the energy levels of the hydrogen atom immersed in a magnetic field are modified by the term $\gamma\hbar m B_z$. This term is either added to or subtracted from $E_n$ depending on whether the magnetic quantum number $m$ is positive or negative. In many situations it is convenient to consider it by itself and call it simply the *energy of a magnetic moment of quantum number $m$* in an applied field $B_z$.

As the magnetic term depends linearly on the applied field, it is instructive to plot the modified energy levels as a function of $B_z$. This is done schematically in Fig. 15.1 for some of the lower states of hydrogen. From Fig. 15.1 we see that the effect of the magnetic field is *to split the energy levels*. The ground state is not split because its magnetic quantum number vanishes. The splitting of energy levels results in the *splitting of spectral lines*, an effect first observed by Zeeman.

Hence the phenomenon is frequently referred to as the Zeeman effect, and the magnetic energy term is called the *Zeeman energy*.

At this point it should be noted that the application of a magnetic field to hydrogen atoms has the effect of partially *lifting the degeneracy of its energy levels* (see p. 209). Now, there are eigenfunctions belonging to distinct energy levels, as identified by the quantum number $m$, in addition to $n$. However, there still remain many degenerate states corresponding to distinct values of the quantum number $l$.

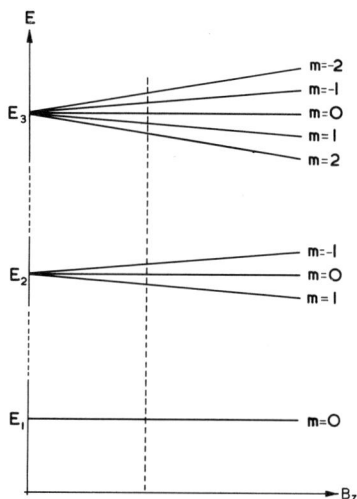

Fig. 15.1.

A calculation of the order of magnitude of the Zeeman splittings for magnetic fields easily realizable in the laboratory shows that they are quite small compared with the spacing between the original levels $E_n$. The magnetic field can therefore be said to have only a *small perturbing effect*. As such it is typical of many effects which are encountered in atomic physics. However, in most cases a theoretical evaluation is quite complicated. The reason for the complication lies in the fact that the small perturbing term does not have simultaneous eigenstates with the basic Hamiltonian to which it is added. This makes it impossible to find the modification of the basic energy level scheme by the simple method that leads up to eqn (15.1.5). Instead recourse must be had to special methods called *perturbation theory*, a subject which is beyond the scope of the present treatment.

The foregoing account of the Zeeman effect does not explain spectroscopic experiments completely. In particular it was found quite early that spectroscopic lines are split even in the absence of a magnetic field. Such observations led to the postulate that the electron has an intrinsic spin magnetic moment in addition to its orbital magnetic moment. Besides causing splitting in the absence of a magnetic field, or zero field splitting, the electron spin introduces additional complications into the Zeeman effect proper, giving rise to the term "anomalous Zeeman effect".

## 15.2 Spin-Orbit Coupling

Since the orbiting electron is a small magnet, it provides a magnetic field, however small. The spin magnetic moment is effectively immersed in this field, and has potential energy due to it. In classical magnetostatics this type of potential is called the mutual potential of two magnetic moments and is given by the expression

$$W = \frac{\mathbf{m}_1 \cdot \mathbf{m}_2}{r^3} - \frac{3(\mathbf{m}_1 \cdot \mathbf{r})(\mathbf{m}_2 \cdot \mathbf{r})}{r^5}. \tag{15.2.1}$$

In quantum mechanics this type of energy is referred to as the *spin-orbit coupling or interaction*. Since the spin magnetic moment cannot be explained on classical principles but is based on relativistic effects (see p. 221) it cannot be taken for granted that its interaction energy with the orbit will assume the classical form given above. In fact, Dirac's theory applied to an electron in a central field leads to the expression

$$\text{Spin-orbit coupling} = \xi(r)\,\mathbf{L}\cdot\mathbf{S}. \tag{15.2.2}$$

This term is added to the basic Hamiltonian of the hydrogen atom to explain the splitting of spectral lines in the absence of an externally applied magnetic field. Unfortunately the modifications of the basic energy level scheme cannot be evaluated by the simple method applied to the Zeeman effect in the preceding section. The reason for this difficulty lies in the fact that the *eigenstates of the hydrogen atom* worked out in foregoing chapters *are not simultaneously eigenstates of the spin-orbit coupling operator* eqn (15.2.2).

To prove that this is so, all we must do is to show that the operator (15.2.2) does not commute with one of the operators listed in eqn (14.4.6). Let us first consider the scalar product $\mathbf{L} \cdot \mathbf{S}$ and form its commutator with $L_z$

$$[L_z, \mathbf{L}\cdot\mathbf{S}] = [L_z\,\mathbf{L}\cdot\mathbf{S} - \mathbf{L}\cdot\mathbf{S}L_z]. \tag{15.2.3}$$

Remembering that $\mathbf{L}$ and $\mathbf{S}$ commute as explained on p. 223, we find after some manipulation that eqn (15.2.3) reduces to the form

$$[L_z, \mathbf{L}\cdot\mathbf{S}] = (L_z L_x - L_x L_z)\,S_x + (L_z L_y - L_y L_z)\,S_y. \tag{15.2.4}$$

By the results established on p. 220 we see that the above expression does not reduce to zero. Hence the operator $\mathbf{L} \cdot \mathbf{S}$ does not commute with $L_z$ and cannot share simultaneous eigenstates with it. This result applies *a fortiori* to the whole operator (15.2.2), as the reader can verify by writing out its commutator with $L_z$.

As a consequence of these complications it is not possible to find the effect of the spin-orbit interaction on the basic energy level scheme of the hydrogen atom by the simple method applied to the Zeeman energy. The principle followed in cases like this is to find a *new set of states which are simultaneously eigenstates of the basic Hamiltonian and of the new operator,* but not necessarily of the other operators used so far. Having found such a set of states the modified energy levels are evaluated by the method used in the preceding section.

In the case of the spin-orbit interaction it is found that the basic levels of the hydrogen atom are split by amounts small compared with the spacing of the levels. It should be noted, that only the excited levels are split. The ground level has no orbital moment, hence there can be no spin-orbit interaction energy and no splitting.

### 15.3 The Principle of Superposition

So far we have considered only some special aspects of the quantum mechanics of the linear oscillator and the hydrogen atom. Having eliminated the time variable by the procedure of Section 11.2, we were limited to the study of the instantaneous state of the oscillator. As it happens this limitation is not very restrictive, because the basic principles of quantum theory can be learned within it. More important is the limitation we have accepted by studying only the individual eigenstates of the Hamiltonian operator, or solutions of Schroedinger's equation, eqn (11.2.2). It is at this point that we propose to make extensions.

At first *we postulate that any solution of Schroedinger's equation, which conforms with the requisite boundary conditions, corresponds to a state of the dynamical system under consideration.*

Secondly, we recall from the theory of linear differential equations that any linear superposition of known solutions is also a solution of Schroedinger's equation. Hence, *any linear superposition of eigenstates of the Hamiltonian is again a state of the dynamical system in question.* However, the new state is no longer an eigenstate of energy, as we shall soon verify on an example. Instead the new state contains *admixtures* of the several eigenstates of energy which enter into it on superposition. Further *we assume that this new state is an eigenstate of some other dynamical variable*, which it may or may not be desirable or easy to find in specific cases.

The foregoing statements, which together make up the *principle of superposition of states*, are peculiar to quantum mechanics. Nothing analogous exists in classical theory, therefore no purpose is served by trying to construct classical pictures. In one way or other they would be misleading. The best one can do is to become accustomed to the new principles by continued study of examples both theoretical and experimental. It is this that we shall attempt to do next, at least in part, on the examples of the linear oscillator and hydrogen atom.

In preceding chapters we found an infinite sequence of solutions of Schroedinger's equation, each associated with a discrete characteristic number or set of numbers, the corresponding eigenvalues. Other solutions are obtained by writing down summations over all the eigenfunctions, each multiplied by a numerical coefficient, as follows

$$\psi = \sum_n a_n u_n. \tag{15.3.1}$$

Here $\psi$ denotes a new state of the oscillator or hydrogen atom, consisting of a mixture of all the eigenstates of energy. The coefficients $a_n$ could be given any (complex) values if eqn (15.3.1) were to be considered merely a general solution of Schroedinger's equation. However, in our case $\psi$ is to be given the same interpretation as the original eigenstates $u_n$: $|\psi|^2$ is to be the position probability density in the state $\psi$. Hence $\psi$ must be normalized according to the same principle as the individual $u_n$-s were normalized.

$$\int \psi^*(x)\,\psi(x)\,dx = 1. \tag{15.3.2}$$

Here the variable $x$ represents either the coordinate $x$ of the linear oscillator or the set of space coordinates $(r, \theta, \varphi)$ in abbreviated form. The integration symbol and the differential element $dx$ are given a similarly general interpretation and adapted as necessary in special cases. The above condition imposes a restriction on the coefficients $a_n$ which is elaborated by substituting the summation (15.3.1) into the integral (15.3.2)

$$\int \sum_n a_n^* u_n^* \sum_m a_m u_m \, dx = 1.$$

Interchanging the sequence of the summation and integration symbols and taking the constant coefficients outside the integral we obtain

$$\sum_n \sum_m a_n^* a_m \int u_n^* u_m \, dx = 1.$$

Whence by the orthogonality conditions applicable to the states $u_n$

$$\sum_n |a_n|^2 = 1. \tag{15.3.3}$$

Equation (15.3.3) places only a general restriction on the coefficients of expansion. It is possible to write down many distinct summations over the eigen states $u_n$, not all of them infinite. Each such superposition is a solution of Schroedinger's equation, and hence a state of the dynamical system in question.

The physical significance attributed to eqn (15.3.3) will be brought out on the example of the linear oscillator. At the same time we shall consider the relation between the principle of superposition and the matrices of operators discussed in Chapter 12.

As our example we take two states of the linear oscillator made up of a superposition of the two lowest energy eigenstates. Equation (15.3.1) reduces in this case to the expressions:

$$\psi_a = a_0 u_0 + a_1 u_1$$
$$\psi_b = b_0 u_0 + b_1 u_1$$

(15.3.4)

where
$$|a_0|^2 + |a_1|^2 = 1, \quad |b_0|^2 + |b_1|^2 = 1.$$

(Note that the states $\psi_a$ and $\psi_b$ are not orthogonal.)

Next let us compute the matrix elements of the Hamiltonian with respect to the states $\psi_a$ and $\psi_b$.

$$H_{aa} = \int_{-\infty}^{\infty} \psi_a^* H \psi_a \, dx$$

$$= \int_{-\infty}^{\infty} (a_0^* u_0^* + a_1^* u_1^*) H(a_0 u_0 + a_1 u_1) \, dx$$

$$H_{aa} = |a_0|^2 E_0 + |a_1|^2 E_1.$$

In the above calculation both the eigenvalue equation for $H$ and the orthogonality properties of the $u_n$ were utilized. The remaining three matrix elements of $H$ are evaluated in the same way. The results are

$$\begin{array}{c} \qquad\qquad \psi_a \qquad\qquad\qquad\qquad \psi_b \\ \begin{array}{c} \psi_a \\ \psi_b \end{array} \left[ \begin{array}{cc} |a_0|^2 E_0 + |a_1|^2 E_1 & a_0^* b_0 E_0 + a_1^* b_1 E_1 \\ b_0^* a_0 E_0 + b_1^* a_1 E_1 & |b_0|^2 E_0 + |b_1|^2 E_1 \end{array} \right]. \end{array}$$

(15.3.5)

The first point to note about this matrix is that it is non-diagonal. Hence, by the postulates explained in Section 12.3 the states $\psi_a$ and $\psi_b$ are not eigen-states of energy of the oscillator. This being so the diagonal elements of (15.3.5) give averages of many measurements of energy in the states $\psi_a$ or $\psi_b$, according to the principle explained on p. 194.

At this point we go beyond the postulates of Section 12.3 and make assumptions which will help interpret the physical significance of the states $\psi_a$ and $\psi_b$ and the coefficients of expansion $a_n$ or $b_n$. *It is assumed that an attempt to measure the energy of the oscillator in one of the states $\psi$ forces the oscillator into one of the energy eigenstates $u_n$ admixed into* $\psi$, and that the measurement yields the corresponding eigenvalue $E_n$. Furthermore, *the probability that a particular value $E_n$ will be registered in an experiment is given by the corresponding coefficient $|a_n|^2$*.

Thus, if in the above example the energy of the oscillator is repeatedly measured while in the state $\psi_a$, values equal to either $E_0$ or $E_1$ will be registered. If the experiment is repeated many times, the two values of energy will be registed with frequencies proportional to the coefficients $|a_0|^2$ and $|a_1|^2$. This interpretation of the coefficients of expansion eqn (15.3.1) justifies the term used to refer to them: *probability amplitudes*.

The principle of superposition can be very neatly expressed in terms of Dirac's notation. The expansion eqn (15.3.1) then assumes the form:

$$|\psi\rangle = \sum_n a_n |n\rangle \tag{15.3.6}$$

where $a_n$ are the probability amplitudes, and $|n\rangle$ denotes the state $u_n$.

The condition that $\sum_n |a_n|^2 = 1$ is obtained with the help of eqn (15.3.2) as follows

$$\begin{aligned}
1 = \langle \psi | \psi \rangle &= \sum_n a_n^* \langle n | \sum_m a_m | m \rangle \\
&= \sum_n \sum_m a_n^* a_m \langle n | m \rangle \\
&= \sum_n \sum_m a_n^* a_m \, \delta_{nm} \\
&= \sum_n |a_n|^2 = 1 \tag{15.3.7}
\end{aligned}$$

by the orthogonality condition eqn (12.4.3).

The steps leading up to eqn (15.3.7) should be carefully traced by reference to the integral expressions leading up to eqn (12.4.3).

So far we have considered one aspect of the principle of superposition: the creation of new unknown states as mixtures of known eigenstates. The coefficients of linear combination appeared as arbitrary constants subject only to the general restriction on their magnitudes given by eqn (15.3.3).

The converse interpretation starts with a *given function* $\psi$ which is presumed to be a state of the system under consideration. The question is then posed whether $\psi$ can be expanded in terms of the known eigenstates $u_n$ and thus expressed as a mixture of these states. We assume that this is, in fact, possible and write

$$\psi = \sum_n a_n u_n.$$

It now remains to evaluate the $a_n$ from a knowledge of $\psi$ and $u_n$. We premultiply both sides by $u_r^*$ and integrate.

$$\int u_r^* \psi dx = \int u_r^* \sum_n a_n u_n \, dx$$

$$= \sum_n a_n \int u_r^* u_n \, dx$$

$$= \sum_n a_n \delta_{rn}$$

$$\int u_r^* \psi dx = a_r$$

where the orthogonality and normalization of the $u_n$ has been utilized. The coefficients of expansion $a_n$ can thus be determined by the above procedure and the state $\psi$ is expressed in terms of the $u_n$. The foregoing method should be familiar to the student on the example of the Fourier series expansion of a function.

Dirac's notation lends itself particularly well to the foregoing calculation of the coefficients of expansion. We write

$$|\psi\rangle = \sum_n a_n |n\rangle$$

$$\langle r|\psi\rangle = \langle r| \sum_n a_n |n\rangle$$

$$= \sum_n a_n \langle r|n\rangle$$

$$= \sum_n a_n \delta_{rn}$$

$$\langle r|\psi\rangle = a_r. \tag{15.3.8}$$

Here again each step should be carefully compared with the corresponding integral expression.

## Chapter 16

# Many-Electron Atoms

### 16.1 The Equation of Motion of a System of many Particles

Before we approach the problem of the many-electron atom a basic question must first be raised and answered. The question is: how does one set up a quantum mechanical equation of motion of a system of many particles.

It should be remembered that so far we have considered systems consisting of one particle only. The linear oscillator was a typical example of such a system, while the hydrogen atom, although consisting of two particles, was simplified by the assumption that the nucleus is merely a fixed source of a radial field of force.

Schroedinger's equation of motion was introduced in Chapter 1 in a form suitable for application to a single-particle system only. We must now extend that equation so that it can be applied to systems of many particles. As a basis of this extension we take the fact that Schroedinger's equation is the eigenvalue equation of the Hamiltonian, or energy operator, of the one-particle system under consideration.

To set up the equation of motion of a system of $N$ particles *all contributions to the energy of all the particles are added together to form a mammoth Hamiltonian*. The equation of motion then takes the form of the eigenvalue equation

$$\mathcal{H}\,\psi = \mathcal{E}\,\psi \tag{16.1.1}$$

where $\mathcal{H}$ is the Hamiltonian, $\mathcal{E}$ is its eigenvalue, which we now write in curly form to distinguish it from the one-particle case, and $\psi = \psi(\mathbf{r}, 0)$ is a time-independent eigenfunction to be found together with $\mathcal{E}$ by solving the eigenvalue problem. The variable $\mathbf{r}$ represents all space coordinates required to determine the position of the system.

In general the Hamiltonian of eqn (16.1.1) will contain both potential and kinetic energy terms. The former will have the form of functions of position, while the latter will be expressed as differential operators following the rules laid down in Section 12.2.

Equation (16.1.1) is analogous to eqn (11.2.5) in that the time variable has already been separated. In its time-dependent form Schroedinger's equation of the many particle system reads

$$\mathscr{H}\,\psi = i\hbar\,\frac{\partial \psi}{\partial t}$$

where $\qquad\qquad\qquad \psi = \psi(\mathbf{r}, t).$ $\qquad\qquad\qquad$ (16.1.2)

The observations on the axiomatic nature of Schroedinger's equation as introduced in Chapter 11 apply to its extension to many-particle systems. The ultimate justification for the approach outlined in this section is that it yields results which agree with experiments.

## 16.2  The Hamiltonian of the Many-electron Atom

To formulate the Hamiltonian of an atom of atomic number $Z$, containing $N$ electrons, we start with a preliminary simplification. As in the case of the hydrogen atom we assume that practically the whole mass of the atom is concentrated in the nucleus, and that the latter may be considered a fixed source of a central electrostatic field of force in which the electrons perform orbital motions.

The Hamiltonian of the atom will consist of the following terms.

(i)  Kinetic energy of the electrons:

$$\sum_{i=1}^{N} \frac{p_i^2}{2m_e} = \sum_{i=1}^{N} \frac{-\hbar^2}{2m_e}\,\nabla_i^2 \qquad\qquad (16.2.1)$$

where the suffix $i$ labels the individual electron, $p_i$ is the momentum of the $i$-th electron, and the operator $\nabla_i^2$ operates on the coordinates of the $i$-th electron only.

(ii)  Potential energy of the electrons in the attractive field of force of the nucleus:

$$\sum_{i=1}^{N} -\frac{Ze^2}{4\pi\epsilon_0 r_i} \qquad\qquad (16.2.2)$$

where $r_i$ is the radial distance of the $i$-th electron from the nucleus.

(iii)  Potential energy of all electron pairs due to their mutual repulsion:

$$\sum_{i>j=1}^{N} \frac{e^2}{4\pi\epsilon_0 |\mathbf{r}_i - \mathbf{r}_j|} \qquad\qquad (16.2.3)$$

where $|\mathbf{r}_i - \mathbf{r}_j|$ is the distance between the $i$-th and the $j$-th electrons and the inequality $i > j$ rules out any possibility that an electron pair might be included twice in the summation.

In addition to the three contributions listed above further terms would have to appear in a really general Hamiltonian of the atom. There would be the spin-orbit interaction of each electron, the spin-spin and orbit-orbit interactions of all electron pairs, and the interaction between the magnetic moments (orbital and spin) of the electrons.

A problem of this generality has never been solved. In fact, the inclusion of the three terms, listed explicitly above, results in an eigenvalue equation which has not been solved analytically. The difficulty is due to the mutual repulsion of the electron pairs (iii), which prevents variables from being separable in Schroedinger's equation. It is, however, possible to separate variables and obtain explicit solutions, if only the first two terms are retained in the Hamiltonian. These solutions yield a number of useful results regarding the structure of atoms and they form the basis of various approximation methods which can be applied to allow for the remaining effects.

### 16.3 The Central Field Approximation

Neglecting all contributions to the Hamiltonian of the many-electron atom except the kinetic energy terms and the potential energy due to the attraction of the nucleus, we obtain an approximate form of the equation of motion. The Hamiltonian of the system is written as follows:

$$\mathcal{H} = \sum_{i=1}^{N} \left( -\frac{\hbar^2}{2m_e} \nabla_i^2 - \frac{Ze^2}{4\pi\epsilon_0 r_i} \right) = \sum_{i=1}^{N} \mathcal{H}_i \qquad (16.3.1)$$

where

$$\mathcal{H}_i = -\frac{\hbar^2}{2m_e} \nabla_i^2 - \frac{Ze^2}{4\pi\epsilon_0 r_i} \qquad (16.3.2)$$

and $N$ = number of electrons in the atom.

Using this Hamiltonian Schroedinger's equation in $3N$ independent variables is written down as follows

$$\mathcal{H} \psi(\mathbf{r}_i \ldots \mathbf{r}_n) = \mathcal{E} \psi(\mathbf{r}_i \ldots \mathbf{r}_n)$$

$$\sum_{i=1}^{N} \left( -\frac{\hbar^2}{2m_e} \nabla_i^2 - \frac{Ze^2}{4\pi\epsilon_0 r_i} \right) \psi(\mathbf{r}_i \ldots \mathbf{r}_n) = \mathcal{E} \psi(\mathbf{r}_i \ldots \mathbf{r}_n)$$

$$\sum_{i=1}^{N} \mathcal{H}_i \psi = \mathcal{E} \psi. \qquad (16.3.3)$$

In the above equation the vector $\mathbf{r}_i$ defines the position of the $i$-th electron. It is operated upon only by the term $\mathcal{H}_i$ in the Hamiltonian. The Coordinates of each electron are to be considered a *separate three-dimensional space*, while the *space of all N electrons*, included in the above equation of motion, has $3N$ *dimensions*. The operators which appear in eqn (16.3.3) operate only on three-dimensional spaces to which they belong.

To illustrate the foregoing general remarks let us write out in full the equation of motion for the helium atom which has two electrons, $N = 2$.

$$\left\{ -\frac{\hbar^2}{2m_e}(\nabla_1{}^2 + \nabla_2{}^2) - \frac{Ze^2}{4\pi\epsilon_0}\left(\frac{1}{r_1} + \frac{1}{r_2}\right) \right\} \psi(r_1, \theta_1, \phi_1, r_2, \theta_2, \phi_2)$$

$$= \mathscr{E}\psi(r_1, \theta_1, \phi_1, r_2, \theta_2, \phi_2)$$

$$(\mathscr{H}_1 + \mathscr{H}_2)\psi(r_1, r_2) = \mathscr{E}\psi(r_1, r_2). \tag{16.3.4}$$

The general atomic number $Z$ has been retained in the above equation in order to preserve its identity and enable us to extend the results to the general case of $N$ electrons.

Equation (16.3.4) can be solved by the method of separation of variables. To this end we assume that the wave function $\psi$ can be put in the form

$$\psi(r_1, r_2) = u(r_1)u(r_2) \tag{16.3.5}$$

substitution into eqn (16.3.4) yields

$$(\mathscr{H}_1 + \mathscr{H}_2)u(r_1)u(r_2) = \mathscr{E}u(r_1)u(r_2)$$

$$u(r_2)\mathscr{H}_1 u(r_1) + u(r_1)\mathscr{H}_2 u(r_2) = \mathscr{E}u(r_1)u(r_2). \tag{16.3.6}$$

At this point the method of operating on the wave functions should be noted. Comparison with eqn (16.3.4) will disclose that eqn (16.3.6) follows from the partial differential nature of the operators $\mathscr{H}_1$ and $\mathscr{H}_2$. However, the foregoing is a typical example of the behaviour of any linear operators, not necessarily those expressible as partial differential coefficients.

Continuing now with the solution of eqn (16.3.6) we divide throughout by $\psi$.

$$\frac{\mathscr{H}_1 u(r_1)}{u(r_1)} + \frac{\mathscr{H}_2 u(r_2)}{u(r_2)} = \mathscr{E}. \tag{16.3.7}$$

This can be rewritten in the form

$$\frac{\mathscr{H}_1 u(r_1)}{u(r_1)} = -\frac{\mathscr{H}_2 u(r_2)}{u(r_2)} + \mathscr{E} = E^{(1)}$$

where $E^{(1)}$ is a new constant of separation. The coordinates of the two electrons are now separated, and eqn (16.3.7) is replaced by two differential equations as follows

$$\mathscr{H}_1\, u(\mathbf{r}_1) = E^{(1)}\, u(\mathbf{r}_1)$$

$$\mathscr{H}_2\, u(\mathbf{r}_2) = E^{(2)}\, u(\mathbf{r}_2). \tag{16.3.8}$$

The second constant of separation is defined by $E^{(2)} = \mathscr{E} - E^{(1)}$ that is by the relation

$$\mathscr{E} = E^{(1)} + E^{(2)} \tag{16.3.9}$$

where the superscripts in brackets serve as identification labels.

The equation of motion of the helium atom (16.3.4) has thus been separated into two equations of the hydrogen type as can be seen by comparing eqn (16.3.8) with eqn (13.1.3). Hence, the methods of dealing with the hydrogen atom, explained in Chapter 13, and all the results obtained there are directly applicable in this case.

The general case of an atom of $N$ electrons is treated by a direct extension of the foregoing argument. The wave function is expressed as a product of $N$ factors where each factor $u(\mathbf{r}_i)$ is a function of the coordinates of one electron only. Similarly the eigenvalue $\mathscr{E}$ is split into $N$ terms $E^{(i)}$. The resulting $N$ hydrogen-like wave equations are solved by the methods of Chapter 13 and the solutions are of the same form as those found there for the hydrogen atom.

Hence by an extension of eqn (16.3.5) the wave function of the many-electron atom are obtained in the form

$$\psi(\mathbf{r}_1 \dots \mathbf{r}_N) = u(\mathbf{r}_1)\, u(\mathbf{r}_2) \dots u(r_N). \tag{16.3.10}$$

The corresponding eigenvalues are obtained as a generalization of eqn (16.3.9).

$$\mathscr{E} = \sum_{p=1}^{N} E^{(p)}. \tag{16.3.11}$$

The simplifying assumptions made at the opening of this section have enabled us to reduce the problem of the many-electron atom to the familiar case of the hydrogen atom using the method of separation of variables. The results obtained in Chapter 13 can thus be applied immediately to the study of the energy eigen states of complex atoms. The approximation which permits this approach is usually referred to as the *central field approximation*.

### 16.4 The Eigenstates and Eigenvalues of Many-Electron Atoms—the Exclusion Principle

We shall now proceed to consider the eigenstates and eigenvalues of many-electron atoms and the way they lead to the classification of elements in the periodic table.

To begin with we define a number of terms which will help us find our way among the multiplicity of quantum numbers, eigenvalues and eigenfunctions.

Every one of the hydrogen-like wave functions $u(\mathbf{r}_i)$ is identified by four quantum numbers (including spin), which we shall call an *individual set of quantum numbers*, to be denoted by the single symbol $a^k$.

$$nlmm_s \to a^k.$$

Each wave function of the complete atom of $N$ electrons will be characterized by $N$ individual sets of quantum numbers listed side by side. This *complete set of quantum numbers* will be denoted by $A$.

$$a^1 a^2 \ldots a^N \to A.$$

An *individual wave function* associated with the $i$-th electron and belonging to the set of quantum numbers $a^k$ will be denoted by the symbol

$$u_k(\mathbf{r}_i).$$

A *complete wave function* of the atom belonging to the set of quantum numbers $A$ will have the form

$$\psi_A = \prod_{i=1}^{N} u_k(\mathbf{r}_i). \tag{16.4.1}$$

As an example of the foregoing definitions let us write out in full some eigenstates of the helium atom.

$$\psi_A = u_{100\frac{1}{2}}(\mathbf{r}_1)\, u_{100-\frac{1}{2}}(\mathbf{r}_2)$$

$$\psi_B = u_{200\frac{1}{2}}(\mathbf{r}_1)\, u_{100\frac{1}{2}}(\mathbf{r}_2). \tag{16.4.2}$$

Here the individual sets of quantum numbers, belonging to the individual electron wave functions, have been written out in full. The values must be assigned in accordance with the restrictions laid down on Chapter 13.

The energy levels of the helium atom, associated with the states $\psi_A$ and $\psi_B$, are

$$\mathcal{E}_A = E_1^{(1)} + E_1^{(2)}$$

$$\mathcal{E}_B = E_2^{(1)} + E_1^{(2)} \tag{16.4.3}$$

in accordance with eqn (16.3.11). The labelling of the individual electron eigenvalues follows the scheme $E_n^{(p)}$, where $n$ is the principal quantum number of an individual set, while the superscript refers to the $(p)$-th electron. In the case of atoms having $N$ electrons the energy levels are given by summations over individual energy eigenvalues of the form (16.3.11).

Let us now turn our attention to the individual sets of quantum numbers used in eqn (16.4.2). Their choice is not entirely arbitrary in the sense that certain lists of quantum numbers may not correspond to atomic states which actually exist. To ensure that they do, the allocation of quantum numbers to the individual electrons must be made in accordance with *Pauli's Exclusion Principle*. This important postulate of quantum theory demands that *no two electrons in a many-electron system may share the same individual set of four quantum numbers*. In other words no two electrons may occupy the same individual state. The exclusion principle applies not only to many-electron atoms but to any dynamical system which includes a number of electrons as constituent parts. A solid metal or semiconductor specimen is an example of such a system, to be discussed at length in subsequent chapters.

The helium states given above illustrate the distinction between the *ground state* and *an excited state* of an atom. As in the case of the hydrogen atom the ground state is the state of lowest energy. In the central field approximation the ground state coincides with the lowest possible set of individual quantum numbers, compatible with the exclusion principle. However, we shall see presently that this does not apply to all atoms.

To specify the state of an atom it is necessary to list a complete set of quantum numbers $A$, for all the constituent electrons. An atomic state defined in this way is frequently referred to as an *electron configuration*. It will be clear from the foregoing that the listing of the individual sets of quantum numbers for atoms containing tens of electrons is a lengthy procedure. For this reason various abbreviated methods have been devised, one of which, the spectroscopic method, will be explained now.

The principal quantum number $n$ is given as a number 1, 2, 3 etc. The angular-momentum quantum number $l$ is specified by letters according to the following code.

$$\text{value of } l \qquad 0 \quad 1 \quad 2 \quad 3 \quad 4 \quad 5$$

$$\text{letter symbol} \qquad s \quad p \quad d \quad f \quad g \quad h \tag{16.4.4}$$

The magnetic quantum numbers $m$ and $m_s$ are not stated explicitly. Instead the number of electrons having a given value of $n$ and $l$ is given as a superscript. Thus the statement $(2p)^2$ means that we are dealing with electrons for which $n = 2, l = 1$, and that there are two of them.

As an example let us write down the complete set of quantum numbers, or electron configuration, for the ground state of the carbon atom ($Z = 6, N = 6$).

$$
\begin{array}{cccccc}
a^1 & a^2 & a^3 & a^4 & a^5 & a^6 \\
A \rightarrow \underbrace{(100 - \tfrac{1}{2})(100\tfrac{1}{2})}_{(1s)^2} & & \underbrace{(200 - \tfrac{1}{2})(200\tfrac{1}{2})}_{(2s)^2} & & \underbrace{(210 - \tfrac{1}{2})(210\tfrac{1}{2})}_{(2p)^2}.
\end{array}
$$

Using spectroscopic symbols only, the electron configuration assumes the short form

$$(1s)^2 \, (2s)^2 \, (2p)^2. \tag{16.4.5}$$

From the above expression it can be seen that the atomic electrons fall naturally into groups. Thus it is customary to speak of $s$-electrons, $p$-electrons, etc. The word *shell of electrons* is frequently used to refer to such a group. Hence one can speak of $s$-shells, $p$-shells, $d$-shells, etc. A word of warning regarding the terminology of shells. Sometimes the word shell is used to denote the whole group of electrons having the same quantum number $n$, while the word subshell is used for electrons having the same quantum number $l$. Despite this ambiguity there is little risk of confusion, as the context usually makes the position clear.

The spectroscopic symbols for electron configurations contain no explicit information regarding magnetic quantum numbers. In the case of shells which are full such information would be redundant. By a *full shell* we mean a group of electrons which have occupied, subject to the exclusion principle, all values of the magnetic quantum numbers within the limitations stated in Chapter 13. Thus $1s$ and $2s$ shells of the carbon configuration (16.4.5) are full, and it is unnecessary to state the magnetic quantum numbers because $m = 0$ by implication, and $m_s$ can only have the values $\pm \frac{1}{2}$.

The position is not quite so clear as regards the partially filled $2p$ shell. The maximum number of electrons which could be accommodated in this shell is 6 but in actual fact there are only 2. Although these have been allocated the orbital quantum number $m = 0$ and the spin quantum numbers $m_s = \pm \frac{1}{2}$, this step was quite arbitrary and there is no reason why some other quantum numbers should not have been used. This question will be discussed further in subsequent sections. In the meantime we note that the ground state configuration of an atom may contain imcomplete shells, and that the electrons residing in such shells are frequently referred to as *unpaired electrons.*

## 16.5 The Periodic Table of Elements

The central field approximation and the classification of atomic states derived from it forms the theoretical basis of the periodic table of elements.

At first the elements are arranged according to ascending atomic numbers as shown in Table 16.1, in which the first column lists the element, the second column gives its atomic number, and the remaining columns give the ground state electron configuration in spectroscopic symbols. Although the table may be familiar, it is nevertheless desirable to pause and study it with reference to the theoretical ideas explained in preceding sections.

For elements of low atomic number, up to argon, the electron configurations develop predictably as expected on the basis of the central field approximation. The ground states, that is states of lowest energy, correspond to lowest possible sets of quantum numbers. However, starting with potassium irregularities set in. Instead of occupying shells of lowest quantum numbers, electrons make their

## Table 16.1

Ground state electron configurations of the elements

| Values of $n, l$ | | 1, 0 | 2, 0 | 2, 1 | 3, 0 | 3, 1 | 3, 2 | 4, 0 | 4, 1 | 4, 2 | 4, 3 |
|---|---|---|---|---|---|---|---|---|---|---|---|
| Spectral Notation | | $1s$ | $2s$ | $2p$ | $3s$ | $3p$ | $3d$ | $4s$ | $4p$ | $4d$ | $4f$ |
| H | 1 | 1 | | | | | | | | | |
| He | 2 | 2 | | | | | | | | | |
| Li | 3 | 2 | 1 | | | | | | | | |
| Be | 4 | 2 | 2 | | | | | | | | |
| B | 5 | 2 | 2 | 1 | | | | | | | |
| C | 6 | 2 | 2 | 2 | | | | | | | |
| N | 7 | 2 | 2 | 3 | | | | | | | |
| O | 8 | 2 | 2 | 4 | | | | | | | |
| F | 9 | 2 | 2 | 5 | | | | | | | |
| Ne | 10 | 2 | 2 | 6 | | | | | | | |
| Na | 11 | | | | 1 | | | | | | |
| Mg | 12 | | | | 2 | | | | | | |
| Al | 13 | Neon | | | 2 | 1 | | | | | |
| Si | 14 | Configuration | | | 2 | 2 | | | | | |
| P | 15 | 10 Electrons. | | | 2 | 3 | | | | | |
| S | 16 | | | | 2 | 4 | | | | | |
| Cl | 17 | | | | 2 | 5 | | | | | |
| A | 18 | | | | 2 | 6 | | | | | |
| K | 19 | | | | | | | 1 | | | |
| Ca | 20 | | | | | | | 2 | | | |
| Sc | 21 | | | | | | 1 | 2 | | | |
| Ti | 22 | | | | | | 2 | 2 | | | |
| V | 23 | | | | | | 3 | 2 | | | |
| Cr | 24 | | Argon | | | | 5 | 1 | | | |
| Mn | 25 | | Configuration. | | | | 5 | 2 | | | |
| Fe | 26 | | | | | | 6 | 2 | | | |
| Co | 27 | | | | | | 7 | 2 | | | |
| Ni | 28 | | | | | | 8 | 2 | | | |
| Cu | 29 | | 18 Electrons. | | | | 10 | 1 | | | |
| Zn | 30 | | | | | | 10 | 2 | | | |
| Ga | 31 | | | | | | 10 | 2 | 1 | | |
| Ge | 32 | | | | | | 10 | 2 | 2 | | |
| As | 33 | | | | | | 10 | 2 | 3 | | |
| Se | 34 | | | | | | 10 | 2 | 4 | | |
| Br | 35 | | | | | | 10 | 2 | 5 | | |
| Kr | 36 | | | | | | 10 | 2 | 6 | | |

**Table 16.1**–*cont.*

| Values of $n, l$ | | 4, 0 | 4, 1 | 4, 2 | 4, 3 | 5, 0 | 5, 1 | 5, 2 | 5, 3 | 5, 4 | 6, 0 |
|---|---|---|---|---|---|---|---|---|---|---|---|
| Spectral Notation | | $4s$ | $4p$ | $4d$ | $4f$ | $5s$ | $5p$ | $5d$ | $5f$ | $5g$ | $6s$ |
| Rb | 37 | | | | | 1 | | | | | |
| Sr | 38 | | | | | 2 | | | | | |
| Y | 39 | | | 1 | | 2 | | | | | |
| Zr | 40 | Krypton | | 2 | | 2 | | | | | |
| Nb | 41 | Config- | | 4 | | 1 | | | | | |
| Mo | 42 | uration. | | 5 | | 1 | | | | | |
| Ma | 43 | 36 | | 6 | | 1 | | | | | |
| Ru | 44 | Electrons. | | 7 | | 1 | | | | | |
| Rh | 45 | | | 8 | | 1 | | | | | |
| Pd | 46 | | | 10 | | 1 | | | | | |
| Ag | 47 | | | | | 1 | | | | | |
| Cd | 48 | | | | | 2 | | | | | |
| In | 49 | Palladium | | | | 2 | 1 | | | | |
| Sn | 50 | Configuration. | | | | 2 | 2 | | | | |
| Sb | 51 | 46 | | | | 2 | 3 | | | | |
| Te | 52 | Electrons. | | | | 2 | 4 | | | | |
| I | 53 | | | | | 2 | 5 | | | | |
| Xe | 54 | | | | | 2 | 6 | | | | |
| Cs | 55 | | | | | 2 | 6 | | | | 1 |
| Ba | 56 | | | | | 2 | 6 | | | | 2 |
| La | 57 | | | | | 2 | 6 | 1 | | | 2 |
| Ce | 58 | | | | 1 | 2 | 6 | 1 | | | 2 |
| Pr | 59 | | | | 2 | 2 | 6 | 1 | | | 2 |
| Nd | 60 | | | | 3 | 2 | 6 | 1 | | | 2 |
| Il | 61 | | | | 4 | 2 | 6 | 1 | | | 2 |
| Sm | 62 | | | | 5 | 2 | 6 | 1 | | | 2 |
| Eu | 63 | Shells | | | 6 | 2 | 6 | 1 | | | 2 |
| Gd | 64 | $1s$ to $4d$ | | | 7 | 2 | 6 | 1 | | | 2 |
| Tb | 65 | 46 Electrons. | | | 8 | 2 | 6 | 1 | | | 2 |
| Dy | 66 | | | | 9 | 2 | 6 | 1 | | | 2 |
| Ho | 67 | | | | 10 | 2 | 6 | 1 | | | 2 |
| Er | 68 | | | | 11 | 2 | 6 | 1 | | | 2 |
| Tm | 69 | | | | 13 | 2 | 6 | 0 | | | 2 |
| Yb | 70 | | | | 14 | 2 | 6 | 0 | | | 2 |
| Lu | 71 | | | | 14 | 2 | 6 | 1 | | | 2 |

**Table 16.1**—*cont.*

| Values of $n, l$ | | 5, 0 | 5, 1 | 5, 2 | 5, 3 | 5, 4 | 6, 0 | 6, 1 | 6, 2 | 6, 3 | 6, 4 | 6, 5 | 7, 0 |
|---|---|---|---|---|---|---|---|---|---|---|---|---|---|
| Spectral Notation | | 5s | 5p | 5d | 5f | 5g | 6s | 6p | 6d | 6f | 6g | 6h | 7s |
| Hf | 72 | | | 2 | | | 2 | | | | | | |
| Ta | 73 | | | 3 | | | 2 | | | | | | |
| W | 74 | Shells | | 4 | | | 2 | | | | | | |
| Re | 75 | 1s to 5p | | 5 | | | 2 | | | | | | |
| Os | 76 | 68 Electrons. | | 6 | | | 2 | | | | | | |
| Ir | 77 | | | 7 | | | 2 | | | | | | |
| Pt | 78 | | | 9 | | | 1 | | | | | | |
| Au | 79 | | | 10 | | | 1 | | | | | | |
| Hg | 80 | | | | | | 2 | | | | | | |
| Tl | 81 | | | | | | 2 | 1 | | | | | |
| Pb | 82 | Shells | | | | | 2 | 2 | | | | | |
| Bi | 83 | 1s to 5d | | | | | 2 | 3 | | | | | |
| Po | 84 | 78 Electrons. | | | | | 2 | 4 | | | | | |
| At | 85 | | | | | | 2 | 5 | | | | | |
| Rn | 86 | | | | | | 2 | 6 | | | | | |
| Fa | 87 | | | | | | | | | | | | 1 |
| Ra | 88 | | | | | | | | | | | | 2 |
| Ac | 89 | Radon | | | | | | | 1 | | | | 2 |
| Th | 90 | Configuration. | | | | | | | 2 | | | | 2 |
| Pa | 91 | 86 Electrons. | | | | | | | 3 | | | | 2 |
| U | 92 | | | | | | | | 4 | | | | 2 |

appearance in states having higher quantum numbers. The electron configurations shown in Table 16.1 are based on experimental evidence which was used to correct and supplement the central field approximation. On this basis we must conclude that the lowest energy state of an atom such as potassium demands that the unpaired electron, which is outside the $3p$ shell, be accommodated in the $4s$ shell instead of the $3d$ shell. This is the first illustration of the shortcomings of the central field approximation.

Looking further down the table we note that in the case of cesium and barium as many as 3 shells are left vacant by the ground state configuration. The rare earth elements which follow have complete $5s$, $5p$ and $6s$ shells, while the $4f$ shell fills up in steps as the atomic number increases. All these are examples of the inadequacy of the central field approximation in predicting the ground states of many-electron atoms.

Let us now turn to the periodicity of the table of elements. This is due to the cyclical appearance of sets of completely filled electron shells in the case of the inert gases, helium, neon, argon etc. Table 16.2 shows the periods in the customary way, by listing the inert gases on the extreme right, while the alkali elements are listed on the left. The latter are characterized by a single $s$-electron

Perio
Atomic Number

| Group | I. A. | B. | II. A. | B. | III. A. | B. | IV. A. | B. | V. A. |
|---|---|---|---|---|---|---|---|---|---|
| Period 1 | Hydrogen, H. 1. $1\cdot008$ | | | | | | | | |
| Period 2 | Lithium, Li. 3. $6\cdot94$ | | Beryllium, Be. 4. $9\cdot0$ | | Boron, B. 5. $10\cdot8$ | | Carbon, C. 6. $12\cdot00$ | | Nitro 7. $1\cdot$ |
| Period 3 | Sodium, Na. 11. $23\cdot00$ | | Magnesium, Mg. 12. $24\cdot32$ | | Aluminium, Al. 13. $26\cdot97$ | | Silicon, Si. 14. $28\cdot06$ | | Phos P. 15. |
| Period 4 | Potassium, K. 19. $39\cdot10$ | Copper, Cu 29. $63\cdot57$ | Calcium, Ca. 20. $40\cdot08$ | Zinc, Zn. 30. $65\cdot38$ | Scandium, Sc. 21. $45\cdot1$ | Gallium, Ga. 31. $69\cdot72$ | Titanium, Ti. 22. $47\cdot9$ | Germanium, Ge. 32. $72\cdot6$ | Vanadium, V. 23. $50\cdot95$ Arsen 33. |
| Period 5 | Rubidium, Rb. 37. $35\cdot44$ | Silver, Ag. 47. $107\cdot88$ | Strontium, Sr. 38. $87\cdot63$ | Cadmium, Cd. 48. $112\cdot41$ | Yttrium, Yt. 39. $88\cdot92$ | Indium, In. 49. $114\cdot8$ | Zirconium, Zr. 40. $91\cdot2$ | Tin, Sn. 50. $118\cdot7$ | Niobium, Nb. 41. $93\cdot3$ Antin 51. |
| Period 6 | Caesium, Cs. 55. $132\cdot91$ | Gold, Au. 79. $197\cdot2$ | Barium, Ba. 56. $137\cdot36$ | Mercury, Hg. 80. $200\cdot61$ | Lanthanum, La. 57. $138\cdot9$ Cerium, Ce. 58. $140\cdot13$ Praseodymium, Pr. 59. $140\cdot92$ Neodymium, Nd. 60. $144\cdot27$ Illinium, Il. 61. Samarium, Sm. 62. $150\cdot43$ Europium, Eu. 63. $152\cdot0$ Gadolinium, Gd. 64. $157\cdot3$ Terbium, Tb. 65. $159\cdot8$ Dysprosium, Ds. 66. $162\cdot46$ Holmium, Ho. 67. $163\cdot5$ Erbium, Er. 68. $167\cdot64$ Thuilum, Tm. 69. $169\cdot4$ Ytterbium, Yb. 70. $173\cdot04$ Lutecium, Lu. 71. $175\cdot0$ | Thallium, Tl. 81. $204\cdot39$ | Hafnium, Hf. 72. $178\cdot6$ | Lead, Pb. 82. $207\cdot22$ | Tantalum, Ta. 73. $131\cdot4$ Bismu 83. |
| Period 7 | 87.— | | Radium, Ra. 88. $225\cdot97$ | | Actinium, Ac. 89. | | Thorium, Th. 90. $232\cdot12$ | | Protoactinium, Pa. 91. |

nents
c Weights thus, *6·94*

| VI. | | VII. | | VIII. | | | 9. |
|---|---|---|---|---|---|---|---|
| | B. | A. | B. | | | | |
| | | | | | | | Helium, He. 2. *4·00* |
| | Oxygen, O. 8. *16·00* | | Fluorine, F. 9. *19·0* | | | | Neon, Ne. 10. *23·2* |
| | Sulphur, S. 16. *32·06* | | Chlorine, Cl. 17. *35·46* | | | | Argon, A. 18. *39·94* |
| nium, Cr. 52·0 | Selenium, Se. 34. *79·92* | Manganese, Mn. 25. *54·93* | Bromine, Br. 35. *79·92* | Iron, Fe. 26. *55·84* | Cobalt, Co. 27. *58·94* | Nickel, Ni. 28. *58·69* | Krypton, Kr. 36. *82·9* |
| denum, Mo. 6·0 | Tellurium, Te. 52. *127·6* | Masurium, Ma. 43. | Iodine, I. 53. *126·92* | Ruthenium, Ru. 44. *101·7* | Rhodium, Rn. 45. *102·9* | Palladium, Pd. 46. *106·7* | Xenon, Xe. 54. *131·3* |
| en, W. 84·0 | Polonium, Po. 84. | Rhenium, Re. 75. *186·31* | 85. | Osmium, Os. 76. *191·5* | Iridium, Ir. 77. *193·1* | Platinum, Pt. 78. *195·23* | Emanation, Em. 86. *222·0* |
| m, U. 38·14 | | Neptunium, Np. 93. *239* | | Plutonium, Pa. 94. *239·14* | | | |

outside a set of filled shells. Between these extremes various possibilities of partial shell occupation can be seen.

As emphasized above, Tables 16.1 and 16.2 list the ground state configurations of the elements. In an *excited atom* the configuration will correspond to a higher energy state. Here again it should be borne in mind that the central field approximation will not predict correctly the higher energy states except in the case of the smallest atoms. There is, however, an abundance of spectroscopic data from which the excited configuration can be deduced.

The electron configurations of ionized atoms will differ from the ground state in either having some of the outermost electrons removed (usually down to a core of complete shells) or having some added (usually to complete a shell). In either case a net electrical charge appears, and the energy level is considerably above the ground state of the atom.

## 16.6  Addition of Angular Momenta

As can be seen from the preceding sections theoretical attempts to work out the states of many-electron atoms become very complicated if the Hamiltonian is not drastically simplified, or else yield only a limited amount of information. In such circumstances much knowledge regarding the structure of atoms has been obtained by experimental methods. In many cases these methods rely on the fact that most isolated atoms possess a resultant angular momentum and magnetic moment derived both from electron orbits and spins. Moreover the magnetic properties of materials in bulk are ultimately traceable to the resultant magnetic moments of individual atoms. The object of the present section is to explain how resultant angular momenta are formed.

To begin with we consider the simplest case of two angular momenta. An example of this type has already been encountered in connection with the hydrogen atom. In an excited state the hydrogen atom has both an orbital and a spin angular momentum, and the problem of the spin-orbit interaction, introduced in Section 15.2 can be reduced to the problem of finding a resultant.

The problem of adding two angular momenta, which we shall denote by the general symbols $\mathbf{J}_1$ and $\mathbf{J}_2$, presents itself whenever the associated magnetic moments interact, that is whenever one is affected by the magnetic field of the other. The interaction causes the moments, and hence the angular momenta, to assume relative orientations which minimize their mutual energy and hence minimize the energy level of the atom or other dynamical system to which they belong. Such *interacting angular momenta* are said to be coupled. As we shall see below the coupling causes the formation of a *resultant angular momentum,* according to the rules of vector addition.

For the purpose of the present argument we isolate two eigenstates $| j_1 m_1 \rangle$ and $| j_2 m_2 \rangle$ which characterize the angular momenta $\mathbf{J}_1$ and $\mathbf{J}_2$. They may be thought of as the angular momenta of the two individual electrons within the helium atom, or indeed two electrons within any atom. The quantum numbers $j_1$ and $j_2$ may be either integers or half integers since we include the possibility of both orbital and spin momenta, and the quantum numbers $m_1$ and $m_2$ of their $z$-components are subject to the rules explained in Sections 14.1 and 14.4.

As has been explained in the preceding sections the individual states $|j_1 m_1\rangle$ and $|j_2 m_2\rangle$ are factors in a more complicated wave function which describes the many-electron atom. Hence they appear in the form of a product which can be written in two ways (see Section 13.4)

$$|j_1\, m_1\rangle|j_2\, m_2\rangle = |j_1\, j_2\, m_1\, m_2\rangle. \tag{16.6.1}$$

The common or coupled state of the two angular momenta is thus determined by four quantum numbers.

The question to be answered is whether there are states $|jm\rangle$ *of a single angular momentum* **J** which are an equivalent expression of the states (16.6.1). The short answer to this question is that such states can be found, and that they correspond to what is intuitively understood by a *vector resultant* of the angular momenta. As the analytical derivation of these results is beyond the scope of the present treatment, we shall only state the results regarding the relations between the various quantum numbers in a form in which they will be required in applications. Since the wave functions themselves will not be used there is no need to introduce them in the present context.

The quantum number $j$ of the resultant angular momentum $\mathbf{J} = \mathbf{J}_1 + \mathbf{J}_2$ is obtained by the vector addition rule of Fig. 16.1, in which it is assumed that $j_1 > j_2$. According to this rule $j$ may have any one of the integral or half integral values, as the case may be, in the range from $j_1 + j_2$ down to $j_1 - j_2$. In cases in

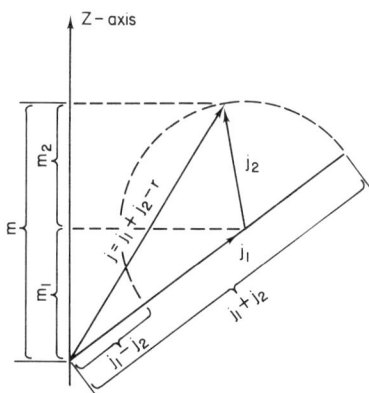

Fig. 16.1. Vector addition of angular momentum quantum numbers.

which $j_2 > j_1$ the latter difference is reversed. Hence the quantum number $j$ of the resultant angular momentum may assume any one of the following values.

$$j = j_1 + j_2, j_1 + j_2 - 1, j_1 + j_2 - 2, \ldots j_1 - j_2 + 1, j_1 - j_2.$$
(16.6.2)

As an example let us consider the resultant angular momentum of a $p$-electrn, which has an orbital angular momentum of quantum number $j_1 = l = 1$ and a spin of quantum number $j_2 = s = \frac{1}{2}$. There are two possible resultants having quantum numbers $j = j_1 \pm j_2 = l \pm s = \frac{3}{2}$ or $\frac{1}{2}$.

Once a resultant angular momentum $j$ has been formed, the quantum numbers $m$ of its $z$-component assume all values between $j$ and $-j$ in steps of unity. Hence

$$m = j, j - 1, j - 2, \ldots -j + 2, -j + 1, -j.$$
(16.6.3)

Since in our example of a $p$-electron two possible resultants exist, we can have two sets of $m$ quantum numbers. In the first case they are

$$m = \tfrac{3}{2}, \tfrac{1}{2}, -\tfrac{1}{2}, -\tfrac{3}{2},$$

while in the second they are

$$m = \tfrac{1}{2}, -\tfrac{1}{2}.$$

An important observation regarding the relationship between the $z$-components of angular momenta should be noted at this point. The $z$-component of the resultant is the algebraic sum of the $z$-components of the individual angular momenta as in every case of vector addition.

$$m = m_1 + m_2.$$
(16.6.4)

This rule is again illustrated by the vector model of Fig. 16.1.

The quantum numbers of the square of the magnitude of the resultant angular momentum have the form $j(j + 1)$ as in the case of individual orbital or spin momenta.

It will be apparent from the foregoing explanations that in cases in which $j_1$ and $j_2$ have values greater than 1 and $\frac{1}{2}$ the various possibilities become quite numerous. Fortunately enough nature itself provides simplification through the exclusion principle and other rules. Before going on to consider these matters, we shall deal briefly with the problem of forming resultants of more than two angular momenta.

Cases of this type are tackled by an extension of the method outlined above. At first the resultant of two angular momenta is formed and then a third one is added on, following the same rules. If there are any more angular momenta, they are added successively by a continuation of the same method. In every possible case the $z$-components may assume any of the values given by eqn (16.6.3).

Since in any but the simplest case the number of possibilities becomes rather bewildering, let us now turn our attention to the simplification introduced into many-electron atoms by the exclusion principle. The most important consequence of this principle is the fact that the *resultant angular momentum of a complete electron shell vanishes*.

This can be demonstrated by reference to eqn (16.6.4) and to the method of assigning $m$ quantum numbers to electrons within a closed shell (see Section 16.4). From the latter it follows that all possible positive and negative values of the $m$ quantum numbers are exhausted by the orbits and spins of electrons forming a complete shell. Hence the algebraic sum of these numbers is zero indicating, according to eqn (16.6.4), that the $z$-component of the resultant angular momentum vanishes. Since the exclusion principle precludes any other value of the $z$-component, we must conclude that the resultant angular momentum of a filled electron shell also vanishes.

The foregoing result makes it possible to neglect filled shells when considering the angular momentum and magnetic moment of many electron atoms. The latter can only reside in incomplete shells. Since the cancellation of the angular momentum of a filled shell can be visualized as the result of pairing of electrons with opposite angular momenta, the electrons in unfilled shells are frequently referred to as *unpaired electrons*.

The angular momenta of unpaired electrons in the ground state of smaller atoms combine according to *Hund's rule*.

This states that as a first step the spin angular momenta are added to form a resultant **S** which assumes the greatest possible value consistent with the exclusion principle. Then the orbital angular momenta are added to form a resultant **L** which also assumes the greatest possible value subject to the exclusion principle. The reason for this procedure is that the energy level of the atom in the resulting configuration is lowest.

The quantum numbers of the resultant orbital and spin momenta are usually denoted by capitals $L$ and $S$ to distinguish them from the individual quantum numbers, $l$ and $s$. The quantum numbers of the squares $|\mathbf{L}|^2$ and $|\mathbf{S}|^2$ follow the pattern of the individual momenta and are given by $L(L + 1)$ and $S(S + 1)$.

The combined orbital and spin angular momenta, **L** and **S**, become coupled according to the scheme outlined above and summarized in Fig. 16.1. However, out of the many possible values of the resultant only two are found in actual cases. In atoms with less than half-filled shells the minimum value applies, $J = L - S$, while in atoms with more than half-filled shells the maximum value applies, $J = L + S$. The quantum numbers of the overall resultant **J** can be denoted by capitals $J$ and $J(J + 1)$ or lower-case letters $j$ and $j(j + 1)$. The quantum number of the $z$-component $J_z$ can likewise be denoted by a capital $M$ or lower case $m$, depending on which symbol happens to be more convenient in a given context. In all cases the actual value of an angular momentum is obtained on multiplication of the quantum number by the constant $\hbar$, or $\hbar^2$ in the case of squared magnitudes.

To summarize, the following equations give the magnitudes of angular momenta in all cases.

$$|\mathbf{L}|^2 = L^2 = \hbar^2 L(L+1)$$
$$|\mathbf{S}|^2 = S^2 = \hbar^2 S(S+1)$$
$$|\mathbf{J}|^2 = J^2 = \hbar^2 J(J+1). \tag{16.6.5}$$

When atoms become excited to energy levels above the ground state the angular momenta of the unpaired electrons still add according to the above rules, but the values of $L$ and $S$ will no longer be the greatest possible. In every case, however, they are combined according to the above rule.

In large atoms the coupling of the angular momenta of unpaired electrons follows a different pattern. At first the spin and orbital momentum of one electron are added, and then the combined momenta of all unpaired electrons form a resultant. This scheme is usually referred to as $j$-$j$ coupling.

Let us now consider a specific example of the foregoing theory, concentrating on one which is found to be of practical importance. The trivalent chromium ion, $Cr^{3+}$, has been extensively investigated by spectroscopic methods and forms the basis of maser and laser amplifiers. A reference to Table 16.1 shows that it has three unpaired electrons in the $3d$ shell, that is $l = 2$. For this reason it is frequently referred to as belonging to the $3d$ group.

To obtain the state of lowest energy Hund's rule demands that the spins of the unpaired electrons should assume orientations yielding the maximum resultant of $S = \frac{3}{2}$. By the exclusion principle the three electrons must, therefore, have different values of the orbital quantum number $m$, and Hund's rule once again dictates that maximum values apply to the lowest energy states. Hence, $m$ = 2, 1, and 0 yielding a sum of 3, which is the value of the resultant orbital angular momentum $L = 3$.

The final step consists of forming the vector resultant of $S$ and $L$ according to Fig. 16.1. Since the $3d$ shell is less than half filled the state of minimum quantum number $L - S = J = \frac{3}{2}$ has the lowest energy and is therefore most likely to be encountered.

Spectroscopists have devised a scheme of symbols to describe the angular momentum states of atoms, which is still widely used. According to this, the lowest energy state of chromium, worked out above, is denoted by $^4F_{3/2}$. The capital letter in the centre denotes the resultant orbital momentum according to the spectroscopic scheme explained on p. 239. Hence $F$ means $L = 3$. The left upper numeral gives the value of the quantity $2S + 1$, while the right suffix gives the value of the quantum number $J$. The scheme can be summarized by the symbol $^{2S+1}L_J$.

### 16.7  Magnetic Moments of Many-Electron Atoms

As in the case of the hydrogen atom the angular momenta of complex atoms give rise to magnetic moments. The presence of the latter introduces considerable variety into the spectra of many-electron atoms and gives rise to the magnetic properties of solids. The complications are due in the first place to interactions between the internal magnetic moments on the pattern of the spin-orbit coupling introduced in Section 15.2. In the second place a variety of splittings is introduced into the spectra by the application of an external magnetic field. These matters will be considered in a later chapter. What should be emphasized now is the fact that the magnetic moments of unpaired electrons are the ultimate source of magnetism. It is for this reason that we shall now consider the relation between the angular momentum and magnetic moment of many-electron atoms.

From eqns (14.2.6) and (14.4.7) we recall that the magnetic moments of an electron orbit and spin are both proportional to their angular momenta but that the constant of proportionality was greater by a factor of 2 in the case of the spin. If the relation is expressed in terms of the Bohr magneton, eqn (14.2.7), we can write the magnitude of the magnetic moment in the form

$$|\mathbf{m}| = g\beta \sqrt{j(j+1)} \qquad (16.7.1)$$

which is simply the square root of eqn (14.2.5) with the generalized symbol $j$ denoting the angular momentum which may be either orbital or spin. The constant of proportionality $g$ is called the *spectroscopic splitting factor*. It assumes the value 1 for an electron orbit and 2 for an electron spin.

When several angular momenta of the same species, say orbital, are added by the rules of the preceding section, the resultant will be associated with a magnetic moment parallel and proportional to it through the same $g$-factor (where $g = 1$).

$$|\mathbf{m}_L| = g\beta \sqrt{L(L+1)} \qquad (16.7.2)$$

However, if two angular momenta of different species are added, their resultant will no longer be parallel to the associated magnetic moment. The position is illustrated by the vector diagram of Fig. 16.2. Two angular momenta $\mathbf{L}$ and $\mathbf{S}$ are added to form the resultant $\mathbf{J}$. The corresponding magnetic moments $\mathbf{m}_L$ and $\mathbf{m}_S$ are added to form the resultant $\mathbf{m}$. The latter is clearly not parallel to the momentum $\mathbf{J}$, but it so happens that this fact is not of great practical consequence. What matters in static magnetism and in spectroscopic experiments is the value of the projection of $\mathbf{m}$ on $\mathbf{J}$ which is denoted by $\mathbf{m}_J$ in Fig. 16.2. The importance of this component lies in the fact that it is simply related to the quantum numbers $J$ of the resultant angular momentum, as indicated in the figure. We shall now derive the expression for $|\mathbf{m}_J|$ on the basis of the vector model, and without recourse to quantum theory, except insofar as we shall use the quantum mechanical expressions for the average values of angular momenta.

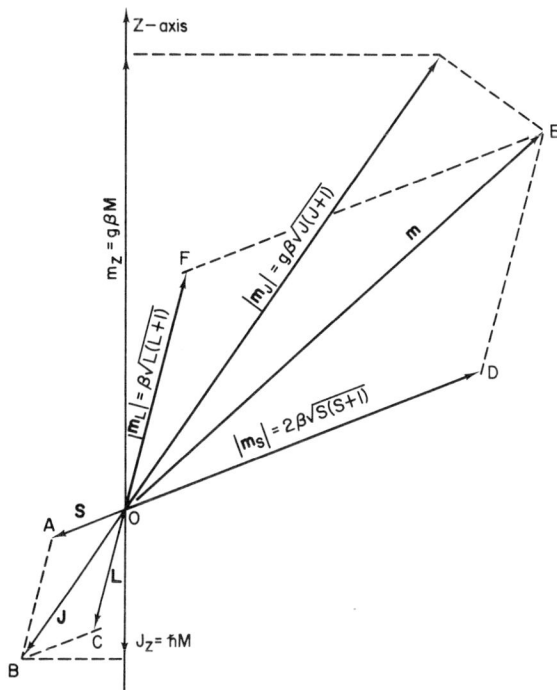

Fig. 16.2. Addition of angular momenta and magnetic moments of different species.

We start with the expression

$$|\mathbf{m}_J| = |\mathbf{m}_L| \cos BOC + |\mathbf{m}_S| \cos AOB. \qquad (16.7.3)$$

To evaluate the cosines we use eqns (16.6.6) and apply the cosine law to the respective triangles.

$$- \cos BOC = \frac{S(S+1) - L(L+1) - J(J+1)}{\sqrt{2\,L(L+1)J(J+1)}}$$

$$- \cos AOB = \frac{L(L+1) - S(S+1) - J(J+1)}{\sqrt{2\,S(S+1)J(J+1)}} \qquad (16.7.4)$$

Substituting these relations into eqn (16.7.3) and using values of $|m_L|$ and $|m_S|$ as marked in Fig. 16.2, we obtain the result

$$|m_J| = g\beta\sqrt{J(J+1)} \qquad (16.7.5)$$

where

$$g = \frac{3}{2} + \frac{S(S+1) - L(L+1)}{2J(J+1)}.$$

The coefficient $g$ is usually referred to as the *Landé g-factor*. It can be seen from the above expression that it may assume values in the range from 1 to 2 depending on the relative admixture of orbital and spin angular momenta in the resultant **J**.

Equation (16.7.5) is of great practical importance because it shows that the component $|m_J|$ of the magnetic moment is related to the quantum number $J$ in exactly the same way as the resultant angular momentum itself. Hence we conclude that the projection of $m_J$ on the $z$-axis is also quantized like the angular momentum. This means that the projection quantum number, which we now denote by $M$, may assume any value between $J$ and $-J$, in steps of unity. The $z$-component of the magnetic moment can now be written in the form

$$m_z = -g\beta M. \qquad (16.7.6)$$

If a magnetic field of induction $B_z$ is applied along the $z$-axis, the energy of the resultant magnetic moment will be

$$W = g\beta B_z M \qquad (16.7.7)$$

according to the principles discussed in Chapter 15, eqn (15.1.5). However, the constant $g$ can now assume values between 1 and 2.

## 16.8 The Stern-Gerlach Experiment

An excellent illustration of the foregoing concepts is provided by the Stern-Gerlach beam splitting experiment. The experimental layout is shown schematically in Fig. 16.3.

A beam of isolated atoms is formed by evaporating a sample of the material under test (say silver) in vacuum. Since the beam is not very well defined as it emerges from the oven, it is passed through collimating slits before being allowed to enter the gap between the pole pieces of a magnet. The pole pieces are especially shaped to provide a magnetic field which is unidirectional along the centre plane $XX$, but highly inhomogeneous in magnitude. As a result of the field inhomogeneity the atoms in the beam experience a force which deflects them according to the magnetic moment which they possess. If the magnetic moment differs from atom to atom the beam will be split.

The motion of the beam is analysed in two stages. In the first place we note that along the centre plane $XX$ the magnetic field has only one component— along the $z$-direction. Hence this is the axis of quantization along which the atomic magnetic moments will have components proportional to the quantum number $M$. By the principles explained in Chapter 14 the average values of the components $J_x$ and $J_y$ of the angular momentum vanish. Hence the corresponding components of the magnetic moment $\mathbf{m}$ also vanish: $m_x = m_y = 0$.

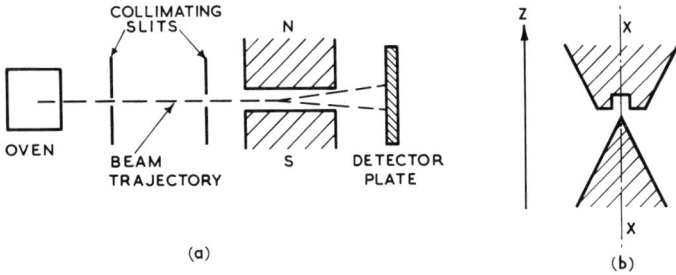

(a)                                                    (b)

**Fig. 16.3.** (a) Schematic layout of the apparatus in the Stern-Gerlach experiment. (b) Cross-sectional view of the magnet.

In the second stage of the analysis we consider the force acting on the magnetic moments in flight. This is given by the expression $\mathbf{F} = (\mathbf{m} \cdot \mathrm{grad})\mathbf{B}$, which in this case reduces to the single component.

$$F_z = m_z \frac{\partial B_z}{\partial z} \qquad (16.8.1)$$

where $$m_z = \langle M | m_z | M \rangle = -g\beta M.$$

From this result we see that the atoms in the beam subject to positive or negative forces depending on the value of the quantum number $M$, and the beam becomes split into as many divergent beams as there are values of $M$. In the case of silver there will be two component beams since there is only one unpaired electron in the $5s$ shell (see Table 16.1), having only a single spin angular momentum. Beams of atoms having greater resultant angular momenta will be split into a greater number of partial beams. E.g., if $J = \frac{3}{2}$ there will be four beams.

In an actual experiment the beams are not observed directly. Instead, deposits of the material under test are found on the detector plate, which is included in the vacuum system containing the oven and the beam trajectory. The deposits will appear as thin films, there being as many patches as there are component beams.

*Chapter 17*

# Electrons in Solids

The subject of electrons in solids was dealt with in Chapter 1 as a preliminary to a discussion of the electrical properties of semiconductors. At that stage every effort was made to simplify the treatment in order to make it accessible to readers with the minimum background. As a result it was not possible to introduce important concepts like the density-of-states function and the Fermi-Dirac distribution on a deductive basis. The aim of this chapter is to correct this deficiency to some extent, taking advantage of quantum mechanical methods introduced in preceding chapters. At the same time the reader is assumed to have some familiarity with the general background material covered in Part I of this book.

## 17.1 The Hamiltonian of Valence Electrons in a Solid

The pool of conduction electrons in a solid body presents a much more complicated example of the motion of many particles than the multielectron atom. In the first place their number is astronomical. Secondly their Hamiltonian is quite complicated, and the associated eigenvalue problem has not been solved to any significant degree of generality. However, some simplified models and a few special cases have been studied extensively, and these provide an adequate theoretical basis for the electrical properties of solids.

To begin with let us state briefly how we visualize the dynamics of electrons in a solid body. Although this matter was dealt with in Chapter 1, it is desirable to review the salient points in terms of quantum mechanical terminology for the purpose of a mathematical formulation.

The solid is pictured as a regular array of atoms of one or more elements. The *nucleus*, together with its complement of *closed electron shells*, forms a hard ionized *core* situated at a lattice site. The cores are assumed to contain all the mass of the solid body and, as a first approximation, to be motionless. Outside the closed shells are to be found the *valence electrons*. Some of these are visualized as being free to move from core to core, in fact throughout the solid

255

body. They move subject to forces of attraction of the stationary ions, of mutual repulsion, and possibly of externally applied forces, e.g. magnetic and electric fields.

To formulate the Hamiltonian of the valence electrons we assume that we have a solid body of volume $V$, containing $N$ valence electrons per unit volume, $NV$ electrons in all. In the absence of externally applied forces (mechanical or electromagnetic) the Hamiltonian will consist of three terms: (i) the kinetic energy of the electrons; (ii) energy due to the attractive potential of the fixed ions, usually called the *crystalline field energy*; (iii) energy due to interaction between the individual valence electrons. Each term will have the form of a summation over all the $NV$ valence electrons.

$$\mathcal{H} = \sum_i \frac{\mathbf{p}_i^2}{2m_e} + \sum_i U(\mathbf{r}_i) + \sum_{i>j} \frac{e^2}{4\pi\epsilon_0 |\mathbf{r}_i - \mathbf{r}_j|}. \qquad (17.1.1)$$

The resulting eigenvalue problem has the form

$$\mathcal{H}\Phi(\mathbf{r}_1 \ldots \mathbf{r}_{NV}) = \mathscr{E}\Phi(\mathbf{r}_1 \ldots \mathbf{r}_{NV}) \qquad (17.1.2)$$

where $\mathbf{r}_i$ is the position vector of the $i$-th electron and $\Phi$ is the electron wave function. No general solution of this eigenvalue problem is available. The main complication is due to the third term of eqn (17.1.1), because it involves the coordinates of two electrons and prevents the method of separation of variables to be applied, as was the case with the multielectron atom. The first step towards a simplification of the problem is, therefore, to drop the interaction term. The variables can be separated at this stage but even so the resulting differential equations are difficult to solve for potentials $U(r_i)$ which have the periodicity of crystal lattices.

As a first approximation the lattice potential is assumed to be constant, and since every potential contains an arbitrary constant, no generality is lost by setting it equal to zero. We shall assume, moreover, that the constant potential prevails only inside the solid body under consideration, and that the electrons are prevented from leaving it by forces present at the boundary. In this way we arrive at the highly idealized concept of a *potential well*. The electrons are free to move inside the well but are unable to leave it.

First we consider in some detail the motion of a single electron in a one-dimensional potential well. Later on we shall extend our considerations to more electrons moving in three dimensions.

### 17.2 Electron Waves in One Dimension

The equation of motion of an electron in a one-dimensional potential well is

$$-\frac{\hbar^2}{2m_e} \frac{d^2\psi}{dx^2} = E\psi \qquad (17.2.1)$$

where $x$ is the distance along the well measured from one of its ends (see Fig. 17.1). For the time being we assume that the potential barriers at $x = 0, L$ are infinitely high, so that the electron cannot penetrate them. It may be helpful to compare the motion of an electron in a potential well with the motion of a billiard ball between the edges of a billiard table.

Equation (17.2.1) has the familiar form of classical equations of wave motion in one dimension. Hence, in this particular case, the motion of an electron derived by quantum mechanical principles is exactly like a classical wave. Let us write down the solution of eqn (17.2.1) which is easily obtained by elementary methods.

$$\psi(x) = A e^{i(2m_e E/\hbar)x} = A e^{ikx}.$$

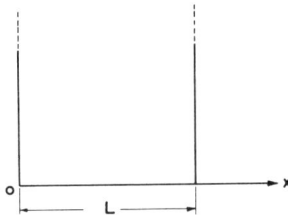

Fig. 17.1.

The constant $k$, which is the *wave number or phase constant* of the electron wave, is an abbreviation for $\sqrt{2m_e E}/\hbar$. The reader may note that the above solution should contain two exponential terms corresponding to the two possible roots of the auxiliary equation. However, it proves more convenient to deal with one term at a time and to generalize at a later stage, when the wave vector is quantized (eqn (17.2.3)). What matters ultimately is that solutions should fit the requisite boundary conditions, the detailed method of obtaining them being of little consequence. The constant of integration $A$ can be determined from the usual normalization condition for wave functions.

$$\int_0^L |\psi(x)|^2 \, dx = 1.$$

Hence

$$A = \frac{1}{\sqrt{L}}. \tag{17.2.2}$$

Before the electron wave function can be written down in its final form certain restrictions must be imposed on the constant $k$ and through it on the constant of separation $E$. The restrictions are introduced via *cyclic or periodic boundary conditions.* Under these conditions it is assumed that the potential well is folded back upon itself and the wave function must remain continuous at the joint. Alternatively the potential well may be visualized as repeating itself along the x-axis indefinitely and the wave function must again remain continuous at the joints. Either of these assumptions leads to the conclusion that the electron wave must contain an integral number of wavelengths within the potential well. For this reason the phase constant k is quantized by the expression

$$k_i = \frac{2\pi}{L} i; \quad \left( L = i\lambda = i\frac{2\pi}{k_i} \right) \qquad (17.2.3)$$

where $i = \ldots -2, -1, 1, 2, \ldots$ (any integer) is a quantum number not to be confused with the imaginary unit $i = \sqrt{-1}$.

By virtue of this relation the constant of separation E assumes the following sequence of values

$$E_k = \frac{\hbar^2 k_i^2}{2m_e} \qquad (17.2.4)$$

where the suffix k emphasizes the functional dependence of E on the wave number. The $E_k$ are, of course, the energy eigenvalues of the electron states in the potential well.

The electron wave functions can now be written in the final form

$$\psi_i(x) = \frac{1}{\sqrt{L}} e^{ik_i x}. \qquad (17.2.5)$$

It is easily verified that the exponentials form an orthogonal set of states.

Let us now reflect on the physical significance of the foregoing results.

First of all the position probability density $| \psi_i(x)|^2$ is constant throughout the potential well. Hence the electron is not confined to any particular location within the box, but is likely to be found anywhere inside it. This result, although obtained on the example of a highly idealized model, is applicable to the motion of the free electrons inside a solid body. The free electrons are to be identified with the conduction electrons in a sample of metal. They are equally likely to be found anywhere inside it, at a given instant of time.

Next let us calculate the matrix elements of momentum of the electron in the potential well by the usual method.

$$\langle i'|p_x|i\rangle = \langle i'| -i\hbar \frac{d}{dx} |i\rangle = \hbar k_i \, \delta_{i'i}. \qquad (17.2.6)$$

The momentum matrix is thus seen to be diagonal. Hence the electron wave functions are eigenstates of momentum, as well as being eigenstates of energy (see p. 194 and p. 218). Recalling the relation between the wave number and wavelength of a wave, the *momentum eigenvalues* $p_i$ can be written in the form

$$p_i = \langle i | p_x | i \rangle = \hbar k_i = \frac{h}{\lambda_i}. \tag{17.2.7}$$

This is de Broglie's expression relating the momentum of particles to the wavelength of the interference phenomenon which they display.

The reader should note the way the differential operator "extracts" the momentum eigenvalue from the exponential wave function.

$$-i\hbar \frac{d}{dx} \left( \frac{1}{\sqrt{L}} e^{ik_i x} \right) = \hbar k_i \left( \frac{1}{\sqrt{L}} e^{ik_i x} \right) = p_i \left( \frac{1}{\sqrt{L}} e^{ik_i x} \right).$$

Straightforward eigenvalue relations of this type were noticed in the early stages of development of quantum theory, and led to the general assumption that components of momentum can always be replaced by differential operators.

It is instructive to compare the momentum and energy eigenvalues of the electron wave as functions of the quantum number $i$.

$$p_i = \hbar k_i = \frac{\hbar}{L} i$$

$$E_k = \frac{p_i}{2m_e} = \frac{\hbar^2}{2m_e} \left( \frac{2\pi}{L} \right)^2 i^2. \tag{17.2.8}$$

From the first equation it is seen that the momentum eigenvalues are evenly spaced at intervals of $\hbar/L$ along the momentum axis, as shown in Fig. 17.2a. This fact is usually stated by saying that the *electron states are evenly distributed in momentum space or k space*. Although the foregoing result has been obtained on the basis of drastic simplification it nevertheless represents a reasonable approximation to states of electron motion in solids. It will be utilized in the derivation of several properties of metals. One more important property of electron momentum should be noted. The spacing of its eigenvalues is inversely proportional to the length $L$ of the potential well. This observation is again generally valid for electron states in solids. It means that in a larger sample the momenta of valence electrons are more closely spaced than in a smaller sample.

It follows from the second of eqn (17.2.8) that the energy eigenvalues do not display the linear regularity of the momentum. However, it is worth noting that the energy-level spacing also decreases with increasing size of the potential well. Again this result is generally valid for electron energies in solid samples. Figure 17.2b shows the energy levels as a function of the electron momentum quantum number. Only one energy level corresponds to the two numerically equal quantum numbers $i$ and $-i$ by virtue of the quadratic relationship between $k_i$

and $E_k$. The negative values of momentum correspond to electrons moving in the negative direction of the $x$-axis with the same energy.

The extension of the foregoing results to more than one electron in the potential well requires a fairly straightforward argument. The eigenvalue equation, eqn (17.1.2) consists of as many terms as there are electrons. The variables are separable by the procedure which was applied to the multielectron atom provided the interaction term is dropped. The resulting eigenfunctions

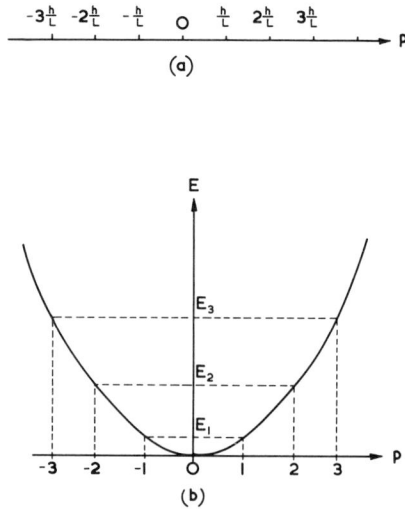

(a)

(b)

Fig. 17.2.

have the form of products of the individual electron eigenfunctions given by eqn (17.2.5). The energy of the system as a whole is the sum of single electron eigenvalues eqn (17.2.4). The question of occupation of the individual electron states is settled by the exclusion principle in conjunction with the fact that each electron state is characterized by two quantum numbers: the *translational* quantum number $k_i$ and the spin quantum number $\pm\frac{1}{2}$. Hence each spatial state, as distinct from a spin state, can be occupied by only two electrons.

At the temperature of $0\,^\circ K$ the system will be in its ground state. All lower levels will be fully occupied up to a height above which there will be no electrons. This highest occupied level is called the *Fermi level*. At temperatures greater than $0\,^\circ K$ some states above the Fermi level will be occupied by electrons, and some states below will be vacant.

The present section has dealt more thoroughly with some of the problems first discussed in Chapters 1 and 3 as a prelude to the treatment of semiconductors. The object here has been to show that the dynamics of electrons in solids can be given a firm theoretical framework based on quantum mechanics. In the next section the discussion will be extended to the more realistic model of electrons in a three-dimensional potential well or box.

## 17.3 Electrons in a Three-Dimensional Potential Well

In this section we consider the motion of electrons in a three-dimensional potential well. This abstraction is rather more difficult to visualize than the one-dimensional or two-dimensional potential well; certainly it cannot be represented graphically. By a three-dimensional potential well we mean a region of space bounded by the planes $x = L_1$, $y = L_2$, $z = L_3$ and the coordinate planes, making a box of volume $V = L_1 L_2 L_3$. Inside the region the electric potential is constant, and we again set it equal to zero. At the boundaries of the region the potential suddenly rises to a very high value, high enough to prevent electrons from escaping. This idealized model yields results which are in good agreement with the electrical behaviour of metals, although it cannot predict the phenomena of thermionic and photo emissions. As it also provides an analytical framework which helps to introduce a number of important concepts, it is worth studying with some care. In the work to follow the reader will find it helpful to visualize the potential well as an actual cube of metal.

The simplified equation of motion for $NV$ electrons has the form

$$\sum_i -\frac{\hbar^2}{2m_e} \nabla_i^2 \, \Phi(\mathbf{r}_1 \ldots \mathbf{r}_{NV}) = \mathscr{E}\Phi(\mathbf{r}_1 \ldots \mathbf{r}_{NV}). \qquad (17.3.1)$$

Separation of variables yields $NV$ single-electron equations of the form

$$-\frac{\hbar^2}{2m_e} \nabla^2 \, \psi(x,y,z) = E\psi(x,y,z). \qquad (17.3.2)$$

Here $x$, $y$ and $z$ are Cartesian coordinations of a representative electron which can in turn be separated. As a final step in the solution of the problem we are left with three ordinary differential equations of the form of eqn (17.2.1). Let

$$\psi(x,y,z) = X(x)\,Y(y)Z(z) \qquad (17.3.3)$$

and substitute this into eqn (17.3.2). Hence

$$-\frac{\hbar^2}{2m_e}\left(\frac{1}{X}\frac{d^2}{dx^2}X + \frac{1}{Y}\frac{d^2}{dy^2}Y + \frac{1}{Z}\frac{d^2}{dz^2}Z\right) = E.$$

The constant $E$ is next split into three parts by the method used on p. 237.

$$E = E_x + E_y + E_z. \qquad (17.3.4)$$

The three ordinary differential equations that now result are of the form

$$-\frac{\hbar^2}{2m_e}\frac{d^2}{dx^2}X = E_x X.$$

By anology with eqns (17.2.1) and (17.2.5) we can write down the solutions immediately

$$X_i(x) = \frac{1}{\sqrt{L_1}}e^{ik_{xi}x}$$

$$Y_i(x) = \frac{1}{\sqrt{L_2}}e^{ik_{yi}y}$$

$$Z_i(x) = \frac{1}{\sqrt{L_3}}e^{ik_{zi}z}.$$

Hence the wave function of an electron has the form

$$\psi_i(\mathbf{r}) = \psi_i(x,y,z) = X_i\,Y_i\,Z_i = \frac{1}{\sqrt{V}}e^{i\mathbf{k}_i\cdot\mathbf{r}} \qquad (17.3.5)$$

where          $V = L_1 L_2 L_3 \qquad \mathbf{k}_i = \hat{i}k_{xi} + \hat{j}k_{yi} + \hat{k}k_{zi}.$

The components of the vectorial phase constant or *wave vector* $k_i$ have eigenvalues of the form given by eqn (17.2.3)

$$\mathbf{k}_i = \hat{i}\frac{2\pi}{L_1}l + \hat{j}\frac{2\pi}{L_2}m + \hat{k}\frac{2\pi}{L_3}n. \qquad (17.3.6)$$

The quantum numbers $l$, $m$ and n are positive or negative integers (except zero), as required by cyclic boundary conditions. The suffix $i$ is an abbreviated symbol denoting a set of values of the quantum numbers $l$, $m$ and $n$.

The constants of separation $E_x$ $E_y$ and $E_z$ are the energy eigenvalues associated with the corresponding directions of electron motion. They can be written down by analogy with eqn (17.2.4)

$$E_{xk} = \frac{\hbar^2 k_{xi}^2}{2m_e}, \; E_{yk} = \frac{\hbar^2 k_{yi}^2}{2m_e}, \; E_{zk} = \frac{\hbar^2 k_{zi}^2}{2m_e}.$$

Hence by eqn (17.3.4) the energy eigenvalues of an individual electron are

$$E_k = \frac{\hbar^2 \mathbf{k}_i^2}{2m_e}. \qquad (17.3.7)$$

This expression gives the allowed energy levels of an electron moving freely inside a three-dimensional potential well.

Reversing further the procedure of separation of variables we find the energy eigenvalues of the system of $NV$ electrons in the potential well by the same method as was applied to the multielectron atom (see p. 237)

$$\mathscr{E} = \sum_k E_k = \sum_i \frac{\hbar^2 \mathbf{k}_i^2}{2m_e}. \tag{17.3.8}$$

The corresponding wave functions have the form

$$\Phi(\mathbf{r}_i \ldots \mathbf{r}_{NV}) = \psi_1(\mathbf{r}_1)\psi_2(\mathbf{r}_2) \ldots \psi_{NV}(\mathbf{r}_{NV}) \tag{17.3.9}$$

where the two sets of subscripts have different meanings. Thus $\psi_i(\mathbf{r}_s)$ denotes the wave function of the $s$-th electron having the quantum numbers $l$, $m$, $n$, represented collectively by the single suffix $i$.

Several comments can be made on the results obtained so far. The states of free electrons in a potential well are best studied with the help of the wave vector $\mathbf{k}$ and its eigenvalues. The wave vector can be represented graphically in three-dimensional rectangular coordinates labelled by the components $k_x$, $k_y$ and $k_z$ of $\mathbf{k}$. This frame of reference is usually referred to as the $\mathbf{k}$-*space* or *momentum space*, since momentum is proportional to $\mathbf{k}$. As the eigenvalues $\mathbf{k}_i$ of the wave vector are related to integral values of the quantum numbers $l$, $m$, and $n$, they are represented in $\mathbf{k}$-space by points situated in the corners of rectangular volume elements as shown in Fig. 17.3. Each eigen value is associated with a single cell, hence it is convenient to visualize each state of motion of the electron as represented by an elementary volume in $\mathbf{k}$-space. We note that points representing eigenstates of electron motion are evenly distributed throughout the volume of $\mathbf{k}$-space. This is completely analogous to electron motion in a one-dimensional potential well which is represented by the one-dimensional $\mathbf{k}$-space of Fig. 17.2(a).

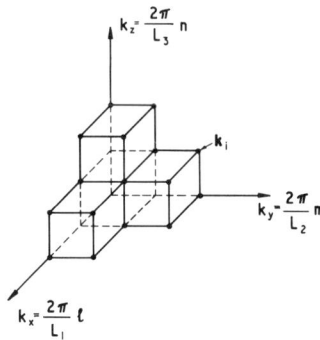

Fig. 17.3.

It will be useful to compute for future reference the volume of an elementary cell in k-space, and then to express the value in terms of momentum. By reference to Fig. 17.3 we see that the volume of a cell is

$$\frac{2\pi}{L_1} \cdot \frac{2\pi}{L_2} \cdot \frac{2\pi}{L_3} = \frac{8\pi^3}{V}. \tag{17.3.10}$$

Taking the reciprocal of the above result we find that there are $V/8\pi^3$ states per unit volume of k-space. This number is proportional to the actual volume of the potential well. *Hence the greater the volume of the potential well the more densely are the states distributed in k-space.* This result is the three-dimensional analogue of the conclusion stated on p. 259 regarding electron motion in one dimension. Electron motion in a well of unit volume $V = 1$ would give rise to $1/8\pi^3$ state points in a unit volume of k-space.

Let us now express the foregoing results in terms of electron momentum. Momentum eigenvalues are given by a relation which is a straightforward extension of eqn (17.2.7), and which can be obtained by the same method.

$$\mathbf{p}_i = \langle \psi_i | p | \psi_i \rangle = \langle \psi_i | -i\hbar \nabla | \psi_i \rangle$$

$$= \hbar \left( \hat{\imath} \frac{2\pi}{L_1} l + \hat{\jmath} \frac{2\pi}{L_2} m + \hat{k} \frac{2\pi}{L_3} n \right)$$

$$= \hbar \mathbf{k}_i. \tag{17.3.11}$$

Hence the coordinates of momentum space are obtained from the coordinates of k-space on multiplication by the constant $\hbar$. States of electron motion are again represented by discrete, evenly spaced points in momentum space, but the volume belonging to a single state-point is now given by the expression

$$\frac{\hbar 2\pi}{L_1} \cdot \frac{\hbar 2\pi}{L_2} \cdot \frac{\hbar 2\pi}{L_3} = \hbar^3 \frac{8\pi^3}{V} = \frac{h^3}{V}. \tag{17.3.12}$$

From this result it follows that *every state of electron motion in a potential well of unit volume V = 1 corresponds to a cell of volume $h^3$ in momentum space.*

The foregoing results will be used in the next section to derive the density-of-states function.

Let us next consider what happens when a large number of electrons are present in the box-shaped potential well. Their distribution over the allowed eigenstates is restricted by the exclusion principle. Bearing in mind that each electronic state is characterized by one of two spin quantum numbers in addition to its k-eigenvalue, every k-state, or *spatial state* of motion, can accommodate two electrons. Since the energy of the states increases with their k values (eqn (17.3.7)), states having low quantum numbers $l$, $m$ and $n$ will be occupied first.

According to eqn (17.3.7) the energy level depends only on the modulus of the wave vector: $E_k \sim |\mathbf{k}|^2$. Hence all states with wave vectors of the same length have the same energy level. This fact is expressed by saying that all states represented by points which lie on a sphere centred on the origin of k-space have the same energy eigenvalues. Thus it is possible to visualize *constant energy spheres* in k-space.

At the temperature of absolute zero all states represented by points within a constant energy sphere will be occupied by electrons, while all states outside will be empty. The corresponding value of the wave vector, $|\mathbf{k}| = k$, defines a particular constant energy sphere, called the *Fermi sphere*. The corresponding energy eigenvalue is referred to as the *Fermi level* of the system.

It is instructive to evaluate the radius of the Fermi sphere in k-space when the number of electrons in the box is large. The radius is determined by the condition that the volume of the sphere must contain half as many k-space cells, eqn (17.3.10), as there are electrons in the potential well. Denoting the radius of the Fermi sphere by $R_F$ we write

$$\frac{\text{volume of Fermi sphere}}{\text{volume of cell in k-space}} = \tfrac{1}{2} \text{ number of electrons in box}$$

$$\frac{\tfrac{4}{3}\pi R_F{}^3}{8\pi^3/V} = \tfrac{1}{2}NV$$

$$R_F = (3\pi^2 N)^{1/3}. \qquad (17.3.12)$$

This value of the wave vector can now be substituted in eqn (17.3.7) to obtain the Fermi level of the system

$$E_F = \frac{\hbar^2 R_F{}^2}{2m_e} = \frac{\hbar^2 (3\pi^2)^{2/3}}{2m_e} N^{2/3}. \qquad (17.3.13)$$

The concept of k-space introduced above on the idealized model of a potential well is applicable to real solids with some reservations regarding details. In particular *constant energy surfaces* will deviate significantly from a spherical shape and the idea of a Fermi sphere must be replaced by that of a *Fermi surface* in k-space. Many properties of solids depend critically on the detailed topology of the Fermi surface. A considerable experimental and theoretical effort is devoted to the study of the Fermi surfaces of various metals and semiconductors.

## 17.4 Density-of-States Function

In this section we consider in greater detail the energy eigenvalues of electrons in a potential well and the distribution of electron states along the energy axis.

To begin with let it be emphasized again that we are always dealing with a very large number of electrons per unit volume, say $N = 10^{23}$ per cm$^3$. On the other hand the volume of the potential well is assumed to be of typical macroscopic size, say $V = 1$ cm$^3$. Hence the spacing of successive energy levels is minute as the reader can verify by substituting some numerical values into eqn (17.3.7). The result thus obtained is truly microscopic as compared with the energy gap between the valence and conduction bands of typical semiconductors.

In the light of this observation it is quite justifiable to consider the energy levels of free electrons in a potential well as forming a *continuum*. Hence expressions which form summations over the energy eigenvalues can be replaced by integrals for purposes of calculation. Although these conclusions have been arrived at on the basis of a highly idealized model, they are applicable to electrons in real solids. Indeed the treatment of semiconductors in Part I was based on these assumptions.

What has been said about the electron energies applies even more to their momentum eigenvalues as represented by points in k-space and hence their wave functions. Only one energy eigenvalue can be derived from a number of distinct values of $\mathbf{k}_i$ by virtue of the dependence of energy on the square $|\mathbf{k}_i|^2$. Furthermore a distinct spatial electronic state or wave function is associated with each distinct value of the wave vector $\mathbf{k}_i$. Finally, by virtue of the spin, two electrons can occupy one spatial state or wave function. All these observations must be borne in mind when the function giving the density of electron states per unit energy range is evaluated.

To derive the density of electron states we consider, to begin with, a potential well of unit volume. According to the results of the preceding section one spatial state corresponds to a volume element $8\pi^3$ of k-space (see eqn (17.3.10) *et seq.*). Hence the number of states in a small range $dk$ of $|\mathbf{k}| = k$ is equal to the number of cells, of volume $8\pi^3$, within a spherical shell of thickness $dk$ in k-space (see Fig. 17.4). We have now dropped the subscript $i$ in $k_i$ to emphasize that we neglect its discrete nature. The number of states within the shell is

$$\frac{\text{volume of shell of } \mathbf{k}\text{-space}}{\text{volume of unit cell in } \mathbf{k}\text{-space}} = \frac{4\pi k^2 \, dk}{8\pi^3}. \qquad (17.4.1)$$

The above equation must be expressed in terms of energy to be of use in applications. The transformation is effected by a change of variable based on eqn (17.3.7). We find

$$k^2 = \frac{2m_e E}{\hbar^2}$$

$$dk = \frac{(2m_e)^{1/2}}{\hbar E^{1/2}} \, dE \qquad (17.4.2)$$

where all suffices have been dropped to emphasize the continuous nature of both $k$ and $E$ in the present context. Substitution into eqn (17.4.1) yields the result

$$\frac{(2m_e)^{3/2}}{4\pi^2\,\hbar^3}\,E^{1/2}\,dE = g(E)\,dE. \qquad (17.4.3)$$

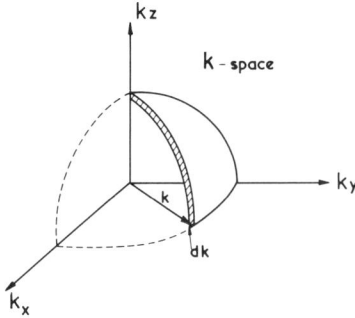

Fig. 17.4.

This equation gives the density-of-states function due to the spatial motion of electrons. In its final form the function must include a factor of 2 to include the effect of spin. Hence

$$g(E)\,dE = \frac{4\pi}{h^3}(2m_e)^{3/2}\,E^{1/2}\,dE. \qquad (17.4.4)$$

This result should be compared with eqn (3.3.1) of Part I. There are two differences. The mass $m_e$ is implied to be the real electron mass while in connection with semiconductors an effective mass was used. The necessity to substitute an effective mass is due to the band nature of allowed energies in real semiconductors and consequently does not appear in the simplified model developed here, which neglects the periodicity of the lattice potential.

The second difference is concerned with the origin of the energy axis. In the present context this is fixed to coincide with the lowest energy level in the potential well. On the other hand the formulae applied to semiconductors in Chapter 3 are expressed in terms of energy differences and thus do not assume any specific origin on the energy axis.

The free electron model developed above applies reasonably well to solids having Fermi surfaces which approach a sphere. This is the case with some metals, particularly alkali metals.

## 17.5 Thermionic Emission Equation

We are now sufficiently equipped to apply the free-electron model to derive the equation for the thermionic current first quoted in Chapter 6. Electrons are able to escape from a potential well, which is to represent a metallic sample, only if the potential barrier at the boundaries of the box is sufficiently low to allow the more energetic electrons to surmount it. The position is represented diagrammatically in Fig. 17.5. At high temperatures some electrons will occupy

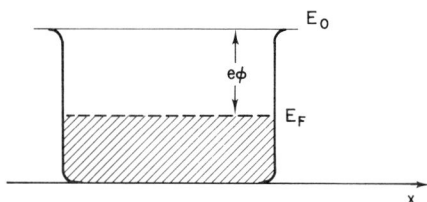

**Fig. 17.5.** Energy relations in thermionic emission.

energy levels well above the Fermi level, indeed some of them will have energies in excess of $E_0$, the *vacuum level* outside the box. It is these last electrons which will contribute to the thermionic current, and our problem is to ascertain their number.

For an electron to be able to escape from the box, that part of its kinetic energy associated with motion along the x-axis must exceed $E_0$. Writing the electron energy according to eqn (17.3.4) in the form

$$E = E_x + E_y + E_z = \frac{\mathbf{p}^2}{2m_3} = \frac{p_x^2 + p_y^2 + p_z^2}{2m_e} \qquad (17.5.1)$$

the following condition must be satisfied if an electron is to be emitted.

$$E_x = \frac{p_x^2}{2m_e} \geqslant E_F + e\phi. \qquad (17.5.2)$$

Since the object of the calculation is to find a current density across the surface of the metal, it is desirable to work in terms of electron velocities and momenta. We consider electrons in momentum space and denote by $n(\mathbf{p})$ the *number of electrons per unit volume of momentum space*. We pick out for attention those electrons which have momenta in the range $\mathbf{p}$ to $\mathbf{p} + d\mathbf{p}$. The rate at which these electrons strike unit area in the yz-plane is

$$v_x\, n(\mathbf{p})\, d\mathbf{p} = \frac{\partial E}{\partial p_x}\, n(\mathbf{p})\, dp_x\, dp_y\, dp_z$$

$$= n(\mathbf{p})\, dE\, dp_y\, dp_z. \qquad (17.5.3)$$

The total number of electrons per unit area of wall able to leave the potential well is obtained by integration of the above expression over the range of $x$-directed energy $E_x$ from $e\phi + E_F$ to $\infty$ and over the full range of the $p_y$ and $p_z$ components of momentum. However, the integration can only be carried out if the number of electrons per unit volume of momentum space, $n(\mathbf{p})$, is expressed in terms of energy. The number of *electron states* per unit volume of momentum space is, from eqn (17.3.12), $1/h^3$ (for a potential well of unit volume). The *actual number of electrons* is obtained on multiplication of this quantity by the Fermi-Dirac distribution.

$$n(\mathbf{p}) = \frac{1}{h^3}\, f(E) = \frac{1}{h^3} \cdot \frac{1}{1 + e^{(E - E_F)/kT}}. \qquad (17.5.4)$$

Hence the number of electrons escaping along the $x$-axis per second

$$= \frac{1}{h^3} \int_{-\infty}^{\infty} \int_{-\infty}^{\infty} \int_{E_x = E_F + e\phi}^{E_x = \infty} \frac{dp_y\, dp_z\, dE}{1 + e^{(E - E_F)/kT}}. \qquad (17.5.5)$$

The integral with respect to $E$ requires a word of comment. The integration is carried out with respect to $E = E_x + E_y + E_z$, but the limits are substituted for $E_x$ only, leaving $E_y + E_z$ in the form $(p_y{}^2 + p_z{}^2)/2m_e$. Hence

$$\int_{E_x = E_F + e\phi}^{E_x = \infty} \frac{dE}{1 + e^{(E - E_F)/kT}} = -kT \left| \log\left(1 + e^{-(E - E_F)/kT}\right) \right|_{E_x = E_F + e\phi}^{E_x = \infty}$$

$$= kT \log\left(1 + e^{-(e\phi + E_y + E_z)/kT}\right).$$

Letting $\dfrac{1}{kT}(e\phi + Ey + E_z) = \dfrac{1}{kT}\{e\phi + (p_y{}^2 + p_z{}^2)/2m_e\} = \theta$

and substituting into eqn (17.5.5) we find

$$\text{No. of electrons} = \frac{kT}{h^3} \int_{-\infty}^{\infty} \int_{-\infty}^{\infty} \log\left(1 + e^{-\theta}\right) dp_y\, dp_z.$$

Under normal conditions $\theta \gg 1$. Hence the log expression can be expanded and the integral simplified.

$$\frac{kT}{h^3}\, e^{-e\phi/kT} \int_{-\infty}^{\infty} \int_{-\infty}^{\infty} e^{-(p_y{}^2 + p_z{}^2)/2m_e kT}\, dp_y\, dp_z.$$

The two repeated integrals have the form of the Gaussian distribution

$$\int_{-\infty}^{\infty} e^{-(1/2)cx^2}\, dx = \sqrt{\frac{2\pi}{c}}.$$

Hence the number of escaping electrons is

$$2\pi m_e (kT)^2\, h^{-3}\, e^{-e\phi/kT}$$

and the thermionic current density is

$$J = 2\pi e m_e (kT)^2\, h^{-3}\, e^{-e\phi/kt}.$$

The last result can be written in the form

$$J = AT^2\, e^{-e\phi/kT} \tag{17.5.6}$$

where the coefficient has the theoretical value $A$ = 120 A/cm$^2$deg$^2$. The variation of the thermionic current with temperature as given by eqn (17.5.6) is well substantiated by experimental data. However, the value of $A$ varies widely for different materials. Thus at one extreme $A$ = 1·5 for a cathode constructed of tungsten (W) covered by a thin surface layer of barium (Ba) and $A$ = 160 for cesium (Cs) at the other extreme.

*Chapter 18*

# Motion of Electron Beams and Individual Electrons— Uncertainty Principle

The motion of electron and other particle beams presents a dynamical problem which is mathematically simpler than any of the quantum mechanical systems considered so far. Moreover it provides many examples which lie on the borderline between classical and quantum physics, and which illustrate very well the conceptual differences between the two disciplines. These attractions are coupled with certain theoretical complications which carry with them risks of confusion. It is for this reason that the discussion of these seemingly simple problems has been delayed. It is hoped that the basic concepts of quantum theory are by now fairly firmly established and that the reader is ready to consider problems which by their very similarity to classical examples may prove confusing at times.

### 18.1  Free Electron Beams and Matter Waves

In this section we consider the motion of electrons which have been formed into a uniform beam by an electron gun and then allowed to travel in a space screened from any electric or magnetic forces. If we neglect mutual repulsion between the electrons we expect, on classical principles, that the electrons will travel along a straight line at a constant velocity, imparted to them by the accelerating voltage of the beam forming arrangement. This view of electron motion is verified experimentally in countless electronic devices whose operation is based on the presence of an electron beam.

In some devices, e.g. a cathode ray tube, the electron beam is deliberately exposed to forces designed to modify its motion. Again, in most cases, the electron paths can be accurately described on the basis of classical principles. However, there are situations in which the electrons behave in a way quite inexplicable in terms of classical concepts. The best example is provided by the diffraction of electrons, which can only be explained on the basis of quantum mechanics. It is this approach which will be dealt with now.

271

Since we assume that the electrons move in a space free of any potential forces and since we neglect their mutual repulsion, the Hamiltonian will contain only the kinetic term. Hence Schroedinger's equation will be of the form

$$-\frac{\hbar^2}{2m_e}\frac{d^2\psi}{dx^2} = E\psi \tag{18.1.1}$$

where we assume that the electrons move along the $x$-axis. Equation (18.1.1) is of the same form as the equation of motion of electrons in a one-dimensional potential well discussed in Section 17.2. The solution will be of the same form as in the latter case.

$$\psi(x) = A\,e^{ikx}; \quad k = \sqrt{\frac{2m_e E}{\hbar^2}}. \tag{18.1.2}$$

The next step in the description of motion is to determine the constants $A$ and $k$ (and hence $E$). In the case of the potential well, $A$ was determined from the normalization of the wave functions and $k$ from the cyclic boundary conditions. None of these procedures can be applied in this case because the dynamics of electrons in the beam do not allow it. The different approach to be adopted is imposed by the physical conditions of the electron beam.

To begin with let us deal with the energy eigen value $E$ and wave number $k$. The energy of each electron in the beam is purely kinetic and is decided by the initial accelerating voltage used to form the beam. Hence

$$E = \tfrac{1}{2}m_e v^2 = eV \tag{18.1.3}$$

where $V$ is the accelerating voltage and $v$ is the velocity of the electrons. Thus we have a case where the *energy eigen values are not discrete but form a continuum*, since the voltage $V$ can be given any value.

The wave number can now be written in the form (from eqns (18.1.3) and (18.1.2))

$$k = \frac{p}{\hbar} \tag{18.1.4}$$

where $p = m_e v$ is the momentum of the electrons. This is again de Broglie's relation, but this time it allows a continuum of wave numbers instead of the discrete values obtained in Section 17.2. It is this continuum of wave numbers or wavelengths which is observed in electron diffraction experiments using uniform beams.

Before attempting to determine the constant of integration $A$ we note that it cannot be fixed by the normalization procedure applied to electron wave functions in the potential well, because we do not wish to assume at this stage that the beam is confined to a limited volume. Indeed we idealize it by assuming that it extends the full length of the x-axis from $-\infty$ to $+\infty$ as is usual when dealing with waves. Since the wave function remains finite over the full extent of the x-axis, an integral of the form $\int_{-\infty}^{\infty} |\psi|^2 dx$ will be infinite. For this reason normalization of the wave function is impossible. It is a general characteristic of *dynamical states associated with continuous sets of eigenvalues* that they *cannot be normalized in the usual way*. The reader is invited to note this fact without a formal proof, which would take up space without significantly illuminating the physics of the subject.

In the circumstances it is customary to leave the constant $A$ undetermined until specific cases make it desirable to assign values to it. This will be illustrated in the sections to follow. It should be remembered however, that the squared modulus of the wave function, $|\psi|^2$, is a real quantity *proportional to the position probability distribution or the charge density in the beam*.

To complete the description of motion of the electron beam let us include the time factor in its wave function. We do this by going back to Section 11.2, where the time-dependent part of Schroedinger's equation was separated and solved. Using eqn (11.2.6) we can write

$$\psi(x,t) = A e^{i\{kx - (E/\hbar)t\}}. \qquad (18.1.5)$$

The above equation represents a plane wave travelling in the positive direction of the x-axis. On changing the sign in front of the term $kx$ a wave travelling in the negative direction is obtained. Since eqn (18.1.5) is of exactly the same form as the equations of electromagnetic waves and elastic waves, it is to be expected that analogous phenomena of reflection, interference and diffraction should be observed on electron beams.

Some examples of these phenomena will be discussed in subsequent sections. In the meantime we note that the application of Schroedinger's equation to uniform beams of any particles yields results similar to the foregoing. Hence we conclude that any material particles are capable of behaving like waves under suitable conditions.

## 18.2  Reflection of Electrons at a Potential Step

In this section we proceed to consider the motion of electron beams through space in which the potential is not constant. As our first example we take a potential step situated at the point $x = 0$. On the left-hand side of this point we set the constant potential arbitrarily to zero (see Fig. 18.1). In this region the equation of motion is of the form of eqn (18.1.1). On the right-hand side the potential has the constant value $U_1$. We write Schroedinger's equation for the electron beam in the form

$$-\frac{\hbar^2}{2m_e}\frac{d^2\psi}{dx^2} = E\psi \text{ for } x < 0$$

$$-\frac{\hbar^2}{2m_e}\frac{d^2\psi}{dx^2} + U_1\psi = E\psi \text{ for } x > 0. \qquad (18.2.1)$$

The motion of the electron beam will be described by the solutions of the above equations subject to the *boundary conditions* that the overall *wave function and its first derivative must be continuous at the potential step* which means that both the probability distribution and momentum remain continuous across the step.

To the left of the potential step we expect electron waves travelling in both directions of the $x$-axis, hence the solution in that region will be of the form

$$\psi(x) = e^{ikx} + R\,e^{-ikx}, \quad (x < 0) \qquad (18.2.2)$$

where we have arbitrarily normalized the wave travelling to the right, or the *incident wave,* by giving it a coefficient equal to unity. The coefficient of the wave travelling to the left is clearly in the nature of a *reflection coefficient* to be determined from the boundary conditions. The wave number in this region is given by eqn (18.1.2) and is completely determined by the accelerating voltage used when the beam is formed.

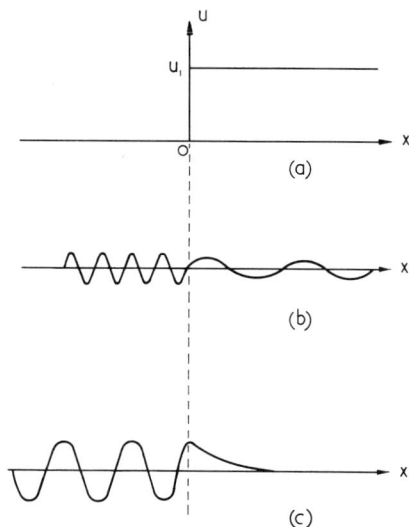

Fig. 18.1.

On the right-hand side of the potential step there is no possibility of a wave travelling to the left since we assume that there are no further obstacles and the beam travels to $x = +\infty$ unimpeded. Hence the wave function in this region will be of the form

$$\psi(x) = T e^{ik_1 x}, \; (x > 0) \tag{18.2.3}$$

where $T$ is in the nature of a *transmission coefficient*, and the wave number is given by the expression

$$k_1 = \left\{ \frac{2m_e(E - U_1)}{h^2} \right\}^{1/2}. \tag{18.2.4}$$

Let us pause to consider $k_1$. If $E > U_1$ the wave number is real and the wave function (18.2.3) has the same oscillatory character as eqn (18.2.2). The electron beam thus retains its wavelike character on mounting the potential step, the only difference between the two regions being the wavelength. Figure 18.1(b) suggests this state of affairs graphically. The increased wavelength implies lower energy of the electrons, which is as it should be, since some of the kinetic energy in the original beam has been spent on surmounting the potential step. This case can be considered as an idealization of the process of thermionic emission. The potential step serves as a model of the potential barrier at the boundary of a metal, and the electrons under consideration correspond to conduction electrons which have sufficient thermal energy to escape.

An analogy of the foregoing phenomenon is provided by electromagnetic waves crossing the boundary between two media. If propagation in the second medium is at all possible, a change of wavelength will generally take place.

An altogether different phenomenon appears where $E < U_1$. In this case $k_1$ is imaginary and the wave function becomes a real exponential as suggested in Fig. 18.1(c). By virtue of the fact that the squared modulus of the wave function $|\psi|^2$ is proportional to the position probability of the electrons, we see that the latter penetrate the potential barrier to a certain extent, but that they are unlikely to be found beyond a distance of the order of $1/k_1$.

In this case an electromagnetic analogy is again readily found. Electromagnetic waves impinging on a metal suffer exponential decay in crossing the surface. Another example is provided by waves travelling along a waveguide whose size is suddenly reduced below the cut-off dimensions. Evanescent modes are set up in the waveguide below cut off, which are characterized by exponential decay.

As the next step we proceed to determine the reflection and transmission coefficients $R$ and $T$ which are in general complex numbers. On the basis of the electromagnetic analogy this is to be expected, since $R$ and $T$ would correspond to voltage or current coefficients, which are complex. It will be recalled, however, that power coefficients are frequently used and that these are related to squared moduli of voltage and current coefficients. In the quantum mechanical case we are interested in probabilities which are defined in terms of squared moduli.

Before applying the boundary conditions at the potential step, it is useful to visualize the problem in terms of three wave functions. The first term of eqn (18.2.2) is the wave function of the incident electrons and its squared modulus has been arbitrarily set to unity. This implies that the modulus of the second term, $|R|^2$, will *indicate the charge density of the reflected beam, as a fraction of the incident electrons.* Similarly the third wave function, eqn (18.2.3), yields the charge density of the transmitted beam through the quantity $|T|^2$.

The continuity of the wave functions and their derivatives at $x = 0$ leads to the equations

$$1 + R = T$$

$$k(1 - R) = k_1 T. \qquad (18.2.5)$$

Solution for $R$ and $T$ yields the results

$$R = \frac{k - k_1}{k + k_1}, T = \frac{2k}{k + k_1}. \qquad (18.2.6)$$

When the potential step $U_1$ is less than the initial kinetic energy $E$, we have $k_1 < k$ from eqn (18.2.4) and $|R|^2 < 1$. Hence the charge density in the reflected beam is lower than in the incident beam as was to be expected. However, under the same conditions $|T|^2 > 1$ and the density of the transmitted beam is greater than the incident beam. This finding is readily explained in the light of the fact that the kinetic energy, and hence the velocity of the transmitted beam is reduced by the effect of the potential step. The electric current carried by the transmitted beam is, of course, smaller than in the incident beam, but this matter will be discussed in a later section.

By analogy with any familiar form of wave motion the incident and reflected electron waves interfere with each other to form a standing wave pattern. More elaborate forms of interference take place when electrons are directed at a crystal lattice. The resulting Bragg diffraction pattern derives from the fact that the wavelengths of electron beams easily formed in the laboratory are of the order of lattice dimensions.

It is rather more difficult to visualize what happens in the case $U_1 > E$. The wave number $k_1$ becomes imaginary and we find

$$|R|^2 = \left| \frac{k - ik_1'}{k + ik_1'} \right|^2 = 1 \qquad (18.2.7)$$

where $k_1 = ik_1'$ with $k_1'$ real.

This result indicates that all electrons impinging on the potential step are ultimately reflected. However, the transmission coefficient in eqn (18.2.3) does not vanish

$$|T|^2 = \left| \frac{2k}{k + ik_1'} \right|^2 > 0. \qquad (18.2.8)$$

Hence it follows that some of the electrons penetrate the potential step, but in the end return to the potential free region, to the left of the origin of coordinates in Fig. 18.1. From the form of the corresponding wave function it further follows that most electrons penetrate depths less than $1/k_1'$, since the probability of finding them beyond that point is very small and decreases exponentially.

The foregoing result suggests an interesting reflection regarding the free electron model of metals described in the preceding chapter. As $U_1 \rightarrow \infty$ the value of $k_1'$ becomes very great (eqn (18.2.4)) and hence $|T|^2 \rightarrow 0$ by eqn (18.2.8). Under these conditions the electrons do not penetrate the potential step at all. However, the potential barrier at the boundary of a metal is of finite height and it appears that some electrons are able to penetrate the space outside and then to return to the metal.

It is interesting to enquire whether some of these electrons could be intercepted by some means, and prevented from returning to the metal. In the following section we shall find that theoretically this is possible. Moreover, the phenomenon can be observed experimentally and it forms the basis of important applications.

### 18.3  Tunnelling of Electrons Through a Potential Barrier

As the next example of electron wave motion we consider the effect of a potential barrier of finite height and length. The $x$-axis is now divided into three regions as shown in Fig. 18.2(a). The equations of motion are

$$-\frac{\hbar^2}{2m_e} \frac{d^2 \psi}{dx^2} = E\psi, \text{ for } x < 0 \text{ and } x > a$$

$$-\frac{\hbar^2}{2m_e} \frac{d^2 \psi}{dx^2} + U_1 \psi = E\psi, \text{ for } 0 < x < a. \qquad (18.3.1)$$

The state of the beam will be described by the solutions of these equations subject again to the boundary conditions that the solutions and their first derivatives must be continuous at the extremities of the potential barrier.

The solutions are of the form

$$\psi(x) = e^{+ikx} + R e^{-ikx}, \quad (x < 0)$$

$$\psi(x) = A e^{+ik_1 x} + B e^{-ik_1 x}, \quad (0 < x < a)$$

$$\psi(x) = T e^{+ikx}, \quad (x > a). \tag{18.3.2}$$

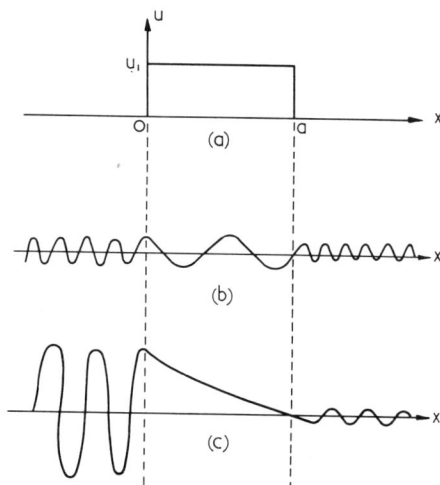

Fig. 18.2.

As in the previous section the incident wave has been arbitrarily normalized to unity. The wave numbers $k$ and $k_1$ are again given by eqn (18.2.4) in which $U_1$ is set to zero to obtain $k$.

Figures 18.2(b) and (c) show graphically the nature of the electron waves along the $x$-axis depending on whether $U_1$ is smaller or greater than $E$. In the former case the electrons retain their wavelike character over the entire length of the $x$-axis. The potential barrier in this case is analogous to a change in the permittivity of a medium still capable of propagating electromagnetic waves. There will, of course, be reflections at the discontinuities, as a result of which the transmitted beam will, in general, have a lower current density than the incident beam, except in the case of a "match".

In the case $E < U_1$ the wave number $k_1$ is imaginary and the electron wave function is a real exponential over the extent of the potential barrier. The probability that an electron may penetrate the barrier and emerge on the other side depends on the length of the barrier $a$. Those electrons, which happen to "tunnel" through the barrier, resume their wavelike motion with the same wave number (the same kinetic energy) as in the first region of the $x$-axis. Since these electrons have the same velocity, their charge density $|T|^2$ is a direct measure of the probability of tunnelling. Let us work it out by applying boundary conditions to eqns (18.3.2). Continuity of the wave function at points $x = 0$ and $x = a$ leads to the equations

$$1 + R = A + B$$

$$Te^{ika} = Ae^{ik_1a} + Be^{-ik_1a}. \tag{18.3.3}$$

Continuity of the derivative leads to the equations

$$k(1 - R) = k_1(A - B)$$

$$kTe^{ika} = k_1(Ae^{ik_1a} - Be^{-ik_1a}). \tag{18.3.4}$$

Solution of the above equations for the transmission coefficient yields the result

$$T = \frac{4kk_1 e^{ika}}{(k_1 + k)^2 e^{ik_1a} - (k_1 - k)^2 e^{-ik_1a}}. \tag{18.3.5}$$

Equation (18.3.5) applies both when the potential barrier is propagating and attenuating. In the latter case it is necessary to substitute $k_1 = ik_1'$ ($k_1'$ real) as in the preceding section. The probability that an electron will "tunnel" through an attenuating barrier is given by $|T|^2$.

$$|T|^2 = \frac{16 k^2 k_1'^2}{(k^2 + k_1'^2)^2} e^{-2k_1'a}. \tag{18.3.6}$$

## 18.4 Expression for Current Flow

The motion of the electron beams discussed in preceding sections constitutes an electric current like the motion of any charged particles. We shall now consider the question of how to formulate an expression for the current flow in terms of the wave functions of charged particles.

The problem is one of formulating a definition which will yield the expected result in familiar cases. It is found that the following expression satisfies the necessary requirements.

$$\mathbf{J} = \frac{q\hbar}{i2m} \left\{ \psi^*(\nabla\psi) - (\nabla\psi)^* \psi \right\} \tag{18.4.1}$$

where $q$ is the electric charge on the particles and $\mathbf{J}$ is *electric current density*.

Referring to the beam of eqn (18.1.2) we note that the electron density is $n = |\psi|^2 = A^2$. Substituting this into the classical formula $\mathbf{J} = -en\mathbf{v}$ the current density can be obtained in terms of the accelerating voltage with the help of eqn

(18.1.3), or in terms of the electron wave number, eqn (18.1.4). The reader will satisfy himself that substitution of the wave function into eqn (18.4.1) yields the same result.

A more general test of the validity of the above definition can be obtained with the help of the equation of charge continuity $\partial \rho / \partial t = -$div $\mathbf{J}$. Remembering that the charge density is $qn = q| \psi|^2$ and substituting in eqn (18.4.1) a verification is obtained.

It is useful to note that without the charge $q$ eqn (18.4.1) gives the flow of particles represented by the wave function $\psi$ as the number of particles crossing unit area per second.

## 18.5  Motion of Charged Particles in Magnetic Fields

Some effects of the magnetic field in the context of quantum mechanics have been dealt with in preceding chapters. In every case so far there was a magnetic moment associated with the dynamical system under consideration and the influence of an applied field could be tackled through the magnetic energy term $-\mathbf{m} \cdot \mathbf{B}$. The reader will be aware that this is not always the case and that a straight beam of charged particles has no magnetic moment (neglecting spin if any).

Nevertheless the motion of charged particles is affected by a magnetic field through the medium of the Lorentz force given by the expression

$$\mathbf{F} = q\mathbf{v} \times \mathbf{B}. \tag{18.5.1}$$

Although this expression forms the basis of any classical treatment of the dynamics of, say, electrons it is not suitable for a quantum mechanical approach since the latter relies on energy expressions which can be included in the Hamiltonion.

The effect of the Lorentz force can be given an energy expression with the help of the Lagrangian formulation of mechanics. Readers interested in the derivation are referred to texts dealing with classical dynamics. In the present context it will be enough to quote the result for future use. The objective is achieved in terms of the magnetic vector potential $\mathbf{A}$, by replacing the momentum $\mathbf{p}$ by the expression $\mathbf{p} - q\mathbf{A}$ or, in operator form, by $(-i\hbar\nabla - q\mathbf{A})$ in expressions for the kinetic energy.

$$\frac{1}{2m}(\mathbf{p} - q\mathbf{A})^2 = \frac{\hbar^2}{2m}(i\nabla + \frac{q}{\hbar}\mathbf{A})^2. \tag{18.5.2}$$

It should be noted that the operators do not generally commute and the expanded form of the bracketed expression will therefore contain four distinct terms.

## 18.6  The Principle of Uncertainty

The motion of electron beams described in preceding sections had two important properties. The momentum of the electrons was precisely defined in eqn (18.1.4) in terms of the accelerating voltage, through eqns (18.1.2) and (18.1.3). On the other hand the position of the electrons was completely undefined since the position probability density $\psi^*\psi = A^2$ was constant, indicating that the electron was equally likely to be found anywhere in the beam.

This observation is a manifestation of the *principle of uncertainty*. The principle lays it down in general terms that certain dynamical variables, referred to as *conjugate coordinates*, cannot be simultaneously defined precisely, either experimentally or theoretically. If one variable can be determined within some limits, the conjugate variably will have an uncertainty extending between corresponding limits. The electrons of Section 18.1 are an extreme example in that their momentum is completely fixed, while their position along the $x$-axis is totally indeterminate. Another pair of conjugate variables is provided by energy and time. Thus if a phenomenon is found to have precisely defined energy its duration in time must be infinite. Examples are provided by atomic electron orbits, whose energy can be fixed precisely only if they persist indefinitely, or again by the electron beam.

Readers familiar with the properties of communication signals will note that the electron beam of eqn (18.1.5) is analogous to a continuous wave signal which has a precisely defined frequency and hence photon energy, according to the relation $E = hf$. If the signal has the form of an individual pulse of finite duration, its frequency will no longer have a single precisely defined value, but will cover a continuous band of values. The relation between the time duration $\Delta t$ and bandwidth $\Delta f$ of a pulse is given by the relation

$$\Delta t . \Delta f \gtrsim 1. \tag{18.6.1}$$

Since $\Delta f = (1/h) \Delta E$, the above relation can be put into the form

$$\Delta t . \Delta E \gtrsim h. \tag{18.6.2}$$

The corresponding relation which is applicable to a position coordinate and the conjugate momentum reads

$$\Delta x . \Delta p \gtrsim h. \tag{18.6.3}$$

The last relation is well exemplified by the linear oscillator. If its position is found by some method to lie within the uncertainty range $\Delta x$, then its momentum can only be determined to within the uncertainty range $\Delta p$ given by eqn (18.6.3).

At this point we recall the statements of Section 14.3 to the effect that commuting variables can be simultaneously measured or calculated with complete precision. This can be expressed by saying that their uncertainties are zero. It is no coincidence that the commutators of such variables are zero. Indeed it can be shown in general that uncertainty products of the form of eqn (18.6.3) have values of the same order of magnitude as the commutators of the corresponding variables. The reader is invited to verify that the commutator of position and momentum of the linear oscillator is indeed numerically equal to $\hbar$.

The foregoing statements can be justified mathematically in terms of Fourier transforms which are found to relate pairs of conjugate variables. Readers interested in the mathematical background of the uncertainty principle are referred to specialist texts on quantum theory.

## Chapter 19

# Statistics of Energy Level Occupation

We have now considered at some length the states of motion of particles and systems of particles on the basis of quantum theory. We have found that the particles can exist in a variety of states or levels, but we have never gone into the question of what states the particles are likely to occupy under actual physical conditions. It is the object of this chapter to consider this problem. We shall find that the occupation of the various states by individual particles or by assemblies of particles is governed by statistical laws, which depend on the nature of the particles themselves, and on the form of the dynamical theory used to describe their motion. Before going into these statistical laws proper, we consider more closely some aspects of quantum states which were only touched upon so far.

## 19.1 Fermions and Bosons

As a start we take a closer look at the exclusion principle as applied previously to electrons, and consider other possibilities of state occupation by particles.

We take as our point of departure two particles, say electrons, enclosed in a box. In the case of electrons the box has the form of a potential well, while in the case of neutral gas molecules it is an actual box. The important property of the system of two particles that concerns us here is their *indistinguishability*. To be able to distinguish the particles one would have to be able to keep track of them, to follow their paths somehow. By virtue of the principle of uncertainty, however, it is impossible to locate the particles precisely, let alone follow their paths. Hence the particles are said to be *non-localized*, a concept most easily appreciated by recalling that their position probability densities (see p. 258) are distributed throughout the box.

The indistinguishability of the particles has important consequences as regards the analytical form of their wave functions. By eqns (17.3.5) and (17.3.9) a state of the two particles can be written in the form:

$$\Phi(\mathbf{r}_1, \mathbf{r}_2) = \psi_i(\mathbf{r}_1)\psi_j(\mathbf{r}_2) \tag{19.1.1}$$

where the letters $i$ and $j$ label the quantum numbers of the individual particle states, while the numbers 1 and 2 are hypothetical tags on the particles themselves. Since the particles cannot, in fact, be distinguished or localized, we can interchange them, or rather interchange their coordinates in eqn (19.1.1), and still have the same state

$$\Phi(\mathbf{r}_2, \mathbf{r}_1) = \psi_i(\mathbf{r}_2)\psi_j(\mathbf{r}_1). \tag{19.1.2}$$

That the above wave functions represent one and the same state is easily seen by considering any quantity having physical significance. Thus the position probability density $|\Phi|^2$ is the same for both, and so is the energy eigenvalue (see eqn (17.3.8)). Taking $|\Phi|^2$ as the basic quantity which has physical significance, we say that it must remain unchanged when the two indistinguishable particles are interchanged. Hence we deduce that the function $\Phi$ must be either *symmetrical* or *antisymmetrical* in the coordinates of the particles. (The terms even or odd parity are also used.) By this we mean that an interchange of the particle coordinates between the individual wave functions either leaves the overall wave function unchanged or at most changes its sign, still leaving the probability density $|\Phi|^2$ unaltered.

The expressions (19.1.1) and (19.1.2) can be combined to form both symmetrical and antisymmetrical wave functions. Thus

$$\Phi_{\text{symm}} = \psi_i(\mathbf{r}_1)\psi_j(\mathbf{r}_2) + \psi_i(\mathbf{r}_2)\psi_j(\mathbf{r}_1)$$

$$\Phi_{\text{anti}} = \psi_i(\mathbf{r}_1)\psi_j(\mathbf{r}_2) - \psi_i(\mathbf{r}_2)\psi_j(\mathbf{r}_1). \tag{19.1.3}$$

Equations (19.1.3) represent states of two particles for which the probability distributions $|\Phi|^2$ remain unaltered when the particles are interchanged. It should be noted, however, that these functions are not normalized. As normalization is irrelevant in the present context, the normalizing coefficients will be omitted for simplicity.

Let us now consider more closely the second of eqns (19.1.3) representing an antisymmetry of two-particle wave functions is an analytical expression of the individual wave functions, that is if we put $i = j$ in defiance of the exclusion principle, the complete antisymmetric wave function vanishes. Hence we see that antisymmetry of two-particle wave functions is an analytical expression of the exclusion principle. We also conclude that *electron states must always be antisymmetric* to satisfy the exclusion principle.

It is easily seen that the antisymmetric wave function of eqns (19.1.3) can be written in the form of a determinant

$$\Phi_{\text{anti}} = \begin{vmatrix} \psi_i(\mathbf{r}_1) & \psi_i(\mathbf{r}_2) \\ \psi_j(\mathbf{r}_1) & \psi_j(\mathbf{r}_2) \end{vmatrix}. \tag{19.1.4}$$

In this form both the antisymmetry of electron states and the exclusion principle are clearly brought out. An interchange of the two particles is equivalent to an interchange of the columns of the determinant which changes its sign. Assigning the same quantum numbers to the individual wave functions, $i = j$, renders the rows of the determinant identical, with the result that the latter vanishes.

The foregoing conclusions regarding the antisymmetry of electron wave functions apply generally to the states of *elementary particles* such as protons, neutrons, etc. and these particles obey the exclusion principle. However, certain *composite particles*, e.g. some gas molecules, have states which are symmetric. Since such states are expressed in the form of the first of eqns (19.1.3) let us consider it in more detail.

The most important point to note about the symmetric state is that it does not vanish when the same quantum numbers are assigned to the individual wave functions, $i = j$. Hence we conclude that two particles of this type can coexist in the same quantum state. In other words the *exclusion principle does not apply to particles whose states are symmetric* in the particle coordinates. In fact, an extension of the foregoing arguments to systems of many particles shows that any number of them can occupy the same state.

Equation (19.1.4) suggests immediately how to extend the concept of antisymmetric states to systems of many elementary particles. Given $n$ elementary particles, say electrons, characterized by individual wave functions of the form $\psi_i(\mathbf{r}_k)$, we obtain an antisymmetric state of the system by forming a determinant of order $n$ as follows

$$\Phi_{\mathrm{anti}} = \begin{vmatrix} \psi_i(\mathbf{r}_1) & \psi_i(\mathbf{r}_2) & \cdots & \psi_i(\mathbf{r}_n) \\ \psi_j(\mathbf{r}_1) & \psi_j(\mathbf{r}_2) & \cdots & \psi_j(\mathbf{r}_n) \\ \cdots\cdots\cdots\cdots\cdots\cdots \\ \psi_p(\mathbf{r}_1) & \psi_p(\mathbf{r}_2) & \cdots & \psi_p(\mathbf{r}_n) \end{vmatrix}. \qquad (19.1.5)$$

The antisymmetry of the wave function and the exclusion principle are embodied in the properties of the determinant. An exchange of two particles is represented by an interchange of two columns of the determinant, and the wave function changes sign. Assigning identical quantum numbers to two individual functions renders two rows of the determinant equal, and the wave function vanishes.

The reader will no doubt note that each term of the determinant (19.1.5) is a wave function of the form of eqn (17.3.9) which satisfies Schroedinger's equation. Given one term of the determinant all the other terms can be obtained from it by permutations of the quantum numbers $i$, $j$, etc. between the individual wave functions. The terms are given positive or negative signs depending on whether they have been generated from the initial term by even or odd permutations.

The above remarks regarding antisymmetric states suggest how to form symmetric states of $n$ particles. Again we start with a single term of the form of eqn (17.3.9) and generate other terms by permutation of the quantum numbers between individual wave functions. However, in contrast with the determinantal states, the different terms are all given a positive sign and are summed as in eqn (19.1.3). Thus

$$\Phi_{\text{symm}} = \sum_P \psi_i(\mathbf{r}_1)\psi_j(\mathbf{r}_2)\dots\psi_p(\mathbf{r}_n) \qquad (19.1.6)$$

where $P$ denotes any one of the permutations of the quantum numbers $i$, $j, \dots, p$. The above state is symmetric, since an exchange of two particles (interchange of two position vectors $\mathbf{r}_s$ and $\mathbf{r}_i$) does not change its sign. Moreover, assigning identical quantum numbers to any of the particles does not make the wave function vanish, that is any number of particles may occupy the same state.

The arguments of this section show that there are two types of particle differing fundamentally in their capacity to fill available states. One type of particle, called a *boson* is capable of crowding into a given state in any numbers whatsoever. The other type, called a *fermion*, is incapable of sharing a state (including a spin state), with another particle. These properties of particles are reflected in the statistical laws governing their distribution among available states. The first type of particle obeys Bose-Einstein statistics, the second Fermi-Dirac statistics. As our main preoccupation is with electrons, which are in the second group, we now go on to derive the Fermi-Dirac distribution of fermions among energy levels.

## 19.2 Possibilities of Quantum State Occupation by Electrons

In an attempt to derive the distribution of fermions over allowed energy levels it is very helpful to keep in mind the example of electrons in a potential well as described in Sections 17.3 and 17.4. In fact throughout the following sections we shall use the terminology of that case although the reasoning and results apply quite generally to any assembly of fermions.

From Section 17.3 we recall two points which will help us greatly. In the first place the energy eigenvalues of electrons in a box are degenerate, that is a number of distinct states belongs to the same energy level. Moreover, the spacing between adjacent energy levels is minute. On the basis of these facts we are quite justified in arranging the electron states in groups, each group being associated with an approximately fixed value of energy. We denote the energy by $E_i$ and the number of distinct electron states (including spin states) associated with it

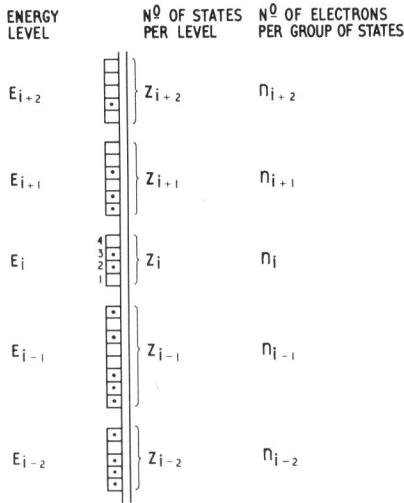

Fig. 19.1.

by $z_i$. We assume that $z_i$ is a number large compared with unity. Further we assume that there are $n_i$ electrons distributed over the group of states $z_i$. Figure 19.1 shows diagrammatically an arrangement of states and electrons which it is useful to keep in mind in the course of the following arguments.

To begin with we consider possibilities of state occupation within one group of states $z_i$. Taking as an example a group of $z_i = 4$ states occupied by $n_i = 2$ electrons, all possible distributions are shown schematically in Fig. 19.2. When making the arrangements of Fig. 19.2, it must be remembered that only one electron can be fitted into a state, and that two electrons cannot be distinguished. The exclusion principle rules out the possibility of both electrons occupying a single state represented by a compartment in Fig. 19.2. Moreover, since the electrons are indistinguishable no new distribution is created by interchanging two black dots between the compartments of Fig. 19.2.

From Fig. 19.2 we see by inspection that the total number of possible distributions of 2 electrons over the 4 individual states is 6. Our problem now is

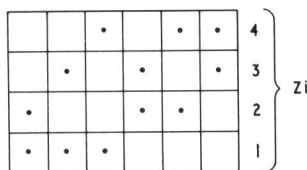

Fig. 19.2.

to formulate a general expression which would yield all possible distributions of $n_i$ electrons over $z_i$ states. At first we note that we could start with any definite distribution, say the first row of Fig. 19.2, and permute the numbered compartments in all possible ways. In this way we would get $z_i!$ arrangements. However, this number would include many distributions which do not differ by virtue of the restrictions we have imposed. Thus, a mere interchange of the empty compartments 3 and 4 would not yield a new distribution, yet it is included in the $z_i!$ permutations. To exclude such possibilities we divide $z_i!$ by the number of permutations of empty states which is $(z_i - n_i)!$. In addition to this limitation we must also exclude permutations which interchange only states occupied by electrons. Thus an interchange of the full compartments 1 and 2 of Fig. 19.2 produces no new distribution by virtue of the indistinguishability of electrons. To allow for these possibilities we divide $z_i!$ by the number of permutations of states occupied by electrons which is $n_i!$. Hence the total number of possible distributions of $n_i$ electrons over $z_i$ states is given by the expression

$$\frac{z_i!}{n_i!(z_i - n_i)!}. \tag{19.2.1}$$

The foregoing equation can be arrived at by a more concise argument. It is sufficient to note that the number of rearrangements of the electrons among the states is equal to the number of ways one can take $n_i$ objects out of a group of $z_i$ objects. This number is equal to the appropriate binomial coefficient and is given by eqn (19.2.1).

Finally, to find the total number of possible distributions for all groups of states indicated in Fig. 19.1, we form a product of expressions like eqn (19.2.1) for all groups.

$$g = \prod_i \frac{z_i!}{n_i!(z_i - n_i)!}. \tag{19.2.2}$$

So far we have been considering the possible distributions of electrons over groups of states $z_i$, each group belonging to one energy level $E_i$. We have tacitly assumed that a fixed number of electrons $n_i$ has been allotted to the corresponding energy level $E_i$ and we have found eqn (19.2.2) as the number of possible rearrangements of electrons subject to this limitation.

However the object of the present exercise is to find the most probable distribution of all the electrons among *the energy levels* in the box, *given that the total amount of energy possessed by the electrons is fixed*. To arrive at this most probable distribution we must consider what happens when the electrons are allowed to wander from energy level to energy level or from one group of states $z_i$ to another group of states $z_j$. To put the problem in mathematical terms we must consider what happens to eqn (19.2.2.) as the numbers $n_i$ of electrons are allowed to vary. It is conceivable that for a particular set of values of $n_i$ the number $g$ of possible arrangements will reach a maximum. As this number $g$ of arrangements is a measure of the probability of finding the corresponding distribution of electrons among the energy levels, we say that the maximum value of $g$ corresponds to the most probable distribution.

In the present context it is necessary to note some terms which are frequently used. Each allowed arrangement of electrons among the states of the system is called a *microstate* of the system. Thus the quantity $g$ of eqn (19.2.2) is the number of possible microstates of the system, given a specific distribution ... $n_{i-1}, n_i, n_{i+1}$ ... of electrons among the energy levels $E_i$. As pointed out above another distribution of electrons among the energy levels may give rise to a greater value of $g$, and hence to a greater number of microstates. If we succeed in finding the distribution which results in the *maximum number of microstates* we shall conclude that this is the most probable distribution in which the system can be found in nature. As we can perform observations and measurements of a macroscopic type on such a system we say that it is in a *macrostate*.

## 19.3 The Fermi-Dirac Distribution

The Fermi-Dirac function gives the most probable distribution of a fixed number $N$ of fermions among a set of energy levels $E_i$, given that the total energy of all the particles is constant and equal to $\mathscr{E}$.

The reasoning that leads to the final result is in three stages.

*Stage (i).* A group of $z_i$ states belonging to one energy level $E_i$ is occupied by $n_i$ electrons. The number of ways in which the electrons can be distributed over this group of states is evaluated. This was done in the preceding section, the result being given by eqn (19.2.1).

*Stage (ii).* All groups of states, associated with all possible energy levels of the system are considered. Each energy level has assigned to it a fixed number of electrons $n_i$, such that $\sum_i n_i = N$ is the total number of electrons. All possible arrangements of electrons among the states are evaluated, subject to this subdivision among energy level. The result is given by eqn (19.2.2).

*Stage (iii).* The subdivision of electrons among the energy levels is varied and the effect of this variation on the quantity $g$ of eqn (19.2.2) is studied. It is expected that for a *certain subdivision or distribution of the electrons among the energy levels* the number of possible arrangements $g$ will reach a maximum. This particular distribution of electrons among energy levels is the most probable and will be found to exist in nature under conditions of equilibrium. This final stage of the reasoning is given in the present section.

The problem we have to solve is to find the maximum value of the quantity

$$g = \prod_i \frac{z_i!}{n_i!(z_i - n_i)!} \tag{19.3.1}$$

considered as a function of the numbers $n_i$ of electrons at the energy levels $E_i$. The numbers $n_i$ are subject to the subsidiary condition that their sum is fixed and equal to the total number of electrons $N$ in the box.

$$\sum_i n_i = N. \tag{19.3.2}$$

In addition we impose the second subsidiary condition that the energy of the system of electrons in the potential well is fixed.

$$\sum_i n_i E_i = \mathscr{E}. \tag{19.3.3}$$

A maximum of eqn (19.3.1) will occur for values of $n_i$ for which the differential $\delta g$ vanishes. However, instead of trying to work out $\delta g$ directly, it proves convenient to first take the natural logarithm of eqn (19.3.1)

$$\log g = \sum_i \log \frac{z_i!}{n_i!(z_i - n_i)!}$$

$$= \sum_i \log z_i! - \sum_i \log n_i! - \sum_i \log (z_i - n_i)!. \tag{19.3.4}$$

To simplify the above expression we use Stirling's formula for the approximate value of a factorial, together with the fact that the number $z_i$ is assumed to be large: $z_i \gg 1$. Hence

$$\log z_i! \cong z_i(\log z_i - 1) \cong z_i \log z_i. \tag{19.3.5}$$

Moreover, we assume that the number $n_i$ is also large, $n_i \gg 1$ and the difference $z_i - n_i \gg 1$ likewise. Hence eqn (19.3.4) assumes the form

$$\log g = \sum_i z_i \log z_i - \sum_i n_i \log n_i - \sum_i (z_i - n_i) \log (z_i - n_i). \tag{19.3.6}$$

As the next step we take the differential of eqn (19.3.6) with respect to the numbers $n_i$, considered as independent variables,

$$\delta(\log g) = \sum_i \frac{\partial(\log g)}{\partial n_i} \delta n_i$$

$$= \sum_i \log \frac{z_i - n_i}{n_i} \delta n_i. \tag{19.3.7}$$

To find the maximum of $g$ it is not enough to equate the above differential to zero, because there are ancillary conditions to satisfy. The latter are included in the problem by Lagrange's method of undetermined multipliers. In this the differentials of the ancillary conditions (19.3.2) and (19.3.3) are multiplied by undetermined constants, added to the differential (19.3.7) of the function under investigation, and the resulting sum is equated to zero. Thus

$$\delta N = \sum_i \delta n_i = 0$$

$$\delta \mathcal{E} = \sum_i E_i \, \delta n_i = 0 \qquad (19.3.8)$$

by virtue of the fact that both $N$ and $\mathcal{E}$ are assumed to be constant. Equations (19.3.8) are next multiplied by unknown constants $-\alpha'$ and $-\beta$, added to eqn (19.3.7), and the sum is equated to zero.

$$\delta \log g - \alpha' \, \delta N - \beta \delta \mathcal{E} = \sum_i \delta n_i \left( \log \frac{z_i - n_i}{n_i} - \alpha' - \beta E_i \right) = 0. \quad (19.3.9)$$

Let us now consider the above summation to see what conditions must be satisfied to ensure that it vanishes. To begin with we observe that the small variations $\delta n_i$ are only effected by shifting some electrons from one energy level to another. If these individual variations are weighted by different numbers, a summation of the form of eqn (19.3.9) cannot, in general, vanish. To ensure its vanishing it is necessary to assume that the bracketed factor remains constant despite variations of the quantities $z_i$, $n_i$ and $E_i$ from one energy level to another. Under this condition the factor can be taken outside the summation symbol, the vanishing of eqn (19.3.9) being then assured by virtue of the first of conditions (19.3.8). On the basis of this argument we can write

$$\log \frac{z_i - n_i}{n_i} - \alpha' - \beta E_i = \text{const.}$$

$$\log \frac{z_i - n_i}{n_i} = \alpha + \beta E_i \qquad (19.3.10)$$

where the value of the bracketed factor of eqn (19.3.9) has now been absorbed in the new unknown constant $a$. From eqn (19.3.10) we can write

$$\frac{z_i - n_i}{n_i} = e^{\alpha + \beta E_i} = A e^{\beta E_i}.$$

After some manipulation we obtain an equation expressing the ratio of the number of electrons $n_i$, to the number of states in a group $z_i$, as a function of the energy level of the group $E_i$.

$$\frac{n_i}{z_i} = \frac{1}{1 + A e^{\beta E_i}}. \qquad (19.3.11)$$

This is the *Fermi-Dirac distribution*, still somewhat indeterminate because of the presence of the unknown constants $A$ and $\beta$.

The constants are determined from physical considerations. $\beta$ is replaced by the statistical definition of temperature which reads

$$\beta = \frac{1}{kT}$$

where $k$ is Boltzmann's constant. It would require a separate proof to show that the temperature defined above is identical with the thermodynamic absolute temperature scale. The Fermi-Dirac distribution can now be written in the form

$$\frac{n_i}{z_i} = \frac{1}{1 + A\,e^{E_i/kT}}. \qquad (19.3.12)$$

The constant $A$ is determined by reference to the Fermi level as introduced in Section 17.3. At $T = 0\,°\mathrm{K}$ all states below the Fermi level $E_F$ are fully occupied so that the ratio $n_i/z_i = 1$, while all states above the Fermi level are empty so that $n_i/z_i = 0$. The function (19.3.12) is capable of expressing this state of affairs if $A$ is given the form $e^{-E_F/kT}$. Hence

$$\frac{n_i}{z_i} = \frac{1}{1 + e^{(E_i - E_F)/kT}}. \qquad (19.3.13)$$

As a final step in the formulation of the Fermi-Dirac function we assume that the energy levels form a continuum which we denote by $E$. The ratio $n_i/z_i$ then becomes a continuous function of energy, usually denoted by the symbol $f(E)$.

$$f(E) = \frac{1}{1 + e^{(E - E_F)/kT}} \qquad (19.3.14)$$

Features of the Fermi-Dirac distribution are discussed at length in Chapter 3. Figure 3.1(b) of that chapter shows the distribution of electrons among the energy levels for several values of the temperature $T$.

## 19.4  The Maxwell-Boltzmann Distribution

In many applications one is concerned with the occupation of states by electrons or other particle systems under conditions of very sparse population of the available states. Situations of this type correspond to conditions high up the "tail" of the Fermi-Dirac distribution of Fig. 3.1(b). In such cases the distribution function can be greatly simplified, as was shown in Chapter 3. The approximate forms of the distribution were then derived in a form adapted for application to semiconductors. As we shall require the distribution function in a more general form for application to paramagnetic materials, we shall now derive it from eqn (19.3.12) which we rewrite in the form

$$\frac{n_i}{z_i - n_i} = \frac{1}{A}\, e^{-E_i/kT}.$$  (19.4.1)

The sparse population of available quantum states means that the number of particles $n_i$ is very much smaller than the number of states $z_i$ at the energy level $E_i$. In that case $n_i$ may be neglected in the denominator of eqn (19.4.1) and the latter simplifies to the form

$$\frac{n_i}{z_i} = \frac{1}{A}\, e^{-E_i/kT}.$$  (19.4.2)

This is the *Maxwell-Boltzmann distribution*, originally established on the basis of classical statistical mechanics. Graphs comparing the exponential distribution (19.4.1) with the original Fermi-Dirac function (19.3.12) are plotted in Fig. 19.3. Both curves are referred to the same temperature, and the constant $A$ is based on the Fermi level introduced in eqn (19.3.13).

The probability distributions discussed above have until now been given the interpretation of population densities. In this one thinks of large numbers $n_i$ of particles populating even larger numbers $z_i$ of quantum states at energy levels $E_i$. It is possible, and indeed desirable, to attach a different physical significance to the probability functions. This is done by assuming that there is only one particle within the system under consideration and that it is open to the particle to occupy any one of the energy levels $E_i$. The probability that the particle will be found at a specified energy level is then given by the exponential probability density of eqn (19.4.2). In Fig. 19.3 this is represented by horizontal lines placed at the height of the energy levels and extending to the exponential curve. As the temperature is reduced the curve approaches the energy axis more rapidly and the probability of finding the particle at the higher energy levels is reduced.

The foregoing concepts are best illustrated on the example of paramagnetic ions, present as trace impurities in diamagnetic host crystals. Since the ions are spaced large distances apart, they do not interact in any way, and each ion can be treated as a separate dynamical system whose energy levels are due primarily to the externally applied magnetic field acting on the magnetic moment of the ion. For a specific applied field the energy levels of an ion may well resemble the scheme of three levels shown in Fig. 20.4. At a definite temperature, say $T_2$, the relative probability of finding the ion in one of the levels is then represented by the corresponding horizontal line.

In spectroscopic experiments to be described in a later chapter the energy level spacing is measured by observing the absorption of electromagnetic waves of the appropriate frequency in a crystal specimen which may contain

$10^{19}$ paramagnetic ions. Since the ions are effectively isolated dynamical systems, by virtue of the relatively large distances separating them, each one of them will occupy one or other of the energy levels indicated in Fig. 20.4 with a probability given by the exponential Maxwell-Boltzmann law. On aggregate the ions in the specimen will distribute themselves among the three energy levels,

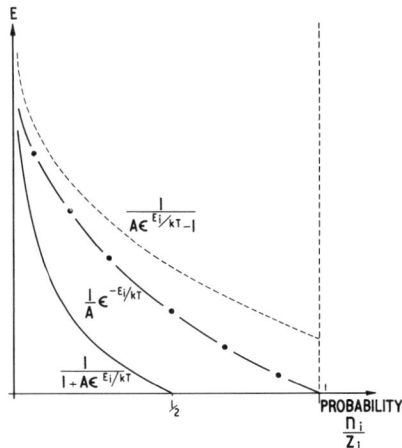

Fig. 19.3.

according to the curves of Fig. 20.4. We are thus led back to the interpretation of the probability distributions as relative population densities, but with a difference. The distinction lies in the fact that here we consider a large number of independent particles, each particle constituting a separate dynamical system, whereas in the original derivation of the Fermi-Dirac law we considered one dynamical system containing many particles.

The foregoing interpretation of the probability of occupation of energy levels invalidates the determination of the constant $A$ in terms of the Fermi level of a potential well. This is so because the Fermi level, as introduced in Section 17.3, is related to the number of electrons in the well, and has no meaning when effectively only one particle is considered as a dynamical system. The question of interpreting and determining the constant $A$ in such cases will be dealt with in connection with paramagnetism when the *state sum or partition function* will be introduced.

Before closing this section on the Maxwell-Boltzmann distribution, it may be appropriate to note that it affords a means of identifying the statistical temperature, defined by eqn (19.3.12), with the thermodynamic absolute temperature scale. As remarked above the Maxwell-Boltzmann law may be established by classical methods. It is then found automatically that the exponential factor depends on the absolute temperature in the same way as eqn (19.4.2) depends on the statistical temperature.

## 19.5 The Bose-Einstein Statistics

To obtain the distribution of bosons among available energy levels we follow exactly the same procedure that led to the Fermi-Dirac distribution for fermions, except that the possible rearrangements of particles among quantum states are much more numerous. This is so because the allowed rearrangements include all possibilities of coexistence of many particles in a quantum state, contrary to the example of fermions discussed in connection with Fig. 19.2.

As the Bose-Einstein distribution will not be applied in later chapters we shall not consider it beyond stating its form

$$\frac{n_i}{z_i} = \frac{1}{A\,e^{E_i/kT} - 1}. \tag{19.5.1}$$

A curve comparing eqn (19.5.1) with the previous probability distributions is shown in Fig. 19.3. From this it is clear that at high energy levels, where the population of states is sparse, the distributions do not differ significantly. Moreover, examples of bosons are provided by gas molecules which are invariably encountered under conditions corresponding to the high "tail" of the distributions. For this reason they are treated in terms of the simpler Maxwell-Boltzmann distribution.

# Part III

*Chapter 20*

# Paramagnetism and the Maser

## 20.1 Macroscopic Description of Magnetic Materials

The electromagnetic description of magnetic materials is carried out in terms of three basic vector quantities: *the magnetic field* (or magnetizing force) **H**, *the magnetic flux density or induction* **B**, and the *magnetization* **M**. The magnetic vectors are related by the following equation (in m.k.s. units)

$$\mathbf{B} = \mu_0(\mathbf{H} + \mathbf{M}) \tag{20.1.1}$$

where $\mu_0 = 4\pi \times 10^{-7}$ is the permeability of free space. For the purpose of this treatment it is assumed that the physical meaning of the vectors **H** and **B** is understood and our concern will be mainly with the magnetization **M**.

From preceding chapters we recall that atoms having unpaired electrons are capable of having a resultant magnetic moment. Indeed such atoms are effectively minute magnets, which can interact with an applied magnetic field by assuming any one of a set of discrete orientations in relation to the field. As we shall see below the majority of atoms, thus exposed to an applied field, tend to assume a direction approaching that of the field. Hence there will be a net component of the magnetic moment parallel to the field. This net magnetic moment, summed over all magnetic atoms per unit volume of the substance, is called the *magnetization per unit volume* **M**, or briefly the *magnetization* **M**.

The foregoing argument suggests that the magnetization **M** is somehow related to the field **H**. In fact it is customary to express their relation in the form

$$\mathbf{M} = \chi\,\mathbf{H} \tag{20.1.2}$$

where $\chi$ is called the *magnetic susceptibility*. The constitutive equation (20.1.1) can now be written

$$\mathbf{B} = \mu_0\mathbf{H}(1 + \chi) = \mu_0\mu\,\mathbf{H}$$

$$[\mathbf{B} = \mathbf{H}(1 + 4\pi\chi) = \mu\mathbf{H} \text{ in c.g.s. units}] \tag{20.1.3}$$

where $\mu = 1 + \chi$ is the *relative permeability* of the material under consideration.

299

In some cases the susceptibility is constant. Materials for which this holds are said to behave *linearly* in response to an applied magnetic field or magnetizing force. More usually, however, the susceptibility varies with the applied field, and materials for which this holds are said to be *non-linear*. Moreover, the susceptibility may be either positive or negative, with the result that the induction **B** may be greater or smaller than in free space.

The various forms and values which the susceptibility may assume give rise to a classification of magnetic substances into three major groups.

1. *Paramagnetic materials:* the susceptibility is *positive* and constant under normal conditions of temperature and applied field. It is also numerically very small: $\chi = 10^{-5}$.
2. *Ferromagnetic materials:* the susceptibility is *positive* but varies with the applied field. It is usually very large, values of $10^4$ being common for ferrous metals.
3. *Diamagnetic materials:* the susceptibility is *negative* and for some non-ferrous metals quite small, $\chi > -10^{-5}$. Its lowest possible value is $-1$ which applies only to superconductors under somewhat restricted conditions.

The problem of magnetism consists to a large extent of an effort to calculate the susceptibility of materials from their atomic properties. It is tacitly assumed in the above summary that the material under discussion is *isotropic*, that is the permeability is the same for each component of the magnetic vectors. This, however, is not always the case particularly with single crystal samples. In such cases of *anisotropy* the field vectors are related by a matrix whose elements constitute the *permeability tensor*, $[\mu_{ij}]$.

$$
\begin{bmatrix} B_x \\ B_y \\ B_z \end{bmatrix} = \mu_0 \begin{bmatrix} \mu_{11} & \mu_{12} & \mu_{13} \\ \mu_{21} & \mu_{22} & \mu_{23} \\ \mu_{31} & \mu_{32} & \mu_{33} \end{bmatrix} \begin{bmatrix} H_x \\ H_y \\ H_z \end{bmatrix}. \tag{20.1.4}
$$

*A susceptibility tensor* is defined on the basis of eqn (20.1.2). Substitution into (20.1.3), conceived as a matrix equation, then yields the relation

$$
[\mu_{ij}] = I + [\chi_{ij}] \tag{20.1.5}
$$

where $I$ is the unit matrix.

A comparison of eqns (20.1.4) and (20.1.5) with eqn (20.1.3) will disclose that an isotropic substance can be thought of as having a diagonal permeability and susceptibility matrix, the diagonal elements being all equal.

## 20.2 Paramagnetic Energy Levels

As a first step we take up the case of paramagnetic substances in the presence of d.c. fields. The static susceptibility can be calculated relatively easily, on the basis of quantum mechanics, for various materials of practical significance.

To begin with let us consider more closely, in terms of atomic and molecular structure, what is meant by a paramagnetic substance.

A paramagnetic material takes the form of a crystal which on the whole is entirely non-magnetic. Thus, alumina, $Al_2O_3$, is non-magnetic in that no atoms have unpaired electrons with a resultant angular momentum and magnetic moment. A crystal of this type becomes paramagnetic if some of the non-magnetic atoms are replaced by atoms or ions having unpaired electrons. For example some of the aluminium atoms in $Al_2O_3$ may be replaced by trivalent chromium ions $Cr^{3+}$. The latter have three 3d electrons which form a resultant magnetic moment, having quantum number $j = \frac{3}{2}$, in the environment of the lattice. It is emphasized that such paramagnetic ions are relatively far apart, say 100 atomic diameters, and that they are unable to interact with each other magnetically. Hence, *each ion can be treated in isolation from the others*, subject only to the influence of the otherwise non-magnetic crystal lattice and an externally applied magnetic field.

Neglecting for the moment the effect of the lattice, the enegy levels of an atom would be like those derived for the hydrogen atom in Chapter 15. For example, if the quantum number is $j = \frac{3}{2}$ in the ground state, the latter will be split into four levels by the application of a magnetic field, as shown in Fig. 20.1. The levels are in the form of straight lines emanating from a point at zero applied field. Moreover as shown in Chapter 14 the energy levels are identifiable

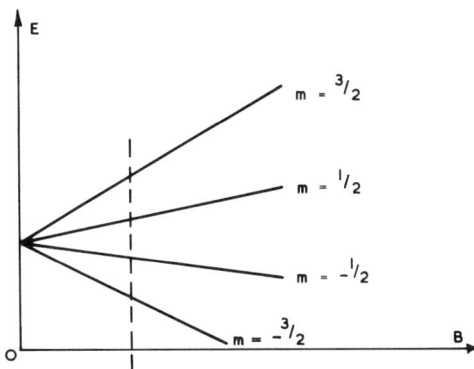

**Fig. 20.1.** Magnetic energy levels of an isolated ion having a resultant angular momentum of quantum number $j = 3/2$.

with states of angular momentum since the corresponding operators commute. As a result it is possible to say that each level is populated by ions, all having the same component of magnetic moment along the direction of the applied field.

The position is generally much more complicated when the effect of the crystal lattice is taken into account. In the first place the interaction of the magnetic ion with its non-magnetic neighbours may split the ground state into two or possibly more levels *in the absence of a magnetic field.* Secondly, the energy levels cease to be straight lines. Accompanying these modifications of the energy level scheme is a more subtle change regarding the classification of the levels. It is no longer possible to identify them with states of angular momentum. Rather, each energy eigenstate must be expressed as a linear mixture of angular momentum states in accordance with the principle of superposition. Hence there is no longer a unique value of the magnetic moment associated with each energy level, a fact which makes the calculation of magnetization more difficult.

In the next section we shall evaluate the susceptibility of a paramagnetic material of the simplest type, leaving aside for the time being more complicated cases.

## 20.3 Static Paramagnetic Susceptibility

The application of a d.c. magnetic field to a simple paramagnetic substance separates the magnetic ions between the several energy levels as shown in Fig. 20.1. Whatever net component of magnetization may appear parallel to the applied field will clearly depend on the relative populations of the energy levels. The paramagnetic ions are assumed to be sufficiently far apart to be out of range of mutual magnetic interaction or coupling. However, they are thermally interconnected through the medium of the crystal lattice and, if left to themselves, they remain in an equilibrium condition just like the molecules of a quantity of gas. Hence they will be distributed among their energy levels according to the Maxwell-Boltzmann exponential law.

$$\rho_E = A\,e^{-E/kT} \tag{20.3.1}$$

where $\rho_E$ is the population of level $E$ and $A$ is a constant still to be determined. The distribution is shown diagrammatically in Fig. 20.2 where the level populations are plotted along the horizontal coordinate. The applied field is assumed fixed as indicated by the dotted line in Fig. 20.1.

Figure 20.2 makes it apparent that there will be a preponderance of ions having components of magnetic moment parallel to the applied field, as compared with those which are antiparallel. The resulting net magnetization can be calculated as follows.

The energy of the ions having magnetic quantum number $m$ is (see eqn 16.7.7)

$$E = g\beta Bm \qquad (20.3.2)$$

where $B$ is the applied magnetic field, assumed to be in the $z$-direction. Hence the population of this level, denoted by $\rho_m$, assumes the form

$$\rho_m = A e^{-g\beta Bm/kT} . \qquad (20.3.3)$$

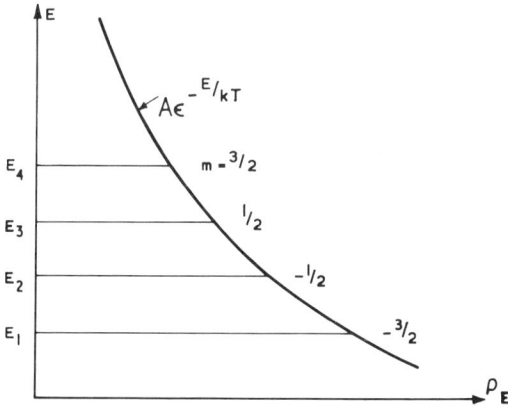

**Fig. 20.2.** Relative level populations of isolated paramagnetic ions having angular momentum quantum number $j = 3/2$.

The sum of the level populations equals the total number $N$ of paramagnetic ions per unit volume of the substance which is assumed to be known in the present context.

$$N = A \sum_{m=-j}^{j} e^{-g\beta Bm/kT} . \qquad (20.3.4)$$

The summation in the above equation is known as the *state sum* or *partition function*. It enables the constant $A$, so far unknown, to be determined.

$$A = \frac{N}{\displaystyle\sum_{m=-j}^{j} e^{-g\beta Bm/kT}} . \qquad (20.3.5)$$

Now come the decisive steps in the calculation of magnetization. At first the magnetic moment per unit volume *due to the population of a single level* is evaluated with the help of eqn (20.3.3).

$$-g\beta m\rho_m = -g\beta mA e^{-g\beta Bm/kT} \qquad (20.3.6)$$

where $-g\beta m$ is the component, along the applied field, of the magnetic moment of a single ion having quantum number $m$ (see eqn (16.7.6)). Next the *net magnetization* is obtained by forming the algebraic sum of all terms like eqn (20.3.6) and substituting for $A$ from eqn (20.3.5)

$$M = Ng\beta \frac{\sum\limits_{m=-j}^{j} -m\,\mathrm{e}^{-g\beta Bm/kT}}{\sum\limits_{m=-j}^{j} \mathrm{e}^{-g\beta Bm/kT}}. \tag{20.3.7}$$

Since the magnetization $M$ is parallel to the field $B$, the vector notation is dropped for convenience.

It is sometimes convenient for experimental purposes *to define an average magnetic moment per ion*, $\mu_{\mathrm{av}}$, as follows

$$N\mu_{\mathrm{av}} = M. \tag{20.3.8}$$

In general this will differ from any of the quantized moments $g\beta m$.

Before calculating the susceptibility and analysing it, we make some algebraic changes in eqn (20.3.7) which is somewhat awkward to handle in its present form.

$$M = N(jg\beta) \frac{\sum\limits_{m=-j}^{j} \frac{m}{j}\,\mathrm{e}^{(m/j)(jg\beta B/kT)}}{\sum\limits_{m=-j}^{j} \mathrm{e}^{(m/j)(jg\beta B/kT)}} \tag{20.3.9}$$

where the negative sign in front of the quantum number $m$ has been omitted since the summations cover all values of $m$ in any case. Noting that the numerator is the derivative of the denominator we write the state sum in the following form

$$f_j(x) = \sum_{m=-j}^{j} \mathrm{e}^{(m/j)x}$$

$$= \frac{\sinh\left(1 + \dfrac{1}{2j}\right)x}{\sinh \dfrac{1}{2j}\,x} \tag{20.3.10}$$

where                    $x = jg\beta B/kT = jg\beta\mu_0 H/kT.$

The fraction in eqn (20.3.9) can now be written in the form

$$L_j(x) = \frac{d}{dx}[\log f_j(x)] = \frac{\sum \frac{m}{j} e^{(m/j)x}}{\sum e^{(m/j)x}}$$

$$= \left(1 + \frac{1}{2j}\right) \coth\left(1 + \frac{1}{2j}\right) x - \frac{1}{2j} \coth \frac{1}{2j} x \qquad (20.3.11)$$

The function $L_j(x)$ is called the *Brillouin function*. It is used to write the magnetization in the compact form

$$M = N(jg\beta)L_j(x). \qquad (20.3.12)$$

Further discussion of the magnetization is carried out in terms of the Brillouin function, the coefficient preceding the latter being a constant of a given material. The general form of this function is illustrated in Fig. 20.3 from which

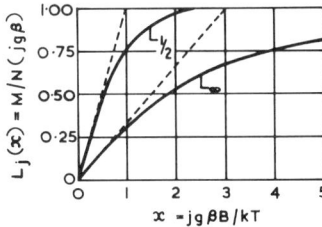

Fig. 20.3. The Brillouin function, $L_j(x)$, for $j = \frac{1}{2}$ and $j \to \infty$. The magentization is proportional to the curves shown.

it can be seen that there is a significant linear section near the origin. This corresponds to applied fields readily realizable in the laboratory at normal temperatures, and suggests that a linear approximation may be useful. A series expansion of eqn (20.3.11) for $x \ll 1$ and retention of the linear term yields the result

$$M = \frac{1}{3}\left(1 + \frac{1}{j}\right) N \frac{(jg\beta)^2}{kT} \mu_0 H \qquad (20.3.13)$$

where $B$ has been replaced by $\mu_0 H$. Expressing the magnetization in terms of the susceptibility as in eqn (20.1.2) one finds

$$\chi = \frac{1}{3}\left(1 + \frac{1}{j}\right) N \frac{\mu_0(jg\beta)^2}{kT} . \qquad (20.3.14)$$

Hence, under these conditions the susceptibility is constant at a fixed temperature. Substitution of typical values for the various constants will satisfy the reader that the susceptibility is also small, corresponding to the conventional case of paramagnetism.

Considering now the susceptibility of eqn (20.3.14) as a function of temperature we see that it is of the form

$$\chi = \frac{C}{T} \tag{20.3.15}$$

where $C = \frac{1}{3}(1 + 1/j)N(jg\beta)^2 \; \mu_0/k$. The above relation is usually referred to as *Curie's law*.

The foregoing statistical calculation of the susceptibility is based on a weighted average of the magnetic moments, parallel to the applied field, associated with the energy levels of the paramagnetic ions. It is instructive at this stage to pause and consider the graphical interpretation of the approximation which led to eqns (20.3.13) and (20.3.14). Figure 20.4(a) depicts the position at constant temperature. The applied field is sufficiently low for the energy levels to be spread over a small range of the population distribution. The exponential Maxwell-Boltzmann law can be approximated by a segment of a straight line. The applied field is assumed to be limited to values within this linear

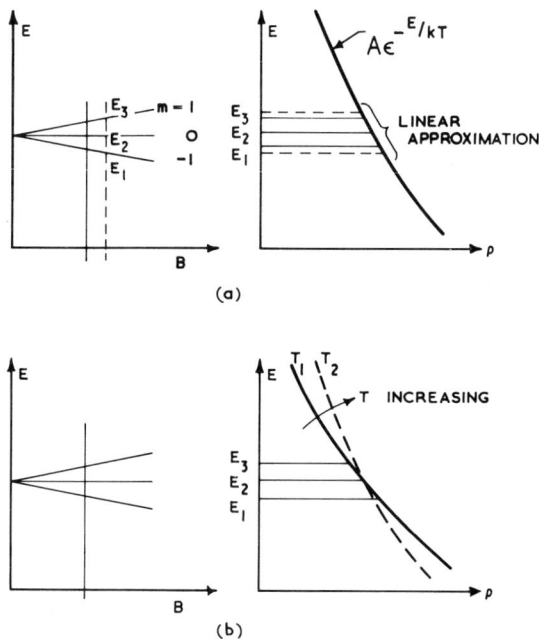

(a)

(b)

Fig. 20.4. (a) Statistical interpretation of the linear variation of magnetization with applied field. (b) Statistical interpretation of Curie's law.

approximation. The effect of temperature variations is represented by Fig. 20.4(b). Changing the temperature is equivalent to tilting the population curve. The assumption of the linear approximation leads to Curie's law.

Once the applied field is sufficiently increased, while the temperature is decreased, the magnetization no longer increases linearly and the susceptibility ceases to be constant. The *saturation magnetization*, which is ultimately reached, is obtained from eqn (20.3.12) on letting $x \to \infty$

$$M_s = Njg\beta. \tag{20.3.16}$$

The average magnetic moment per atom, as defined in eqn (20.3.8), now assumes the value

$$\mu_{av} = \frac{M_s}{N} = jg\beta. \tag{20.3.17}$$

The above results imply that all the ions contribute the maximum possible component of their magnetic moment, $jg\beta$, towards the magnetization. This means that all the ions occupy the lowest energy level, corresponding to the magnetic quantum number $m = -j$. Figure 20.5 shows graphically the approach to saturation. The energy levels are spread over a wide range of the exponential population curve, the lowest having by far the greatest population. In the limit $x \to \infty$ only the lowest level is populated.

A notable feature of eqn (20.3.16) is the fact that the saturation magnetization is proportional to the quantum number $j$ which characterizes the *maximum component of the atomic magnetic moment along the applied field*. This observation should be contrasted with the *magnitude* of the magnetic moment which is proportional to $\sqrt{j(j+1)}$.

The theoretical value of the saturation magnetization, eqn (20.3.16), can be compared with experimental results which provide a verification of the principal predictions of quantum theory regarding the angular momentum of atoms having unpaired electrons. However, when a comparison of theory and

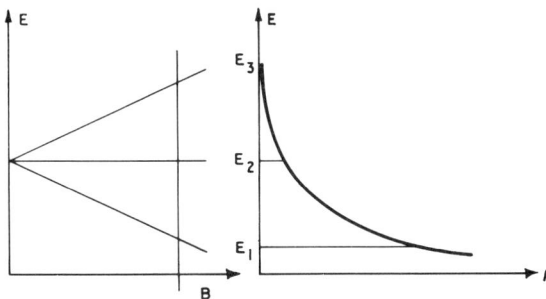

**Fig. 20.5.** Populations of the paramagnetic energy levels on approach to saturation.

experiment is attempted, the simplifying assumptions must be taken into account. In particular it must be remembered that the influence of electrostatic lattice forces on the paramagnetic ions has been neglected. This is most likely to be justified in the case of the rare earth elements whose unpaired electrons are located in inner shells, screened from the lattice by outer shells, which are complete (see Table 16.1 of electron configurations). Under such conditions the

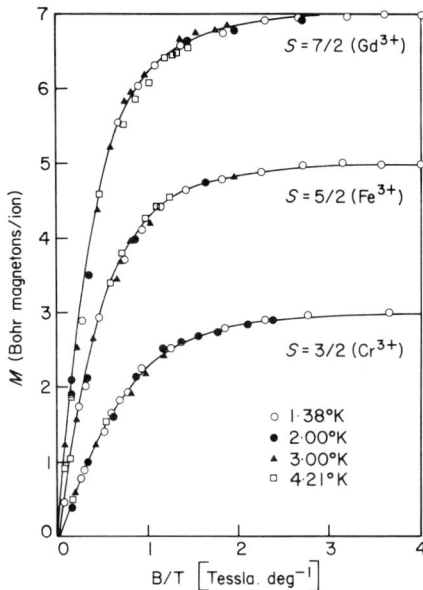

**Fig. 20.6.** Paramagnetic saturation in high fields and low temperatures. (After W. E. Henry, *Phys. Rev.*, 88, 559, 1952).

angular momenta arrange themselves according to Hund's rules (see Section 16.6). The trivalent gadolinium ion, $Gd^{3+}$, provides a good example. Ionization removes the two $6s$ electrons and one $5d$ electron leaving seven unpaired electrons in the $4f$ shell. After positive spins have been allocated to these electrons, yielding the maximum resultant $S = \frac{7}{2}$, the exclusion principle permits only the values of $l = 3, 2, 1, 0, -1, -2, -3$, which yield a zero resultant. Hence the overall resultant angular momentum is a pure spin $J = S = \frac{7}{2}$ and the corresponding magnetic moment per ion is $jg\beta = 7\beta$ by eqn (20.3.17), or 7 Bohr magnetons. This result is well substantiated by the experimental data for $Gd^{3+}$, reproduced in Fig. 20.6.

The example of $Gd^{3+}$ also demonstrates that the spectroscopic splitting factor is $g = 2$ for spin angular momenta. In the case of ions having resultant angular momenta which are composed of both spins and orbits the value of $g$ is expected to be less than 2. Now, it is shown in Section 16.6 that $Cr^{3+}$ has a mixed resultant, whence it would follow that its $g$ value is $g < 2$, and the saturation magnetic moment is $\frac{3}{2}g\beta < 3\beta$. A reference to the experimental data in Fig. 20.6 shows that $Cr^{3+}$ has a resultant moment of 3 Bohr magnetons, contrary to expectations. This is caused by the electrostatic forces of the crystal lattice which interfere with the orbital motions of the outermost $3d$ electrons forcing them to form a zero resultant, $L = 0$, and leaving only the net spin $S = \frac{3}{2}$. This effect is usually referred to as *orbital quenching*. Similar complications apply to most paramagnetic ions having unpaired electrons in the outermost shell. Although the resultant angular momentum and the $g$-value may be evaluated for the isolated ion, it does not follow that the results will apply when the ion is placed in the environment of a crystal lattice. Interactions due to the neighbouring ions may drastically alter the results. Finally it should be noted that although most atoms and ions have a magnetic moment when isolated, this is no longer the case after they combine to form gas molecules, liquids or solids. The various chemical bonding arrangements usually involve the cancellation of the magnetic moments of neighbouring atoms due to exchange forces (see Chapter 21).

Before closing this section it is emphasized that the results obtained apply only to the simplest materials in which the paramagnetic ions behave as if they were isolated. Paramagnetic solids which have more complicated energy levels still have a magnetization and susceptibility, which can be calculated by the approach used above, but eqn (20.3.2) for the energy level spacing no longer applies. Since the energy level configuration varies with the relative orientation of the applied field and the crystal system in many materials, it is to be expected that the susceptibility will be anisotropic in such cases.

## 20.4  The Effect of Time Varying Fields on Energy Level Populations

So far we have discussed only *static paramagnetism*, that is the effect of a d.c. magnetic field on the magnetization of paramagnetic materials. In so doing we have used results, established earlier, regarding the effect of externally applied d.c. magnetic fields on the energy levels of atoms (see Chapters 15 and 16). The basic fact arrived at in this connection was the *modification of atomic energy levels by the applied field*, which took the form of energy level splitting.

In the present and following sections we shall consider some *dynamic aspects of paramagnetic materials*. By this we mean the *effect of a.c. applied fields or electromagnetic radiation* on the populations of paramagnetic ions. As a preliminary to this subject we first introduce some basic concepts and facts regarding the effect of time varying fields and potentials on dynamical systems in the context of quantum mechanics. The discussion will be heuristic in approach since a theoretical treatment would require a mathematical background which is beyond the scope of the present book.

The presence of electromagnetic radiation means the application of electric and magnetic fields of the form $B(t) = B \sin \omega t$ where $B$ is the amplitude of, say, a magnetic field. For the moment we leave aside the question of how atomic systems can be exposed to such influences and go on to consider their effect. At first sight it might appear obvious to refer back to the time-dependent form of Schroedinger's equation (see Chapter 11), to include in the potential a time-dependent term, and to hope that solutions can be obtained which will provide an answer to our problem. However, such an approach is mathematically very difficult and all the results obtained to date have been arrived at by other methods called *perturbation theory*.

All perturbation methods start with solutions of the time-independent Schroedinger equation and seek to include additional effects by approximation. If the *additional force is time independent*, e.g. a d.c. electric or magnetic field, its effect is to modify the basic energy level scheme. A simple example of this kind is the splitting of the energy levels of the hydrogen atom by a magnetic field, calculated in Chapter 15 (without recourse to perturbation theory). However, if the *additional force is time dependent* its effect is interpreted as *causing transitions between existing energy levels*. Hence, if the time-dependent force is applied to a population of atoms occupying several energy levels, it will tend to cause changes in the population distribution. Time-dependent perturbation theory makes it possible to ascertain the conditions under which radiation can affect atomic systems, and to evaluate the probabilities of transition between energy levels. As it is beyond the scope of the present treatment to introduce perturbation theory we shall only state the most important results and explain their physical significance.

The first result of a perturbation calculation relates the frequency of the incident radiation to the spacing between levels through the well-known formula

$$E_j - E_i = hf_{ij}. \tag{20.4.1}$$

This means that the a.c. field must have the frequency $f_{ij}$ if it is to cause transitions between the levels $E_i$ and $E_j$. Another interpretation of eqn (20.4.1) is that the quanta of incident radiation or photons must have energy $hf_{ij}$ if they are to induce transitions between the given energy levels.

However, the above condition is not sufficient to ensure that an interaction takes place because the corresponding transition may have zero *transition probability*. This is the second principal result of a perturbation calculation: it yields a numerical value for the probability that the incident radiation will cause transitions between levels. *The transition probability is defined as the fraction of all atoms occupying an energy level, transferred to another level by the incident radiation, per unit time.* Moreover, the *transition probability is the same for upward and downward transitions between a pair of levels*. Thus, referring to eqn (20.4.1), radiation of frequency $f_{ij}$ will transfer the same fraction of atoms from level $i$ to level $j$ and vice versa, per unit time.

Two factors enter into the transition probability. (i) The transition probability is proportional to the intensity of the incident radiation or, to use the terminology of microwaves, is proportional to the energy stored by the electromagnetic wave per unit volume. This in turn is proportional to the square of the wave amplitude. (ii) Secondly the transition probability is related to the matrix element of the time-varying potential with respect to the states between which transitions are to take place. For many pairs of states such matrix elements vanish, which means that the *transitions are forbidden.*

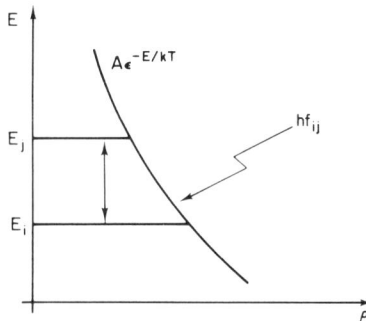

Fig. 20.7.

Using the above concepts let us consider the effect of electromagnetic radiation on a population of atoms, distributed over a number of energy levels, of which we pick out two for special consideration, say $E_i$ and $E_j$. To be specific we assume that the atoms are sparsely spread over space as in a gas and therefore do not interact with each other. In the absence of interaction they are distributed over their energy levels according to the Maxwell-Boltzmann law shown in Fig. 20.7. Assuming that there is finite transition probability between the two levels, radiation of frequency $f_{ij}$ will cause transitions in both directions: upwards and downwards. In every upward transition by an atom one photon of energy, $hf_{ij}$, is absorbed from the electromagnetic wave. In every downward transition one photon is given up to the wave. The net result is *absorption of energy* from the wave because the population of the lower level is greater, and the transition probability is the same both ways. This absorption forms the basis of most spectroscopic experiments, particularly those made at microwave frequencies.

The foregoing argument assumes that the level populations remain substantially undisturbed by the action of the radiation. This assumption is valid in most spectroscopic measurements by virtue of *relaxation processes.* The latter tend to restore level populations to their equilibrium distributions given by the

Maxwell-Boltzmann law. Thus any excess atoms in level $E_j$ will tend to drop spontaneously to level $E_i$, but they will do so at a finite rate, called the *relaxation rate.*

Under specially arranged conditions the level populations may indeed be substantially affected by electromagnetic radiation. Moreover, absorption of radiation may be replaced by amplification as a result of *stimulated emission.* Such conditions are set up by design in maser and laser amplifiers and will be discussed at a later stage.

### 20.5 Spectroscopy of Paramagnetic Solids

In the present section we shall consider some spectroscopic properties of paramagnetic substances, that is transitions between the energy levels of magnetic moments immersed in a d.c. magnetic field. The transitions are caused by a.c. fields or electromagnetic waves of frequency satisfying eqn (20.4.1). In the first place we reiterate that absorption can only be observed between levels for which finite transition probabilities exist. There is a general selection rule, applicable to angular momentum states, which predicts that transitions are only allowed between levels whose magnetic quantum number $m$ differs by 0 or $\pm 1$

$$\Delta m = 0, \pm 1. \tag{20.5.1}$$

This result can be obtained by a perturbation calculation but in the present context we shall only quote and apply it.

Transitions subject to the above rule can be observed in two ways.

(i) At frequencies in the optical range, Zeeman effect of the form discussed in Chapter 15 can be measured. The observed splitting of spectral lines is due to transitions between the split ground and excited states of atoms.

(ii) Transitions between the paramagnetic levels of the split ground state are observed at much lower frequencies, usually in the microwave band. Only this case will be dealt with in the present section.

Let us first consider the simplest case of isolated paramagnetic ions. The energy-level scheme is shown in Fig. 20.1. As explained at the close of Section 20.3 such behaviour is exhibited by salts of the rare earths, in which the unpaired electrons are screened from the crystalline fields by the outer shell of electrons. At any specific value of the d.c. magnetic field (marked by the dotted line) transitions are allowed only between adjacent pairs of levels according to the above selection rule. However, if there are more than 2 levels, several transitions will be taking place simultaneously.

A more complicated example is provided by the trivalent chromium ion, $Cr^{3+}$, when embedded in a non-magnetic host crystal lattice. Since the unpaired electrons are in the outermost shell, we must expect that the energy level scheme will be complicated by crystalline field effects. Leaving aside the more difficult question of how the energy level scheme can be evaluated in this case, we draw on information available elsewhere and reproduce the energy levels of chromium in ruby in, Fig. 20.8, for one particular orientation of the applied field and the crystal lattice.

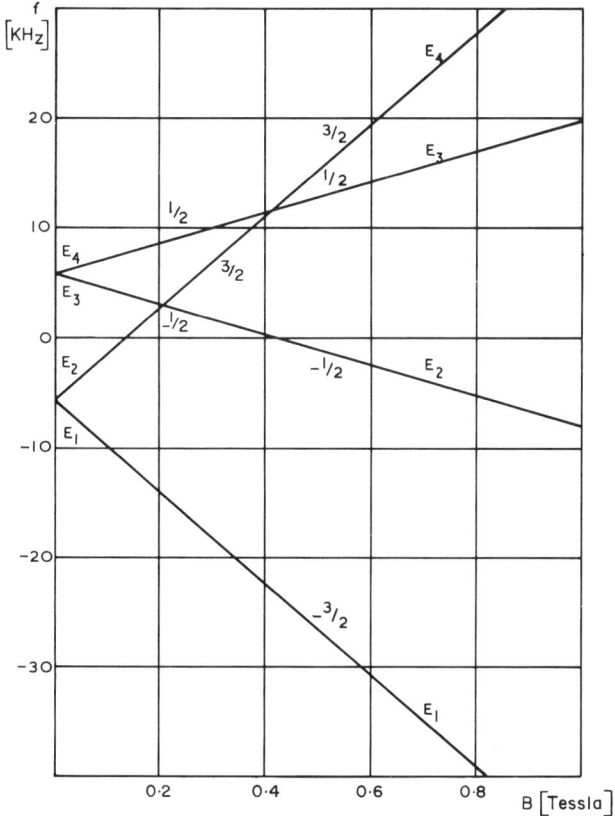

**Fig. 20.8.** Energy levels of $Cr^{3+}$ in ruby for the orientation $\theta = 0°$.

It was shown in Section 16.6 that the ground state of the $Cr^{3+}$ ion is of the type $^4F_{\frac{3}{2}}$ yielding a four-fold splitting in the presence of a d.c. applied field. When the ion is immersed in the alumina ($Al_2O_3$) lattice the crystalline field causes a *zero field splitting* as shown. The detailed form of the energy levels varies with orientation, which is specified by the angle $\theta$ between the $c$-axis of the hexagonal crystal system and the applied field. At the orientation $\theta = 0$, shown in Fig. 20.8, the energy eigenstates are identifiable with angular momentum eigenstates as indicated. While the energy levels are always labelled consecutively from lower to higher energies, angular momentum eigenstates correspond to straight lines labelled by the magnetic quantum numbers. The energy axis is calibrated in frequencies, as this happens to be the most convenient measure for spectroscopic experiments in the microwave band. As an examples of eqn (20.5.1) the reader should satisfy himself that transitions are allowed between levels 2 and 3 at fields in excess of about 4kGauss (0·4 T), but are forbidden below this value.

For energy level spacings corresponding to microwave frequencies, the actual population differences are very small at room temperatures and absorption of the test signal or radiation is frequently beyond the sensitivity of available

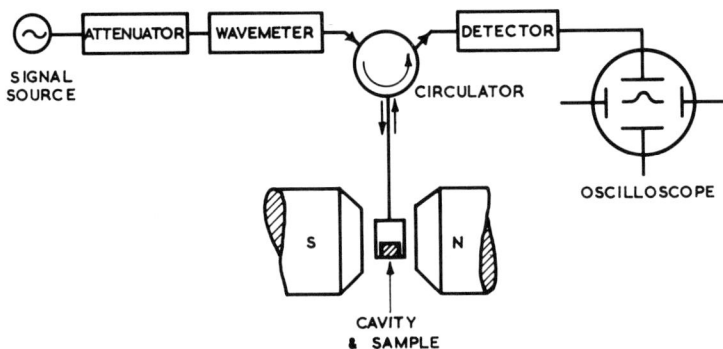

**Fig. 20.9.** Basic elements of a microwave spectrometer.

equipment. For this reason most measurements are carried out at low temperatures, e.g. 77 °K or 4·2 °K (liquid nitrogen or liquid helium temperatures).

A block diagram of a *microwave spectrometer* is shown in Fig. 20.9. Only the most important parts of the equipment are included. Normally various additions are made, designed to increase the sensitivity of the instrument. Electromagnetic radiation is generated by the signal source and transmitted by waveguide to a cavity containing a paramagnetic sample. On the way the signal level is set to the desired value by the attenuator, and the frequency is checked with a wavemeter. The circulator is a device used to separate the reflected wave from the incident wave, as indicated by arrows in the diagram. The radiation reflected by the cavity is passed through a detector which extracts the envelope of the microwave signal for display on the oscilloscope. The signal frequency is adjusted to coincide with the resonant frequency of the cavity. All other frequencies are reflected at the cavity input port, and are thus unable to interact with the sample. The magnetic field provided by an electromagnet is adjustable in such a way that the spacing between a pair of energy levels of interest corresponds to the radiation frequency according to eqn (20.4.1).

Spectroscopic absorption in the sample increases the effective cavity losses and changes the reflection coefficient. The signal displayed on the oscilloscope is that reflected by the cavity at its resonant frequency. Since it would be cumbersome to adjust manually the frequency of the signal source and the applied magnetic field, to observe a desired allowed transition, it is usual to superimpose a small modulation on one of these variables. In practice the signal frequency of a klystron is easily swept over a useful range by modulating the reflector voltage. Equally, it is feasible to modulate the magnetic field with the help of auxiliary coils mounted on the pole pieces. In either case the signal reflected by the cavity is displayed as a function of the modulation. Absorption appears as a dip or hump on the display.

### 20.6  Stimulated Emission and the Three-Level Maser

As suggested at the close of Section 20.4 the process of absorption can be reversed and amplification of the incident radiation can take place as a result of *stimulated emission*. In the present section we shall discuss under what conditions this can happen.

As explained on p. 310 the incident radiation can cause both upward and downward transitions. What is more, the *transition probability for each process is the same*. Absorption was seen to be due to the greater population of the lower level under equilibrium conditions. The same statistical considerations lead us to believe that *net emission of radiation will be observed if the population of the higher level is arranged to be greater*. Such stimulated emission of radiation will have the effect of amplifying the incident signal. Hence the word MASER which means "microwave amplification by stimulated emission of radiation". The LASER is based on the same principle but refers to light amplification.

To achieve amplification the equilibrium populations of the energy levels must be sufficiently disturbed to cause *population inversion* between two levels. There are several ways of inverting populations, the most important of them rests on the *principle of pumping*, applied to a system of three energy levels. To explain this we refer to Fig. 20.10.

Given three levels, their equilibrium populations follow an exponential law as shown by the dotted curve. As a first step a very strong radiation, called the *pumping signal*, is applied to the sample. The pumping signal has the frequency

$$f_{13} = \frac{E_3 - E_1}{h}.$$  (20.6.1)

It must be strong enough to *saturate or equalize the populations* of levels 1 and 3 as indicated by the vertical chain line. As a result the populations of levels 2 and 3 or 1 and 2 are inverted as indicated in Fig. 20.10(a) and (b) respectively. The application of a low level signal will stimulate downward transitions between the inverted levels. The quanta of radiation thus emitted are added to the incident signal which swells in magnitude and is therefore amplified. The amplified signal must be at one of the two frequencies

$$f_{23} = \frac{E_3 - E_2}{h}; \quad f_{12} = \frac{E_2 - E_1}{h}. \qquad (20.6.2)$$

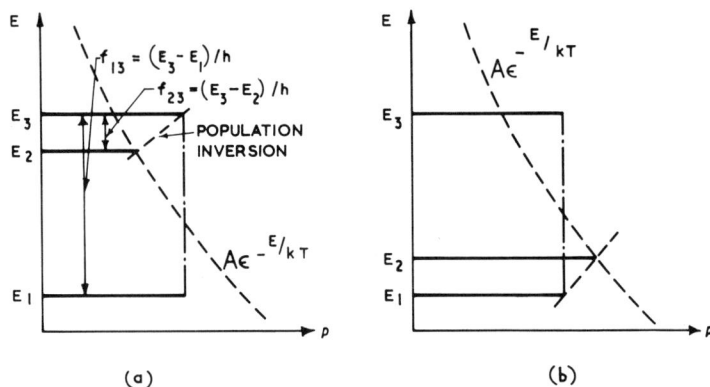

Fig. 20.10. Principle of the three-level maser.

Amplification by stimulated emission in a three-level system can only be realized if several other conditions are satisfied. In the first place the relaxation rates between the various energy levels impose various restrictions. Thus it is desirable that relaxation across the pump transition, from level $E_3$ to $E_1$, should be as slow as possible to facilitate effective pumping.

A similar consideration applies to the signal transition levels $E_3$ and $E_2$ or $E_2$ and $E_1$ in Fig. 20.10(a) and (b) respectively. Relaxation at a high rate between these pairs of levels would obliterate population inversion and frustrate the pumping effort. On the other hand relaxation between levels $E_2$ and $E_1$ of case (a) should be as fast as possible to enable the ions stimulated from level $E_3$ to $E_2$ to vacate the latter rapidly. A similar consideration applies to levels $E_3$ and $E_2$ of case (b).

In practice it must be expected that the relaxation process will prevent the level populations from assuming the precise values shown in Fig. 20.10. In particular the population of level $E_2$ will exceed somewhat the equilibrium value given by the Maxwell-Boltzmann law. However, it is possible to set up rate equations, which take into account both relaxation rates and radiation intensities, and to predict the actual populations with sufficient accuracy for practical purposes.

So far it has been tacitly assumed that transitions are allowed between any pair of levels of a three-level system. In fact this is not always the case. An examination of Fig. 20.8 in conjunction with the selection rule eqn (20.5.1) will disclose that above a field of about 4kGauss it is impossible to find a set of three levels satisfying this requirement. Indeed, the condition of significant transition probabilities coupled with satisfactory relaxation rates is satisfied in

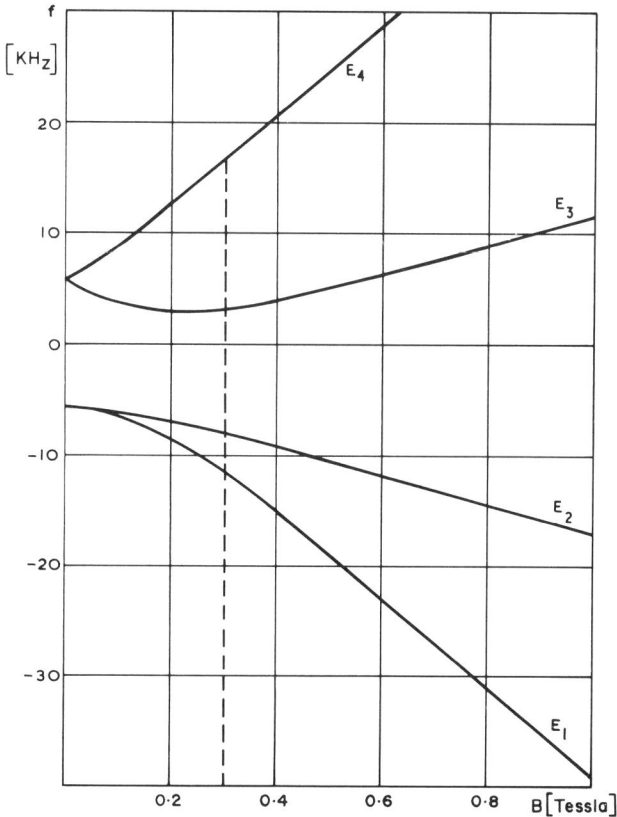

Fig. 20.11. Energy levels of $Cr^{3+}$ in ruby for the orientation $\theta = 90°$.

relatively few cases. Hence the choice of usable maser substances is rather restricted.

The most suitable maser known at present is ruby, used in orientations other than $\theta = 0°$. The $\theta = 90°$ orientation which has been used in many masers, has the energy levels shown in Fig. 20.11. For this orientation it is not possible to identify energy eigenstates with eigenstates of angular momentum. By the principle of superposition (see Section 15.3) the energy levels contain in general admixtures of all or some of the angular momentum states according to the equation.

$$|n\rangle = a|-\tfrac{3}{2}\rangle + b|-\tfrac{1}{2}\rangle + c|\tfrac{1}{2}\rangle + d|\tfrac{3}{2}\rangle \qquad (20.6.3)$$

where $1|n>$ represents any one of the energy levels and $a$, $b$, $c$ and $d$ are constants, some of which may be zero. An application of the selection rule $\Delta m = 0, \pm 1$ to the energy levels now shows that transitions between any of them are allowed. Hence any set of three levels can provide the basis for a maser

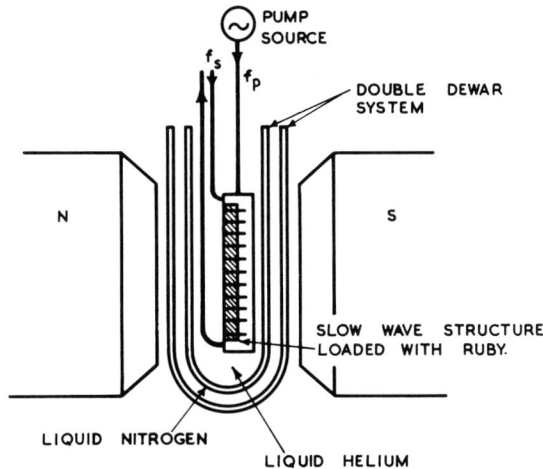

Fig. 20.12. Schematic diagram showing the essential elements of a travelling wave maser. $f_s$ = signal frequency. $f_p$ = pump frequency.

amplifier, giving a range of choice as regards signal and pump frequencies and the necessary d.c. magnetic fields.

A maser can be realized either in the form of a cavity device or a travelling-wave maser. A cavity maser consists of the same basic elements as the spectrometer in Fig. 20.7 but the cavity must be designed to be resonant of both the signal and pump frequencies. A separate source of pump power must be provided and applied to the cavity via a waveguide.

The cavity maser finds little application because it has a narrow bandwidth and unstable gain, features which make it unsuitable for use in communications and radar systems. These shortcomings are absent in the travelling wave maser which is shown diagrammatically in Fig. 20.12.

In a travelling wave maser the interaction between electromagnetic waves and a paramagnetic crystal takes place in a *slow wave structure*. The latter is a periodic structure in which the group velocity of the wave is very much less than the velocity of light, $c$, in free space. Structures for travelling wave masers are usually designed to have a group velocity of the order of $c/100$. As a result there is high concentration of stored energy, or a high intensity of radiation. This is necessary to increase transition probabilities and hence to increase the amount of stimulated radiation. The slow wave structure can be designed in a variety of forms, one of which is illustrated in Fig. 20.13. This comb structure forms the basis of masers which are used as the first stage of amplification in ground receivers of satellite communication systems. Fig. 20.13 omits several features of an actual amplifier for the sake of simplicity.

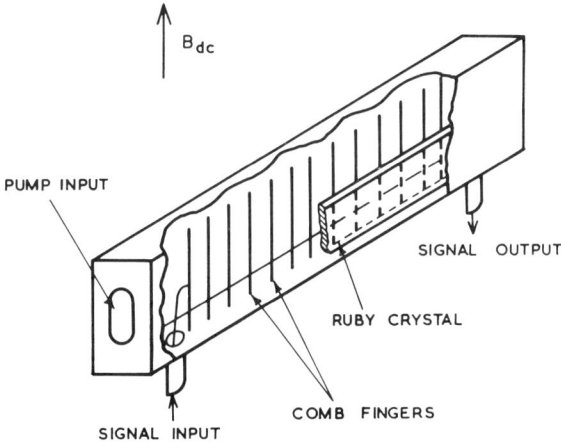

**Fig. 20.13.** Diagrammatic view of the comb structure maser.

Such maser amplifiers are invariably operated at temperatures of liquid helium for the following reasons: (i) to make the population differences between the paramagnetic levels as great as possible, (ii) to reduce relaxation rates, (iii) to ensure a low noise figure. The last consideration makes the maser the ultimate in low noise amplification. Equivalent noise temperatures of about 3 °K have been realized.

The noise generated inside a maser amplifier is greatly influenced by *spontaneous emission*. This is a radiative process which takes place in addition to stimulated emission. However, whereas stimulated emission is *coherent*, that is it forms part of the well-defined stimulating wave, spontaneous emission is *incoherent*. This means that it appears in the form of a jumble of irregular wavelets, and in effect constitutes noise. Spontaneous emission of radiation is the result of spontaneous downward transitions, which are part of the general tendency of the system to maintain equilibrium level populations. Interested readers are referred to specialized texts for further information.

The use of paramagnetic crystals in masers makes it necessary to provide a d.c. magnetic field of the correct magnitude and direction to secure the requisite spacing of energy levels. Although in Fig. 20.12 a magnet outside the low temperature system is shown for clarity, superconducting magnets inside the liquid helium bath are used in practice.

*Chapter 21*

# Ferromagnetism

## 21.1 Exchange Forces as the Basis of Ferromagnetism

As in paramagnetic substances the magnetic properties of ferromagnetic materials are derived from the magnetic moments of individual atoms. The principal difference between the two types of magnetism results from the fact that a large proportion of the atoms, if not all, in a ferromagnet have resultant magnetic moments. This is to be contrasted with the sparsely spread magnetic atoms or ions in a paramagnetic crystal, which are too far apart to interact with each other magnetically. In a ferromagnetic material the direct interaction of the magnetic moments is the principal source of its properties. Before discussing possible interactions we note, as a preliminary, that it is the *spin magnetic moment* which plays the dominant role in Ferromagnetism. This conclusion is based on experimental evidence, some of which will be discussed in subsequent sections.

There are two types of magnetic interaction in which neighbouring spins may be involved: (i) the classical dipolar interaction; (ii) the quantum mechanical exchange interaction introduced in the appendix to this chapter. Early attempts by Weiss to explain ferromagnetism on the basis of the dipolar interaction failed, and it was not until the advent of quantum theory that the exchange interaction was recognized by Heisenberg as the proper basis.

The exchange energy can be written in the form

$$W_e = -2J\mathbf{S}_i . \mathbf{S}_j \qquad (21.1.1)$$

where $J$ is the exchange integral computed from the electronic wave functions of two neighbouring atoms, and $\mathbf{S}_i$ and $\mathbf{S}_j$ are their resultant spin vectors. An outstanding property of the exchange energy is its rapid decrease as the spacing between the $i$-th and $j$-th atoms is increased. Indeed, in many situations it is accurate enough to assume that it vanishes except for nearest neighbours in a lattice. This is because the radial parts of the electronic wave functions contain exponential factors which decrease very rapidly with distance (see Section 13.1).

At the same time, the magnitude of the exchange energy for nearest neighbours is much greater than the dipolar energy, although the latter decreases relatively slowly with the dipole spacing—in fact as $1/r^3$.

The exchange energy of eqn (21.1.1) also depends on the relative orientation of the resultant spin vectors and assumes a minimum value when the latter are parallel, provided that *J is positive.* Hence, the *exchange interaction will tend to align the magnetic moments of neighbouring atoms.* This is the characteristic property of *ferromagnetic* materials. In many cases the *exchange integral J is,* however, *negative.* In such substances, *the effect of the exchange interaction is to antialign neighbouring spins,* a property called *antiferromagnetism.*

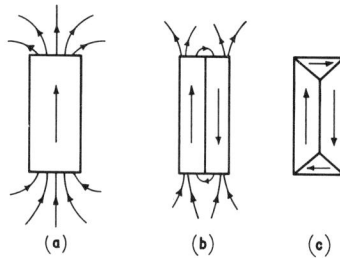

Fig. 21.1. Magnetic domains.

For the moment, we shall be concerned with ferromagnetism only, and we note that the tendency of spins to align themselves must result in *spontaneous local magnetization* of the material. Indeed, if no other forces were present, one would have to conclude that an entire macroscopic sample would be uniformly magnetized as shown in Fig. 21.1(a). However, in this condition the sample energy would be increased by the substantial amount stored in the resulting external magnetic field, which would more than counterbalance the minimum of the exchange energy. Hence the tendency will be to minimize the external field by breaking up the magnetization into *domains.* The simple arrangement of Fig. 21.1(b) will have a magnetostatic energy less than half that of Fig. 21.1(a). In Fig. 21.1(c) the domains arrange themselves in such a way that the associated flux closes upon itself. This particular case represents an idealized situation, applicable in two dimensions only, but it does illustrate the general principle.

The breaking up of the magnetization of a macroscopic sample into domains eliminates the magnetostatic energy of the external field. However, some additional energy is involved in forming *domain walls,* which form the transition between groups of spins of different orientation. The nature of these domain walls, their thickness, and the size of the domains, will be discussed after the concept of magnetic anisotropy has been introduced in the next section.

## 21.2  Crystalline Anisotropy

In the discussion of the preceding section it was tacitly assumed that the spins in a ferromagnetic sample are free to align themselves in any direction with equal ease. This is not the case in practice, due to crystalline lattice forces, which will not be uniform in all directions (isotropic), but which will be related to the symmetry of the crystal system (anisotropic). The lattice forces are electrostatic in nature; they cannot, therefore, affect the spins directly. However, they do affect the orbital motion of the outer electrons and make the position of the orbits conform to a minimum energy configuration relative to the lattice. This effect is then transmitted to the spins via the spin-orbit coupling which was introduced in Chapter 15.

As a result of these interactions the spins find it easier to align themselves along some lattice directions rather than others. This phenomenon is called *anisotropy*, or more precisely *crystalline anisotropy*, to distinguish it from other directional effects which are macroscopic in origin and which are related to sample geometry. The extra forces required to direct a group of spins along a *hard direction* imply that such an arrangement involves an additional amount of energy. This is referred to as the *anisotropy energy*. For many purposes, it is quite adequate to assume that a group of spins, arranged in various directions, will have an average anisotropy energy given by a constant $K$ per unit volume.

The change in the direction of magnetization from one domain to another implies that within the walls separating domains there will be spins which have to assume directions associated with high anisotropy energy. This fact must be assumed to be one of the factors influencing the thickness of domain walls and the size of the domains, which must be expected to form in such a way as to minimize the sum of the various contributions to the sample energy.

## 21.3  Magnetic Domains and Domain Walls

To consider the extra energy involved in forming domain walls we assume that the transition between two domains magnetized in opposite directions has the form shown in Fig. 21.2. This arrangement is referred to as a *Bloch wall*; the direction of magnetization rotates gradually about a normal to the plane of the

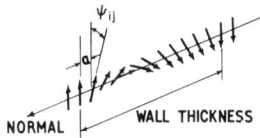

**Fig. 21.2.** Diagrammatic representation of a Bloch wall. The orientation of adjacent spins rotates about a normal to the wall plane.

wall. Other arrangements are possible, e.g. a Néel wall, in which the spins rotate in the plane containing the normal. However, it can be shown that in bulk samples the Bloch wall requires least extra energy, it is therefore more likely to be found.

The additional exchange energy involved in forming a Bloch wall can be estimated as follows. Assuming that the atoms forming the ferromagnetic material all have the same resultant angular momentum $S$, the exchange energy of two neighbouring spins is by eqn (21.1.1)

$$-2JS^2 \cos \psi \qquad (21.3.1)$$

where $\psi$ is the angle between the spins. The total exchange energy of a row of spins forming a Bloch wall is

$$- \sum_{i \,\rangle\, j} 2JS^2 \cos \psi_{ij}. \qquad (21.3.2)$$

If the wall extends over many atoms, say 100, $\psi_{ij}$ is small and it is possible to use the approximation $\cos \psi_{ij} \cong 1 - \frac{1}{2}\psi_{ij}^2$. Equation (21.3.2) can now be rewritten in the form

$$- \sum 2JS^2 + \sum JS^2 \psi_{ij}^2. \qquad (21.3.3)$$

The first term represents the exchange energy of a row of parallel spins. The second term represents the *extra exchange energy of a row of non-parallel spins.*

For a wall separating two antiparallel domains $\Sigma \psi_{ij} = \pi$. If $\psi_{ij}$ is the same for each successive pair and there are $n$ spins in the wall, we can write $\psi_{ij} = \pi/n$. Hence

$$\sum_{i \,\rangle\, j} \psi_{ij}^2 = n \left(\frac{\pi}{n}\right)^2 = \frac{\pi^2}{n}. \qquad (21.3.4)$$

Since $J$ and $S^2$ are constants of a particular material the *extra exchange energy of a row of spins forming a Bloch wall* can now be written in the form

$$JS^2 \frac{\pi^2}{n} \qquad (21.3.5)$$

For purposes of comparison it is desirable to obtain an expression for the energy per unit area of wall rather than per line of spins within the wall. If the lattice constant of the material is $a$, there will be $1/a^2$ rows of spins running in a direction normal to a unit area of wall. Hence the *extra exchange energy per unit area of Bloch wall* is

$$\Delta W_e = JS^2 \frac{\pi^2}{na^2}. \qquad (21.3.6)$$

The above result shows that the extra exchange energy of domain walls decreases as the thickness of the wall increases. On this basis, it appears that there would be a tendency for wall thicknesses to increase, to minimize the energy. However, this tendency is counterbalanced by anisotropy energy. It follows from the discussion of the preceding section that some of the spins in a domain wall must be aligned in directions of hard magnetization, thus causing an increase in anisotropy energy. If this is put approximately equal to a constant $K$ per unit volume, then the *anisotropy energy per unit area of Bloch wall* is

$$W_a = Kan \tag{21.3.7}$$

since $an$ is the volume of a unit area of wall. Hence the anisotropy energy increases with wall thickness and thus constitutes an influence which tends to minimize it.

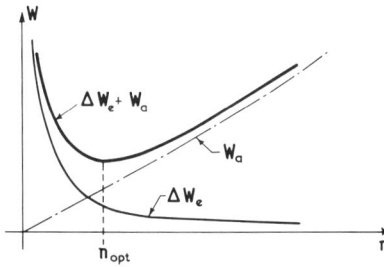

Fig. 21.3. Optimum thickness of a Bloch wall.

The position is illustrated graphically in Fig. 21.3, which shows the two contributions to the wall energy and their sum

$$\Delta W_e + W_a = JS^2 \frac{\pi^2}{a^2 n} + Kan. \tag{21.3.8}$$

It is apparent that an optimum wall thickness exists at which the wall energy has a minimum. A straightforward calculation shows that this is

$$n_{\text{opt}} = \frac{\pi S}{a^{3/2}} \sqrt{\frac{J}{K}}. \tag{21.3.9}$$

Since extra energy is required to form domain walls, it follows that the size of domains will tend to be as large as possible, to keep the volume occupied by walls to a minimum, consistent with an absence of magnetostatic energy. However, under certain circumstances, domains may have to be magnetized in a hard direction, thus increasing the anisotropy energy. This tends to decrease the domain size.

Magnetic domains are readily observed under an optical microscope. Their average diameter is of the order of $10^{-4}$ cm. Domain walls extend typically over

100 atoms (a few hundred Ångstrom units). More refined methods than an optical microscope are therefore required to resolve them and to measure their thickness.

## 21.4 Magnetization of Ferromagnetic Materials

In the foregoing sections we have discussed the state of ferromagnetic materials in the absence of an applied magnetic field. The explanations adduced have shown that substances composed of atoms with resultant magnetic moments become *spontaneously magnetized in domains*. The agency causing the spontaneous alignment of neighbouring spins is the exchange interaction. The uniform spontaneous magnetization of macroscopic samples is prevented by the magnetostatic energy, which would be greater than the energy required to form domain walls.

In this section we consider what happens when magnetic materials are subjected to the action of an *externally applied magnetic field*.

At first we consider polycrystalline samples. It may be assumed that, initially, individual domains will be mostly magnetized in easy directions, resulting in no net magnetization over a sample. The application of a small magnetic field will have the effect of slightly moving domain walls, so that domains magnetized in a direction roughly parallel to the applied field gain in volume at the expense of domains magnetized in other directions.

This process is illustrated diagrammatically in Fig. 21.4(a) and (b), and corresponds to the initial segment of the magnetization curve of Fig. 21.5 marked *OA*. Such movements of the domain walls represent adjustments of the local energy balance in response to an externally applied field, and are *reversible*. The segment *AB* of the magnetization curve represents much more pronounced movements of domain walls resulting in a rapid increase of the net magnetization of the sample. A further increase of the magnetization can only be brought about by rotation of domains away from the easy directions (Fig. 21.4(c)). Although this process adds relatively little to the net magnetization it requires considerable force, as indicated by the relatively high applied fields required to complete it. Finally, the subdivision into domains disappears and the magnetization reaches a saturation value $M_s$. However, it does not follow that, at this stage, the spins are rigidly aligned with the applied field. At temperatures above the absolute zero, thermal lattice vibrations will cause random deviations of individual spins from the minimum energy position of alignment with the field. Consequently, the average component of magnetization along the field will vary with temperature, increasing to a maximum at $0\,°K$. This topic will be considered in greater detail in a later section.

For practical purposes, it is usual to represent the magnetization of a ferromagnetic sample by a graph of $B$ versus $H$. Its relation to the $M$ versus $H$

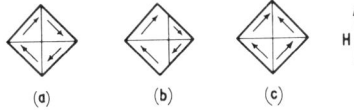

**Fig. 21.4.** Adjustments of ferromagnetic domains in the presence of an applied field $H$.

curve is given by the equation $B = \mu_0(H + M)$. From this, it follows that the saturation condition is represented by a straight line of slope $\mu_0$ and intercept $\mu_0 M_s$, as indicated in Fig. 21.6. In cases of actual ferrous metals, this straight line is not readily perceived because very high values of induction $B$ are reached for very low applied fields. Thus, in iron, saturation is reached at applied fields of about 250 A/m (3 oersted) with a value of $B = 1\cdot5$ T (15,000 Gauss).

It follows from the foregoing discussion that the susceptibility and hence the permeability of a ferromagnetic material vary rapidly with the applied field, as suggested by the dotted line in Fig. 21.6. The maximum value of the permeability is very high as compared with paramagnetic crystals. In iron, it is of the order of $10^4$ and in some special alloys it reaches $10^6$.

The magnetization curves described above apply to polycrystalline samples and represent the average behaviour of many crystallites oriented at random. The magnetization process in such samples can be expected to be the same in any direction, since the easy or hard axes of magnetization of individual crystallites will be evenly distributed over all directions of space.

The properties of single-crystal ferromagnetic samples are in general aniso-tropic. As an example, we now describe the magnetization of single crystals of iron which is body-centred cubic. Easy directions of magnetization lie along cube edges, e.g. [1 0 0]. The corresponding $M$ versus $H$ curve is illustrated in Fig. 21.7. Also shown are curves for face and body diagonals, which are hard directions of magnetization.

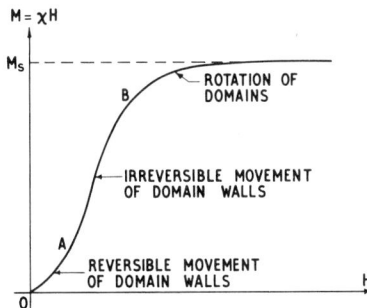

**Fig. 21.5.** Magnetization of a ferromagnetic sample.

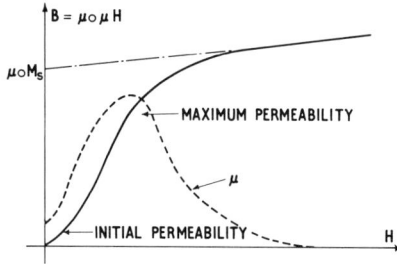

Fig. 21.6. $B/H$ curve of a ferromagnetic sample.

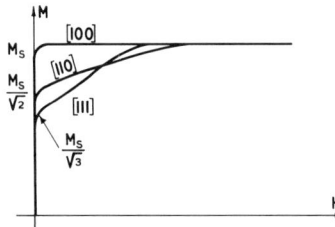

Fig. 21.7. Magnetization of a single crystal sample of iron.

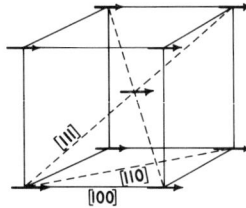

Fig. 21.8. Magnetization of the b.c.c. lattice.

To explain the main features of these curves we refer to Fig. 21.8, which represents the b.c.c. lattice with the magnetic moments of individual atoms aligned along a cube edge. If a field is applied to the sample along a face diagonal, [1 1 0], the spins at first assume the direction of easy magnetization shown. The component of this magnetization along the field is $M_s/\sqrt{2}$, which corresponds to the knee of the appropriate curve of Fig. 21.7. Beyond this point, the magnetization increases entirely as a result of rotation away from the easy direction, until, at a relatively high applied field, all the spins are aligned with the field. A similar argument applies when the field is applied along a body diagonal. Analogous results have also been obtained for other ferromagnetic crystals. For example, in nickel, which forms face-centred cubic crystals, the easy direction is along a body diagonal, [1 1 1], while in cobalt, which is hexagonal, the easy direction is along the $c$-axis, [0 0 1].

The process of magnetization requires energy. The energy is partly stored in the resulting magnetic field and is partly consumed in the process of shifting domain walls irreversibly. The first part of the expended energy is retrieved when the magnetic field is removed from the sample, while the latter part is dissipated in the form of heat in the sample.

It is assumed that the reader is familiar with the expression for the energy of magnetization

$$W = \int_0^{B_1} HdB. \tag{21.4.1}$$

This integral is represented by the area between the $B/H$ curve of Fig. 21.6 and the vertical axis. It is instructive to refer at this point to Fig. 21.7 and to reflect that, according to the above equation, much more energy is required to magnetize a crystal along a hard axis than along an easy axis.

### 21.5 Magnetic Hysteresis

The magnetization curve discussed in the preceding section was only traced for an increasing applied field. However, it was emphasized that the movement of domain walls is to a large extent irreversible. This implies that, when the applied magnetic field is decreased, the *magnetization of the sample will not decrease along the initial curve.* In fact, the magnetization, and the magnetic induction in the sample, will remain above the initial curve, the deviation varying substantially from material to material. This irreversible property of ferromagnetic materials is referred to as *hysteresis* and is of crucial importance in applications. In some situations, it is necessary to minimize hysteresis in order to reduce energy losses, while in others hysteresis is itself utilized.

The irreversibility of magnetization is best displayed by a *hysteresis loop*, which is traced when the applied field is varied from zero up to some maximum value, then back through zero to the corresponding negative value, etc. A typical example is shown in Fig. 21.9. The dotted line in the centre of the loop respresents the *initial magnetization curve* which is obtained when a field is applied to a sample which was completely unmagnetized. When the field is then reduced, the upper line leading to point $B_r$ is traced. $B_r$ is the *remanent flux* present in the sample after the applied field is reduced to zero. To remove the remanent magnetization, associated with the remanent flux, a negative field $H_c$ must be applied to the sample. This is called the *coercive force*. The remaining part of the loop is symmetrical with the first half and follows by analogy. A symmetrical and regular loop is only obtained after the sample under test has been cycled between $+H_m$ and $-H_m$ many times. The first few loops differ from each other in subtle ways, due to the fact that the state of the sample at any given applied field depends on its past history. The principal causes of irreversibility in ferromagnetic materials are anisotropy and "pinning" of domain walls.

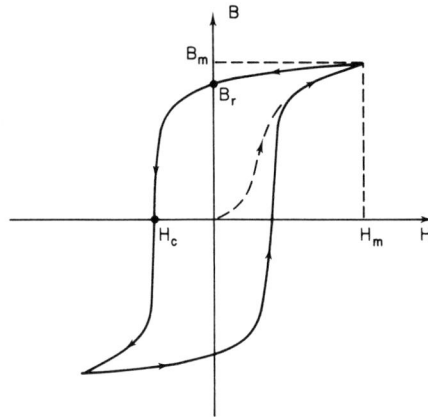

**Fig. 21.9.** Typical hysteresis loop.

Anisotropy forces can be used to account for most of the remanent flux. When the applied field is at its maximum, $H_m$, the magnetization directions of individual domains will be clustered within a relatively small solid angle about the direction of $H_m$. The domains whose easy axes happen to be parallel to $H_m$ will be magnetized in directions coincident with the field. On the other hand, domains whose hard axes are parallel to $H_m$ will be magnetized in directions deviating signficantly from the field. When the applied field is reduced to zero, the domain magnetizations will redistribute themselves in direction. Those magnetized in easy directions will retain their orientations parallel to the field, while those magnetized in hard directions will relax towards the easy directions. The net result will be a remanent magnetization and flux $B_r$ in the sample. It is possible to estimate the remanent magnetization of polycrystalline samples from a knowledge of the crystal structure and the type of magnetic anisotropy which accompanies it. However, such estimates are of small technological signficance, because materials used in practice frequently have additional anisotropic properties caused by particular methods of preparation, e.g., cold rolling. Hence precise values of the remanent magnetization for any specific application are usually obtained by measurement.

As can be seen from Fig. 21.9, it is necessary to apply a magnetic field in the reverse direction in order to reduce the magnetization below the remanent value. The necessity for such a coercive force can be understood in terms of "pinning" of domain walls. Let us consider the application of a field parallel to the spins on one side of a Bloch wall. The spins on the other side of the wall will be antiparallel to the field and their energy will be correspondingly much higher. The spins within the Bloch wall will have energies lying between these two extremes. On application of the field, the arrangement will adjust itself to reduce the extra energy, the adjustment taking the form of a slight rotation of successive spins within the wall. The rotation will have the effect of slightly displacing the Bloch wall to allow the low energy domain to gain in volume at the expense of the high energy domain. If the spins were completely free to rotate within the lattice, such adjustments would take place in response to a minimal applied field and effectively no coercive force would be required to reverse the magnetization. However, small non-magnetic inclusions in the sample will provide regions where segments of domain walls can reside at lower energy than elsewhere. Such segments will become effectively "anchored" in energy minima and it will require a finite external force to dislodge them. This is the *coercive force*. This pinning effect also contributes towards the retention of an established magnetization after the field is removed—that is, it is one of the contributory causes of remanent flux. The structural irregularities which cause pinning need not be strictly non-magnetic. A local change in magnetic properties due to chemical composition or crystallographic irregularities may be enough to cause pinning of domain walls.

It was pointed out at the close of the preceding section that some of the energy expended when a sample is magnetized is irrevocably lost in the form of heat imparted to the sample. This irreversible energy loss can be measured by tracing out a complete hysteresis loop of the material and is therefore called the *hysteresis loss*. Application of eqn (21.4.1) to successive segments of the hysteresis loop of Fig. 21.9 yields the result that the energy loss is equal to the area of the loop

$$W = \oint H dB. \tag{21.5.1}$$

If $B$ and $H$ are expressed in m.k.s. units the energy will be given in joules. If, moreover, the loop is traced out many times per second in an application where the material is subject to an alternating magnetic field, the power loss per second is given by the loop area multiplied by the frequency.

## 21.6  The Molecular or Exchange Field

It was explained in Section 21.4 that the saturation magnetization $M_s$, measured at any given temperature, is rather less than the value to be expected at $0\,^\circ$K, because of the thermal agitation imparted to individual spins by lattice vibrations. The variation of $M_s$ with temperature presents a problem of considerable importance in some applications, e.g., the use of ferrite materials in microwave devices. It is, therefore, essential to have some foreknowledge of the trend to be expected. To this end, we shall now obtain an estimate of the likely effects. In the present section we introduce the concept of the internal field, and in the next we shall consider the changes in the properties of a ferromagnetic material as a function of temperature.

It was argued in Section 21.1 that the exchange interaction between neighbouring magnetic atoms was of a short-range nature, diminishing exponentially with distance. In many cases, it is quite adequate to assume that it is of significant magnitude for nearest neighbours only. Making this assumption for the present purpose, we find that the expression $W = -2J\mathbf{S}_i \cdot \mathbf{S}_j$ need only be summed over a limited number of values of $j$ representing the nearest neighbours of a specific atom $i$. Assuming further, for simplicity, that the $i$-th atom has $z$ neighbours, all equidistant, the exchange energy of the group can be written in the form

$$W = -2J\mathbf{S}_i \cdot \sum_{j=1}^{z} \mathbf{S}_j = -2J\mathbf{S}_i \cdot (z\mathbf{S}_{jav}) \tag{21.6.1}$$

where $\mathbf{S}_{jav} = (\Sigma\mathbf{S}_j)/z$ is the mean value of the neighbouring spins. It should be noted that $\mathbf{S}_{jav}$ is approximately equal to the average of all spins in the sample. Moreover, in the presence of a saturating applied field, this average will vary somewhat with temperature due to lattice vibrations.

As the next step we transform eqn (21.6.1) to show that the exchange effect of the neighbours on the atomic spin $\mathbf{S}_i$ is equivalent to an *internal field* of induction $\mathbf{B}_i$. Since $\mathbf{S}_{jav}$ is the average of all spins in the sample, it can be written in the form

$$\mathbf{S}_{jav} = \frac{\mathbf{M}}{Ng\beta} \tag{21.6.2}$$

where $\mathbf{M}$ is the magnetization and $N$ is the number of spins per unit volume of the material. Using this expression, together with the value of the atomic magnetic moment $\mathbf{m} = g\beta\mathbf{S}_i$, the exchange energy, eqn (21.6.1) can be written in the form

$$W = -2J\left(\frac{\mathbf{m}}{g\beta}\right) \cdot \frac{z\mathbf{M}}{Ng\beta} = -\frac{2Jz}{N(g\beta)^2}\mathbf{m} \cdot \mathbf{M}$$

$$= -\mathbf{m} \cdot \lambda\mathbf{M} = -\mathbf{m} \cdot \mathbf{B}_i. \tag{21.6.3}$$

The above result shows that the exchange energy of the $i$-th atom is equivalent to the energy of its magnetic moment $\mathbf{m}$ immersed in a field of induction $\mathbf{B}_i$, the *internal field of molecular field* of the substance. The internal field is directly related to the magnetization **M** via the *molecular field constant* $\lambda = 2Jz/N(g\beta)^2$.

Theoretical and experimental estimates of the internal field lead to values of the order of $10^3$ T ($10^7$ Gauss). As this field is about 100 times greater than anything obtainable in the laboratory, it will be appreciated that the exchange forces which cause the spontaneous magnetization of domains are enormous.

The possibility that some sort of internal field causes spontaneous magnetization was first suggested by Weiss long before the advent of quantum theory. He thought that this field would be due to the dipolar interaction, but a comparison of theoretical estimates derived on this assumption, with experimental results, disclosed a discrepancy of several orders of magnitude. This discrepancy disappeared when estimates based on the exchange interaction were made by Heisenberg.

## 21.7 Temperature Variation of Magnetization

As suggested before, the properties of a ferromagnetic material vary with temperature. In particular, the saturation magnetization attainable in a given substance decreases with increasing temperature. Moreover, at sufficiently high temperatures, the ferromagnetic properties disappear altogether and the materials acquire paramagnetic properties. Under such conditions, the energy of lattice vibrations is greater than the exchange interaction and it dominates the properties of the material.

To gain some insight into these matters, which are of considerable practical importance, we adopt an approach which might appear rather crude at first sight. However, the results are found to agree remarkably closely with experiments. We go back to the principles of paramagnetism and enquire how the position is modified by the presence of an internal field of the type introduced in the preceding section.

The *total field* acting on a magnetic moment now has the form

$$\mathbf{B} = \mathbf{B}_0 + \mathbf{B}_i = \mathbf{B}_0 + \lambda \mathbf{M} \qquad (21.7.1)$$

where $\mathbf{B}_0$ is the externally applied field, while $\mathbf{B}_i$ is the internal field. At this stage we confine our attention to conditions at high temperatures, when **M** can be expected to be small compared with saturation magnetization, due to disturbances caused by lattice vibrations. We further assume that **M** is related to the *total field* through a susceptibility obeying Curie's law as in paramagnetism, eqn (20.3.15). Hence we can write

$$\mathbf{M} = \chi'\mathbf{H} = \chi'\frac{\mathbf{B}}{\mu_0} = \frac{C}{\mu_0 T}(\mathbf{B}_0 + \lambda\mathbf{M}). \qquad (21.7.2)$$

Solving the above equation for **M** in terms of the *externally applied field* we arrive at the result

$$\mathbf{M} = \frac{C}{T - \lambda C/\mu_0} \frac{\mathbf{B_0}}{\mu_0} = \frac{C}{T - \lambda C/\mu_0} \mathbf{H_0}.$$  (21.7.3)

Hence the susceptibility which relates the magnetization to the *external field* has the form

$$\chi' = \frac{C}{T - \lambda C/\mu_0} = \frac{C}{T - \theta}.$$  (21.7.4)

This result is usually referred to as the *Curie-Weiss law* and $\theta = \lambda C/\mu_0$ is called the *Weiss constant*.

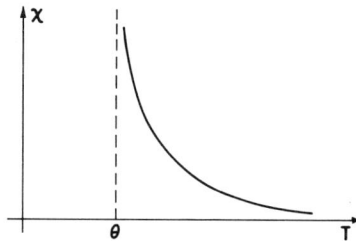

Fig. 21.10. The Curie-Weiss law of ferromagnetic materials.

According to eqn (21.7.4) a ferromagnet has the properties of a paramagnetic material at temperatures $T > \theta$. The Curie-Weiss law is shown diagrammatically in Fig. 21.10. It differs from Curie's law for paramagnetic crystals in that the vertical asymptote is at $T = \theta$ instead of $T = 0\,^\circ$K. The foregoing predictions are well substantiated by experiments. The Weiss constant $\theta$, which can be estimated theoretically from the other constants which enter into it, is of the order of $1000\,^\circ$K for the common ferrous metals. At temperatures in excess of $\theta$, these materials display a constant and quite small susceptibility, similar to the paramagnetic crystals discussed in the preceding chapter.

Let us now turn our attention to temperatures below $\theta$. At $T = \theta$ the law derived above clearly breaks down, since $\chi \to \infty$. This is interpreted as the onset of spontaneous magnetization caused by exchange forces which now dominate the thermal energy of lattice vibrations. Despite this fact, we continue to apply the theory of paramagnetism in the hope that the results will justify the procedure. Equation (20.3.12) is rewritten for the present purpose in the form

$$M = M_{so} L_j(x) = Njg\beta L_j(x)$$

$$x = jg\beta B/kT$$  (21.7.5)

where it should be remembered that $j$ is the angular momentum quantum number, in this chapter denoted by $S$. The coefficient $Njg\beta$ represents the maximum magnetization which the material can have at $0\,°K$. In the present context, it is therefore denoted by $M_{so}$. The symbol $M$ will be presently identified with the saturation magnetization of a ferromagnetic material at temperatures below $\theta$, as discussed in Section 21.4. Using the foregoing symbols we can write

$$\frac{M}{M_{so}} = L_j(x) \tag{21.7.6}$$

$$x = \frac{M_{so}}{NkT}\,B = \frac{M_{so}}{NkT}\cdot(B_0 + \lambda M).$$

The induction $B$ is assumed to include both the applied field and the molecular field, this being again the principal difference in the above equations as between the case of paramagnetism and the present situation. Solving the last expression for $M/M_{so}$, we find

$$\frac{M}{M_{so}} = \frac{NkT}{\lambda M_{so}^2}\,x - \frac{B_0}{\lambda M_{so}}. \tag{21.7.7}$$

Equations (21.7.6) and (21.7.7) can be treated as a pair of simultaneous equations in the unknowns $M/M_{so}$ and $x$. In principle, one could eliminate $x$ and obtain an expression for $M$ as a function of temperature. However, because of the complicated nature of the Brillouin function, $L_j(x)$, this cannot be done analytically and a graphical method must be used. When plotted in $M/M_{so}$ versus $x$ coordinates, the above equations represent the Brillouin function and a straight line respectively. The corresponding graphs are shown diagrammatically in Fig. 21.11, where the straight line occupies several possible positions. At first, we consider the case of no applied field $B_0 = 0$. The straight line representing eqn (21.7.7) passes through the origin and its *slope decreases with decreasing temperature T*.

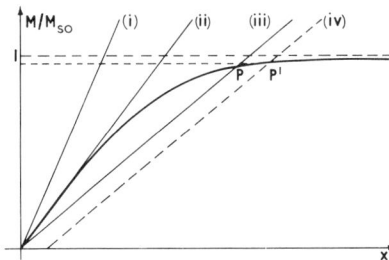

Fig. 21.11. Graphical solution of eqns (21.7.6) and (21.7.7).

(i) In this case, the slope of the line is greater than the slope of the Brillouin function at the origin. The only solution is a state of zero magnetization $M$. This case clearly corresponds to the paramagnetic state of the material in the absence of an applied field.

(ii) The line is tangent to the Brillouin function at the origin and represents a transitional condition which we characterize by the temperature $T_c$. Equating the slope of the line to the slope of $L_j(x)$ at the origin, we find

$$\frac{NkT_c}{\lambda M_{so}^2} = \frac{1}{3}\left(1 + \frac{1}{j}\right)$$

$$T_c = \frac{1}{3}\left(1 + \frac{1}{j}\right)\lambda\frac{N(jg\beta)^2}{k} = \frac{\lambda C}{\mu_0} \qquad (21.7.8)$$

where eqn (20.3.15) has been used to obtain the last expression. $T_c$ is called the Curie temperature. Comparison with eqn (21.7.4) shows it to be identical with the Weiss constant $\theta$ according to the above derivation. However, in practice, it is necessary to make a distinction between these two temperatures in a manner to be explained presently.

(iii) At a temperature below $T_c$ the line intersects the Brillouin function at two points: the origin and point $P$. From what we know about exchange forces and spontaneous magnetization, the only possible stable state corresponds to point $P$ and the finite magnetization indicated on the vertical axis. As the temperature is lowered from the Curie temperature to $0°K$, the point $P$ traces out the Brillouin function. The corresponding values of normalized magnetization can be plotted separately against temperature yielding the curve of Fig. 21.12.

(iv) The effect of an applied field $B_0$ is to displace the straight line slightly to the right, resulting in a magnetization which corresponds to the point $P'$. This increase in magnetization is minute, because the applied fields of the kind obtainable in the laboratory are orders of magnitude smaller than the molecular field $\lambda M$. Hence the curve of Fig. 21.12 can be considered as the *variation of the*

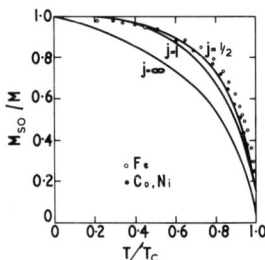

Fig. 21.12. Variation of spontaneous magnetization with temperature. Experimental points after F. Tyler (1931), *Phil. Mag.* 11, 596.

*saturation magnetization with temperature* in a ferromagnetic material, and the symbol $M$ used above can be identified with $M_s$ of Section 21.4.

The graphs of Fig. 21.12 are plotted in normalized coordinates: they are, therefore, in the nature of universal curves applicable to all ferromagnetic materials. As it may not be clear from the foregoing arguments that such a universal relationship does exist, we proceed to derive it now. Combining eqn (21.7.7) $(B_0 = 0)$ with eqn (21.7.8), we find

$$\frac{M}{M_{so}} = \frac{T}{T_c}\left(\frac{j+1}{3j}\right)x. \qquad (21.7.9)$$

To obtain a functional relationship between $M/M_{so}$ and $T/T_c$ it is once again possible in principle to eliminate $x$ with the help of eqn (21.7.6). Hence an equation of the following general form holds

$$\frac{M}{M_{so}} = f\left(\frac{T}{T_c}\right). \qquad (21.7.10)$$

The quantum number $j$ is implicit in the above relation as a parameter. The experimental points shown in Fig. 21.12 lie closer to the curve corresponding to $j = \frac{1}{2}$ than to any other. This is generally accepted as one of the indications that ferromagnetism is due to the spin of electrons rather than to their orbital motions.

It was mentioned before that, in practice, a distinction must be made between the Curie temperature $T_c$ and the Weiss constant $\theta$, despite eqn (21.7.8). Experimentally, $T_c$ is defined as the temperature at which the spontaneous magnetization drops to zero, and $\theta$ as the temperature above which a fixed susceptibility can be measured, following the Curie-Weiss law. It is found that these two types of behaviour are separated by a small temperature interval listed in Table 21.1. In this interval, the susceptibility deviates from the law $1/\chi \sim T - \theta$.

Table 21.1
Magnetic data for some ferromagnetic materials

| Material | $T_c$ (°K) | $\theta$ (°K) | Saturation moment at 0 °K | |
| --- | --- | --- | --- | --- |
| | | | ampere-metre$^2$ per kg | Bohr magnetons per atom |
| Fe | 1043 | 1093 | 221·7 | 2·22 |
| Co | 1403 | 1428 | 162·6 | 1·715 |
| Ni | 631 | 650 | 57·6 | 0·606 |
| Gd | 289 | 302 | | 7·10 |

The last column of Table 21.1 gives the *effective number of Bohr magnetons per atom* of the material. This quantity is obtained by dividing *the experimental* saturation magnetization at $0\,°K$, $M_{so}$, by the number $N$ of atoms per unit volume. Contrary to what could be expected on the basis of the foregoing theory, this number is *not integral*. These results should be taken as an indication that the theory is only approximate. In particular, the implicit assumption that each atomic magnetic moment is localized on a lattice site is subject to strong reservations in the case of metallic ferromagnets, in which the conduction electrons are free to migrate throughout a macroscopic sample. In view of this, it appears that the energy band structure of solids must be taken into account in any explanation of ferromagnetism. However, the foregoing theory helps greatly in coordinating experimental findings and utilizing them in engineering designs.

## Appendix to Chapter 21—The Exchange Interaction

In addition to the classical electrostatic repulsion, two electrons in close proximity will exert forces on each other which have a purely quantum mechanical basis. The origin of these forces is to be found in the interchangeability of the individual electron states discussed in Section 19.1. For this reason the term *exchange interaction* is used to refer to these forces.

To explain the exchange interaction we consider two electrons in close proximity. They may be part of the electron configuration of one atom, or they may be the outermost unpaired electrons of two separate atoms, sufficiently close together for their wave functions to overlap. The second point of view applies to ferromagnetic materials. In either case the electron states are subject to the exclusion principle since they form part of one dynamical system. In consequence the complete wave function of the two electrons must be antisymmetric according to the principles of Section 19.1.

As a first step we proceed to set up the antisymmetric wave functions of the two electrons in the ground state. To satisfy the requirements of antisymmetry we shall use both the spatial wave functions and the spin states. The spatial wave functions are denoted by the symbols $u_i(\mathbf{r}_1)$ and $u_j(\mathbf{r}_2)$, where $i$ and $j$ label individual states while $\mathbf{r}_1$ and $\mathbf{r}_2$ label the first and second electron. Individual spin states are denoted in the present context by the symbols $\alpha$, corresponding to the quantum number $m_s = \frac{1}{2}$, and $\beta$, corresponding to the quantum number $m_s = -\frac{1}{2}$. Thus the symbol $\alpha(1)$ indicates the electron 1 is in the spin state $m_s = \frac{1}{2}$. The symbols $\alpha(2)$, $\beta(1)$ and $\beta(2)$ have analogous meanings. The effect of the spin operators on these states is as laid down in Section 14.4, e.g. $S_z\alpha(2) = m_s\hbar\alpha(2) = \frac{1}{2}\hbar\alpha(2)$ or $S^2\beta(1) = \hbar^2 S(S + 1)\beta(1) = \frac{3}{4}\hbar^2\beta(1)$.

Let us now consider the following states.

$$\psi_1 = \frac{1}{\sqrt{2}}\{u_i(\mathbf{r}_1)u_j(\mathbf{r}_2) + u_i(\mathbf{r}_2)u_j(\mathbf{r}_1)\}\{\alpha(1)\beta(2) - \alpha(2)\beta(1)\}$$

$$\left.\begin{array}{c}\psi_2 \\ \psi_3 \\ \psi_4\end{array}\right\} = \frac{1}{\sqrt{2}}\{u_i(\mathbf{r}_1)u_j(\mathbf{r}_2) - u_i(\mathbf{r}_2)u_j(\mathbf{r}_1)\}\left\{\begin{array}{c}\alpha(1)\alpha(2) \\ \beta(1)\beta(2) \\ \alpha(1)\beta(2) + \alpha(2)\beta(1).\end{array}\right. \qquad (1)$$

In the first place the reader will readily check that all four states are antisymmetric in the two electrons, by interchanging the labels 1 and 2 and verifying that each state changes sign. Secondly by operating on each state with the operator $S^2$ it will be found that the state $\psi_1$ has a zero resultant spin, $S = 0$, while the states $\psi_2$, $\psi_3$ and $\psi_4$ have a resultant spin given by $S = 1$. Thus the individual spins of state $\psi_1$ are antiparallel while those of states $\psi_2$, $\psi_3$, $\psi_4$ are parallel.

We shall now show that the arrangement of the two spins is associated with a difference in the energy level of the corresponding states, by considering the Coulomb interaction $e^2/|\mathbf{r}_1 - \mathbf{r}_2|$ between the electrons. The matrix elements of this interaction with respect to the states listed above are given by the following expressions.

$$E_1 = \int \psi_1^* \frac{e^2}{|\mathbf{r}_1 - \mathbf{r}_2|} \psi_1 \, d\tau$$

$$E_{2,3,4} = \int \psi_{2,3,4}^* \frac{e^2}{|\mathbf{r}_1 - \mathbf{r}_2|} \psi_{2,3,4} \, d\tau \qquad (2)$$

where $E_n$ denotes the average value of the Coulomb energy and $d\tau$ is a volume element in the coordinate space of both electrons. Since the spin states remain unaffected by the operator $e^2/|\mathbf{r}_1 - \mathbf{r}_2|$ they drop out of the integral expressions leaving only the spatial wave functions to be considered. As a result the energy is found to be identical for the three states $\psi_{2,3,4}$ and the following difference can be formed with the help of eqns (1).

$$E_1 - E_{2,3,4} = 2\int u_i^*(\mathbf{r}_1)u_j^*(\mathbf{r}_2) \frac{e^2}{|\mathbf{r}_1 - \mathbf{r}_2|} u_i(\mathbf{r}_2)u_j(\mathbf{r}_1) \, d\tau. \qquad (3)$$

The form of this integral expression should be carefully noted. A straightforward electrostatic interaction between two electrons would be related to their position probability distribution, or charge density, which is given by the expression $|u_i(\mathbf{r}_1)u_j(\mathbf{r}_2)|^2 = u_i^*(\mathbf{r}_1)u_j^*(\mathbf{r}_2)u_i(\mathbf{r}_1)u_j(\mathbf{r}_2)$. However, in eqn (3) the electrons have been *exchanged* between the corresponding wave functions. For this reason the difference in enrgy between the state $\psi_1$ on the one hand, and the states $\psi_{2,3,4}$ on the other, is called the *exchange energy*. It is clearly connected with the parallel and antiparallel arrangement of the spins and can be written in terms of the spin vectors as follows

$$W_e = -2J\mathbf{S}_i \cdot \mathbf{S}_j \tag{4}$$

where $J$ is come constant still to be determined.

Equation (4) yields a maximum value of $J/2$ when the spins are antialigned and a minimum value of $-J/2$ when they are aligned. If this difference in the exchange energy of the two types of state is equated to eqn (3) we can write

$$E_1 - E_{2,3,4} = J. \tag{5}$$

Thus the quantity $J$ can be identified with the integral expression in eqn (3), including the factor 2. $J$ is usually referred to as the *exchanged integral* or *exchange constant*. By virtue of the exponential factor included in the spatial wave functions (see Section 13.2) the value of the exchange integral is very sensitive to the spacing of the two electrons as discussed at length in Section 22.1.

*Chapter 22*

# Ferrimagnetism

The classification of materials given in Section 20.1 envisaged one category whose properties are due to the exchange interaction of closely packed moments. Such materials were called ferromagnetic. However, in Section 21.1 it was pointed out that the exchange interaction could lead to either alignment or antialignment of neighbouring spins, the latter case being referred to as antiferromagnetism. In that discussion it was assumed that all the spins were of the same kind so that an antiferromagnet would have no spontaneous magnetization. In fact there are materials in which the neighbouring spins are not of the same magnitude. Moreover, they may be arranged in various directions, not necessarily antiparallel, and their interaction may be more complicated than the direct exchange envisaged so far. In any case a net magnetization usually appears, resulting in ferromagnetic properties of the type described in the preceding chapter. However, it is frequently necessary to draw a distinction between materials in which neighbouring spins are aligned or antialigned, and the latter are frequently called *ferrimagnetic*. They form the main subject matter of the present chapter although antiferromagnetism is briefly dealt with as an introduction.

## 22.1 Antiferromagnetism

In the preceding chapter materials having a positive exchange constant $J$ have been discussed. In such cases the exchange energy $W = -2J\mathbf{S}_i \cdot \mathbf{S}_j$ assumes a minimum value when the spins are aligned, resulting in a tendency towards spontaneous magnetization. The exchange constant can, however, assume negative values, in which case the exchange energy of the lattice of spins becomes a minimum when neighbouring magnetic moments are *antialigned*, and the material is *antiferromagnetic*.

Antiferromagnetic materials may be visualized as consisting of two *interlocking sublattices*, $A$ and $B$, each spontaneously magnetized in one direction. The possibility is illustrated in Fig. 22.1 on the example of a simple cubic lattice.

If the magnetic moments of the opposing *dipolar lattices* are equal, no net magnetization appears. In such circumstances it is not easy to detect by magnetostatic means whether there are any magnetic moments in the material and other experimental methods must be adopted, e.g. neutron diffraction.

An important indication of the antiferromagnetic nature of a substance is provided by the analogue of the Curie temperature. If the antiferromagnetic

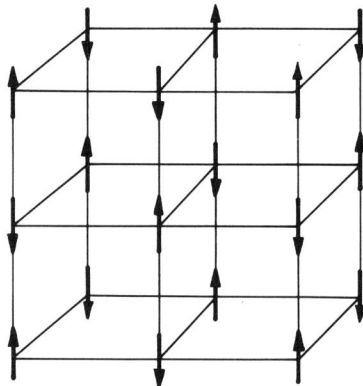

**Fig. 22.1.** Antiferromagnetic arrangement of magnetic moments on two opposing sublattices.

state really exists then at some temperature the exchange interaction should be broken up by thermal vibrations of the lattice and paramagnetism should appear, as it does in ferromagnets above the Curie temperature. This is indeed observed above a point referred to as the *Néel temperature.*

Theoretical estimates of the exchange energy indicate that it is a very sensitive function of the interatomic spacing. For short distances between atoms it is negative, resulting in antiferromagnetism. For greater distances it is positive, leading to ferromagnetism. However, due to the exponential factor contained in the electron wave functions, the exchange energy decreases rapidly to zero for greater atomic spacings. Hence the incidence of ferromagnetism among the *elements* is quite rare. The incidence of ferromagnetism among *compounds* is due to a more elaborate interaction, called *indirect exchange or superexchange,* which will be discussed in the next section. The situation is illustrated diagrammatically in Fig. 22.2. In the antiferromagnetic range one finds the elements Ti, Pt, Cr, Mn together with many others. The ferromagnetic range is occupied by the typical ferrous metals, Co, Ni, Fe and some of their alloys. The ferromagnetic region extends roughly over lattice spacings of 2·5 Å to 3·5 Å. Above that the exchange interaction becomes insignificant. Around the transition point $X$ the exchange integral varies rapidly with interatomic distance and small changes of the latter, due to temperature or alloying, may change the character of a material.

Since antiferromagnetism itself has little practical importance, its discussion will not be continued beyond this point. Instead we shall now go on to introduce the indirect exchange interaction which forms the basis of ferrimagnetism.

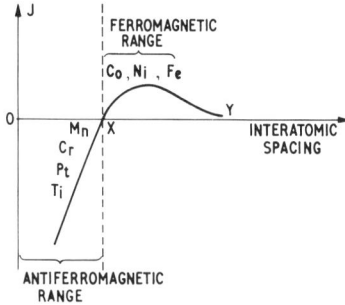

Fig. 22.2. Diagrammatic variation of exchange constant $J$ with interatomic spacing.

## 22.2 The Indirect Exchange Interaction

The basis of the exchange interaction, assumed so far, is the possibility of two electrons, belonging to neighbouring atoms, interchanging their identity. If the spins point in the same direction ferromagnetism follows. If the spins point in opposite directions antiferromagnetism results. Fig. 22.2 shows that for materials with small lattice spacings the latter is to be expected.

In the light of these observations the properties of many substances would be difficult to understand. Neutron diffraction studies indicate, for example, that a strong antiferromagnetic interaction binds *not the nearest neighbours* but the *next* nearest neighbours in MnO. The latter are $Mn^{2+}$ ions separated by 4·4 Å along a cube edge, and a reference to Fig. 22.2, and the discussion accompanying it, indicates that a very weak ferromagnetic interaction is to be expected.

To understand the position various models have been proposed which envisage that *non-magnetic oxygen ions*, $O^{2-}$, serve as intermediaries in an *indirect or superexchange interaction* between two magnetic $Mn^{2+}$ ions. The simplest model involves 4 electrons which can be arranged between the participating ions in two ways shown in Fig. 22.3(a) and (b). It is emphasized that both configurations are compatible with the ground state of the entire arrangement and being at the same energy level are equally probable. The single unpaired $d$-electrons of the manganese ions have been nevertheless effectively exchanged. It is possible to visualize that the exchange involves the oxygen ion through the formation of excited states. The latter are sufficiently low on the energy scale in relation to the ground state to have a significant probability at normal temperatures.

The excited states which can form a transition between the arrangements of Fig. 22.3(a) and (b) are illustrated in Fig. 22.3(c) and (d). One of the $p$-electrons is first transferred to the neighbouring Mn ion as suggested in (c). This is relatively probable because the $p$-orbital wave functions extend along the axis of the $Mn^{2+}-O^{2-}-Mn^{2+}$ arrangement as indicated in the figure. The next stage of the operation is illustrated in Fig. 22.3(d). This is another excited state which

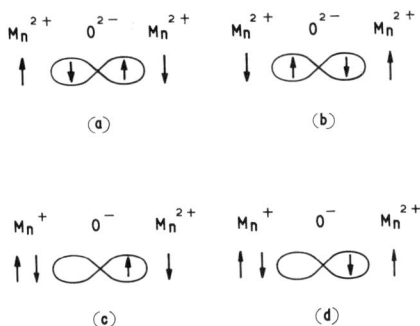

Fig. 22.3. Diagrammatic representation of the indirect exchange interaction involving four electrons.

differs from the preceding states in that the remaining $p$-electron of the $O^-$ ion becomes exchanged with the $d$-electron of the right hand $Mn^{2+}$ ion. The final electron transfer takes place between the left hand $Mn^+$ ion and the $O^-$ ion leading to the alternative ground state (b).

The foregoing sequence of operations demonstrates the possibility of an indirect antiferromagnetic exchange interaction between relatively distant magnetic ions. Other, more complicated, possibilities have been investigated theoretically in connection with other compounds and have been found to be consistent with observed data. However, these matters will not be pursued any further in the present context, since the object is only to show that the properties of the rather complicated compounds discussed in the following sections are not inconsistent with the discussion of Section 22.1. It should be noted however that the superexchange interaction illustrated above on an antiferromagnetic compound can give rise to ferromagnetic properties in the case of sublattices having unequal magnetic moments.

The simplest case arises when two dipolar sublattices have opposing but unequal moments. A net magnetization appears which at saturation equals the difference of the individual sublattice magnetizations as shown symbolically in Fig. 22.4(a). Another possibility is suggested in Fig. 22.4(b). Here one sublattice contains two distinct but equal moments while the other sublattice contains one opposing moment of the same magnitude. Figure 22.4(c) illustrates the possibility of two unequal moments on the first sublattice and one opposing moment on the other. In all cases resultant magnetization exists.

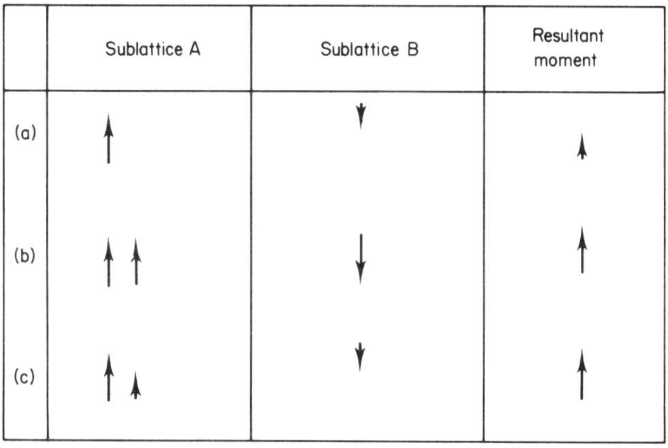

| | Sublattice A | Sublattice B | Resultant moment |
|---|---|---|---|
| (a) | ↑ | ↑ | ↑ |
| (b) | ↑ ↑ | ↓ | ↑ |
| (c) | ↑ ↑ | ↑ | ↑ |

Fig. 22.4. Resultant magnetic moment in a ferrimagnetic material.

The saturation magnetization of materials having this type of structure must be expected to be smaller than that of the ferrous metals for two reasons. Only some of the atoms making up the compound have resultant magnetic moments. Furthermore, only a fraction of these moments will contribute towards a net magnetization.

## 22.3  Ferrites and Garnets

Among magnetic materials whose properties are based on the principles outlined in the preceding section the *ferrites and garnets* are the best known and have found most widespread applications. Perhaps the most important techno-logical property of these substances is the fact that they are insulators. This permits their application in situations where the highly conductive ferrous metals cannot be used. There is nothing logical or systematic about their names; ferrites are a group of chemical compounds, while garnet is a type of crystal structure. The term *ferrites* is used as a generic name to describe insulating magnetic materials particularly those applicable in microwave devices.

Ferrites are compounds of the form $PFe_2O_4$ where P is another metal ion, e.g. Ni, Co, Mn. In the special case of magnetite, $Fe_3O_4$, this additional metal ion is replaced by iron itself. Ferrites form crystals of the *spinel* type. This is a cubic system in which each unit cell contains eight chemical formula units and consequently includes 32 oxygen ions. A cube edge is about 8 Å long.

The magnetic properties of ferrites originate in the metallic ions P and Fe which occupy two distinct crystallographic sites, separated by oxygen ions. Detailed crystallographic studies indicate that there are several possible arrangements involving two metal ions of either type and one oxygen ion situated approximately between them. However, the distances and angles are such that only one of them is favourable to indirect exchange interaction. As it happens this is enough to provide an antiparallel spin configuration on two sublattices, resulting in net magnetization. The saturation magnetization is only a fraction of what is found in ferrous metals, due to the reasons stated at the close of the preceding section.

The *garnet* is a crystal structure shared by many compounds of the form $P_3Q_2R_3O_{12}$, where P, Q, and R are positive ions, frequently, but not necessarily metallic. The basic symmetry of the structure is cubic. A unit cell contains eight formula units, including 96 oxygen ions and measures about 12·5 Å along its edge.

Only some of the garnets have magnetic properties, the most important being *yttrium iron garnet* $Y_3Fe_2Fe_3O_{12}$ or YIG for short. Yttrium enters the compound in the form $Y^{3+}$ and has no resultant magnetic moments. In general there are three distinct crystallographic sites in garnets which can be occupied by metallic ions, but in YIG there are only two, occupied by iron ions. Their configuration indicates a strong superexchange interaction resulting in two antiparallel sublattices with a resultant magnetization.

An extensive family of magnetic garnets is obtained upon replacement of the $Y^{3+}$ ion by trivalent ions of the smaller rare earth atoms (Sm to Lu), the remaining rare earth atoms (La to Pm) being too large to be accommodated within the lattice. The *rare earth iron garnets* have magnetic ions on three distinct sites, resulting in more complicated superexchange interactions. They all exhibit net magnetization.

The ferrites and garnets are usually prepared by growing single crystals or by a sintering process. Crystal growing is a laborious and expensive process and is only used when samples having special properties, e.g. narrow resonance linewidth, are required. *Sintering* is accomplished by mixing simple compounds and elements in powder form and then heating them under pressure. Reactions which take place at the elevated temperatures yield the desired ferrite in the form of a hard ceramic. This can be cut and ground to any desired shape using diamond tools.

The variation of saturation magnetization with temperature in some ferrites exhibits a striking peculiarity referred to as a *compensation point*. At this point the measured magnetization dips to zero and then increases again, contrary to the monotonic behaviour of the ferrous metals, shown in Fig. 21.12. The phenomenon is readily understood in terms of the magnetization of sublattices. Figure 22.5(a) depicts the variation of saturation magnetization with tempera- ture on two sublattices $A$ and $B$, and the resultant, as measured on an actual sample. In this case the behaviour is similar to ferromagnetic material. Figure 22.5(b) illustrates the appearance of a compensation point which is caused by the different shape of the graph of $M_s(T)$ for the individual sublattices. The *measured* resultant magnetization is given by the *modulus* of the difference since the application of an external magnetic field will always orientate the resultant in its own direction.

The graphs of Fig. 22.5 represent only the simplest cases. In particular they assume that there are only two magnetized sublattices and that their Curie temperatures are the same. The rare earth garnets are beyond these simplifi- cations in that they have three magnetic sublattices, with differing Curie temperatures. Consequently their $M_s(T)$ graphs exhibit a variety of unusual features.

The saturation magnetization and its variation with temperature are important design parameters in many ferrite devices. The foregoing discussion is intended to make the reader aware of the general trends to be expected. For quantitative data reference should be made to specialist texts.

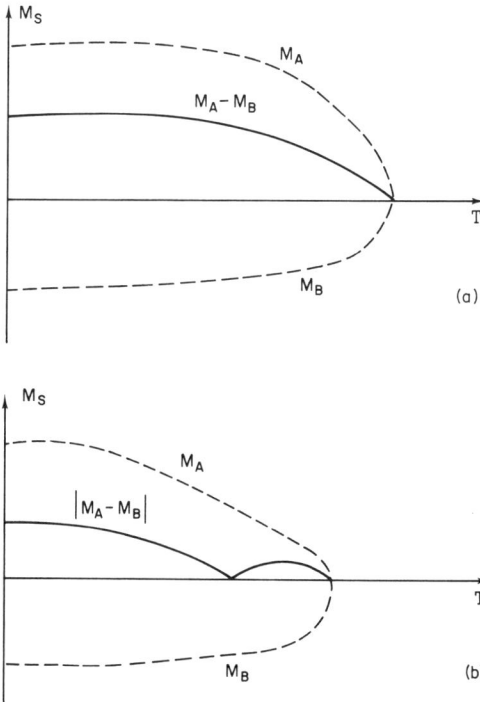

**Fig. 22.5.** Resultant magnetization of two sublattices.

*Chapter 23*

# Ferromagnetic Resonance

### 23.1 Larmor Precession

The concepts of quantized angular momentum and the associated paramagnetic energy levels first discussed in Chapters 14 and 15 are consistent with a classical view of precession of the angular momentum vector about an applied magnetic field **B**. The phenomenon is usually referred to as *Larmor precession.* As it is particularly helpful when dealing with the problem of ferromagnetic resonance we start the present chapter by discussing it briefly.

The effect of a magnetic field of induction **B** on a magnetic moment **m**, is to apply a torque given by the well-known expression

$$\mathbf{T} = \mathbf{m} \times \mathbf{B}. \tag{23.1.1}$$

If the magnetic moment is due to an atomic electron orbit or spin, it is related to the angular momentum through eqn (14.2.3). Hence the torque can be written in the form

$$\mathbf{T} = -\gamma \mathbf{L} \times \mathbf{B} \tag{23.1.2}$$

where $\gamma$ is a *positive* constant.

The application of a torque to an angular momentum vector causes a rate of change according to the classical equation of motion

$$\frac{d\mathbf{L}}{dt} = \dot{\mathbf{L}} = \mathbf{T} = -\gamma \mathbf{L} \times \mathbf{B}. \tag{23.1.3}$$

Assuming that the magnetic field is applied in the $z$-direction, $\mathbf{B} = \hat{k} B_z$, the above equation has the following components

$$\dot{L}_x = -\gamma B_z L_y$$

$$\dot{L}_y = \gamma B_z L_x$$

$$\dot{L}_z = 0. \tag{23.1.4}$$

348

The solution of the above equations is best visualized with the help of Fig. 23.1 which depicts the relative configuration of the various vectors.

According to the last of eqns (23.1.4) the $z$-component of the angular momentum is constant. Hence the orientation of the vector $\mathbf{L}$ is confined to the surface of a cone, consistent with precessional motion about the $z$-axis. This is confirmed by solving the first two of eqns (23.1.4) which is done by

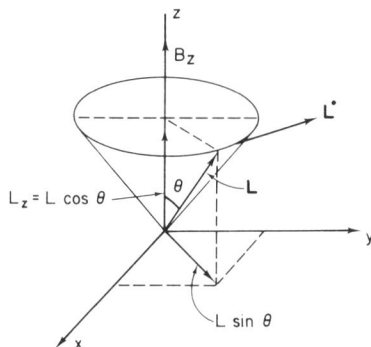

Fig. 23.1. Precession of the angular momentum about a magnetic field.

differentiating them with respect to time, and eliminating the components $L_x$ and $L_y$ in turn. The results have the form

$$\left.\begin{aligned} L_x &= L \sin\theta \cos(\omega_0 t + \epsilon) \\ L_y &= L \sin\theta \sin(\omega_0 t + \epsilon) \end{aligned}\right\} \quad \omega_0 = \gamma B_z$$
$$L_z = L \cos\theta \qquad\qquad\qquad\qquad\qquad (23.1.5)$$

where $\omega_0 = \gamma B_z$ is the precession frequency or *Larmor frequency*. $\theta$ and $\epsilon$ are constants of integration. $\epsilon$ is a phase angle to be determined from any initial conditions, while $\theta$ has the geometrical significance shown in Fig. 23.1. The precessional nature of the motion is further confirmed by noting the direction of the vector $\dot{\mathbf{L}}$. It can be seen directly from eqn (23.1.3) that this is normal to both $\mathbf{L}$ and $\mathbf{B}$. The magnetic moment vector $\mathbf{m}$ describes similar precessional motion, but its direction is opposite to $\mathbf{L}$.

Let us now compare the foregoing results with the properties of angular momentum established on the basis of quantum mechanics in Chapter 14. The $z$-component $L_z$, as given by the diagonal matrix element, is indeed constant in any given eigenstate, and it has the value $m\hbar$. From state to state $L_z$ varies in units of $\hbar$, reflecting the discrete orientation of $\mathbf{L}$ relative to $B_z$. The diagonal matrix elements of $L_x$ and $L_y$ vanish (see examples on Chapter 14) indicating that the average values of these components are zero. This fact is consistent with the steady rotation of the components $L_x$ and $L_y$ about the $z$-axis, yielding a zero average over a length of time.

So far a state of steady precession of the angular momentum in a d.c. magnetic field has been described. A situation of the greatest practical importance arises when an a.c. magnetic field is also present. This question has been dealt with to some extent in Section 20.4 on the basis of quantum mechanics. There it was stated that a.c. fields cause transitions between the energy levels of the quantized values of $L_z$, and in the process quanta of radiation are absorbed, provided the waves are of the right frequency.

The conclusion that energy will be transferred from an a.c. field to the precessing system can be arrived at on the basis of the classical model. A field, circularly polarized in the $xy$-plane, rotating in the direction of $\dot{\mathbf{L}}$, will have the effect of increasing the *angle of precession* $\theta$, by virtue of the additional torque applied to the magnetic moment vector. Provided the frequency of the a.c. field equals the Larmor frequency $\omega_0$, the kinetic energy of the precessing system will be increased. Under such conditions a phenomenon of *forced resonance at the frequency* $\omega_0$ will be observed and energy will be absorbed from the a.c. field. It should be noted that if the sense of polarization is opposite to the direction of precession no interaction will take place. A linearly polarized a.c. field can be decomposed into two contrarotating circularly polarized waves, one of which can cause interaction.

The precessional motion visualized above has been assumed to be free of dissipation due to any internal loss mechanism. Although in actual cases of ferromagnetic resonance this is by no means the case, the idealized model is a useful guide to many phenomena and their applications.

The precession frequency $\omega_0$ lies in the microwave range for d.c. fields of the order of 1 T, as the reader will verify on substitution of typical values in eqn (23.1.5), assuming a simple electron spin or orbit. Some nuclear particles also have a resultant spin angular momentum and magnetic moment. However, their gyromagnetic ratios may be orders of magnitude smaller, resulting in a much lower precession frequency. For example proton resonance appears at radio frequencies, say $10^5$ Hz.

## 23.2 Ferromagnetic Resonance

Ferromagnetic resonance is more conveniently discussed in terms of the magnetic moment and magnetization instead of the angular momentum. For this reason we rewrite eqn (23.1.3) in terms of the magnetic moment by multiplying both sides by $\gamma$ and remembering that $\mathbf{m} = -\gamma \mathbf{L}$.

$$\frac{d\mathbf{m}}{dt} = -\gamma \mathbf{m} \times \mathbf{B}. \qquad (23.2.1)$$

To adapt this equation to the description of spins in a ferromagnetic substance we recall once again that the latter are closely coupled by the exchange or superexchange interaction. Hence the assembly of spins moves in unison, which means that if one spin precesses about a magnetic field, the neighbouring spins do the same. In this way we can visualize the macroscopic magnetization of the material in a state of precession about a d.c. magnetic field. However, this picture represents the behaviour of a ferromagnetic sample, without undue complications, if all the spins are aligned in the same direction along the applied field, a condition realized only if the field is strong enough to remove the domain structure, that is to magnetize the material to saturation. An equation of motion of the magnetization is then readily formulated by forming the vector sum of the individual spin magnetic moments $\mathbf{m}$ over a unit volume of the material. Writing $\mathbf{M} = \Sigma\mathbf{m}$ and assuming that the gyromagnetic ratio $\gamma$ is the same for all $\mathbf{m}$, the equation of motion of the magnetic moment per unit volume, $\mathbf{M}$, has the form

$$\frac{d\mathbf{M}}{dt} = -\gamma\mathbf{M} \times \mathbf{B}, \quad (\omega_0 = \gamma B_z) \tag{23.2.2}$$

where $\mathbf{M}$ is understood to be the saturation magnetization in the present context.

Equation (23.2.2) has the same form as that for a single atomic moment $\mathbf{m}$. The solution has the form of precession of the magnetization vector $\mathbf{M}$ about the magnetic field $\mathbf{B}$ at the frequency $\omega_0 = \gamma B_z$ where $\mathbf{B} = \hat{k}B_z$.

At this point an important qualification must be made regarding the precession frequency $\omega_0$ and the magnetic field which enters into it. The field $\mathbf{B}$ appearing in eqn (23.2.2) is that seen by the individual spins, and this differs in general from the field applied by an external magnet. The difference is due in the first place to demagnetizing effects associated with differing sample shapes. Readers not familiar with this problem should consult a textbook on electromagnetism which deals with potential problems in terms of demagnetizing or depolarizing coefficients. In addition the field to be substituted in eqn (23.2.2) should include the effect of anisotropy forces which also modify the magnetic condition of individual spins. However, these are only apparent in single crystal samples, and since most applications use polycrystalline materials the effect will be neglected in this context. For our present purpose it will be sufficient to state that in isotropic samples of *spherical shape* the value of the magnetic field to be substituted in eqn (23.2.2) happens to be identical with the externally applied field. Hence the precession frequency is evaluated using the *applied* field in the expression $\omega_0 = \gamma B_z$. Qualitatively it is useful to note further that the precession frequency in samples of flat geometry, with a normal applied field, will be *lower* than predicted by eqn (23.2.2). If it is desired to set up precession at a given frequency $\omega_0$ in such samples, the required applied field will be *greater* than predicted by eqn (23.2.2). A quantitative discussion of the equation of motion, including demagnetizing effects, can be found in the references list on page 413.

It now remains to consider the effect of a.c. magnetic fields on the precessing magnetization of a ferromagnetic sample. In the first place it should be noted that the precession discussed above will not continue indefinitely after the application of a d.c. field to a sample. Internal losses will soon damp down any motion in the absence of an a.c. driving force. However the application of a circularly polarized a.c. field of frequency $\omega_0 = \gamma B_z$ will sustain the precession of the magnetization **M** in the manner described at the close of the preceding section. A condition of forced resonance is maintained, provided the sense of rotation of the a.c. field is the same as the precession of **M**. In the process energy is transferred from the a.c. field to the exchange coupled spin system, and hence to the lattice as thermal vibration. If the sense of the circular polarization is contrary to the precessional motion no interaction will take place and the a.c. field or wave will remain substantially unaffected by the spin system. In the case of a linearly polarized wave partial interaction will take place.

The phenomenon of forced resonance described above is usually referred to as *ferromagnetic resonance*. Since it is usually observed and utilized in ferrite materials it is sometimes called *ferrimagnetic resonance*. The resonant frequency $\omega_0$ is usually in the microwave range for d.c. magnetic fields readily available, say up to 1 T . Since the application of a.c. fields in this frequency range to the highly conductive ferrous metals *in bulk* is frustrated by skin effect, their use at microwave frequencies is precluded. However, this restriction does not apply to the insulating ferrites and garnets described in the preceding chapter. For this reason they form the basis of microwave devices.

### 23.3 A.C. Susceptibility

To put the discussion of the preceding section on a quantitative basis the concept of a.c. *susceptibility* is used. The application of an a.c. field to a magnetic material will induce time-dependent changes of the magnetization. The a.c. susceptibility is defined as the ratio of the magnetization to the a.c. field by analogy to eqn (20.1.2). However, since both magnitude and phase differences will enter into the relationship, the definition must be formulated in terms of complex amplitudes (see Chapter 10).

To be specific let the a.c. field be applied in the $x$-direction. We can write

$$H_x(t) = H_x e^{i\omega t} \qquad (23.3.1)$$

where $H_x$ is in general a complex amplitude. The magnetization will vary with time at the same frequency as the applied field and can, therefore, be written in the form

$$M_x(t) = M_x e^{i\omega t} = \chi H_x e^{i\omega t} \tag{23.3.2}$$

where $M_x$ is again a complex amplitude. $\chi$ is the a.c. *susceptibility*. In general it is a complex number which can be written as follows

$$\frac{M_x(t)}{H_x(t)} = \frac{M_x}{H_x} = \chi = \chi' - i\chi'' = |\chi| e^{-i\phi} \tag{23.3.3}$$

where $\phi$ is the phase difference and $|\chi|$ the ratio of real amplitudes. More useful in the present context are the real and imaginary parts of the susceptibility. In the first place it should be noted that they must in no way be considered to be constants of a material. They vary with frequency, with the amplitude of the a.c. field, and with the magnitude of the applied d.c. field, if any. The function $\chi'(\omega)$ is usually referred to as the *dispersion relation*. The *imaginary part* of the *susceptibility*, $\chi''(\omega)$, is a measure of the energy absorbed by the material from the a.c. field. This can be shown by the application of eqn (21.4.1) to a complete cycle of the field given by eqn (23.3.1). Since the proof is an exercise in electromagnetism, it will be omitted here. It follows from the foregoing discussion that the a.c. field will suffer a power loss to the material only if there is a phase difference between the field and the magnetization, accompanying a non-vanishing imaginary susceptibility $\chi''$.

The above definitions apply to one component of the a.c. field and magnetization. Although analogous relations apply to the remaining components of the magnetic vectors, the actual values of the susceptibilities will not in general be the same. This means that the material is anisotropic to a.c. fields, an observation which will be elaborated below.

Let us now apply these concepts to the problem of ferromagnetic resonance. Referring to eqn (23.2.2) we shall substitute the vector sum of the d.c. and a.c. fields

$$\mathbf{B} = \hat{\imath} B_x e^{i\omega t} + \hat{k} B_z \tag{23.3.4}$$

where the amplitude of the a.c. field is assumed to be very much smaller than the d.c. field, $B_x \ll B_z$. Regarding the magnetization we neglect any small effect that the a.c. field may have on the $z$-component by virtue of the fact that the latter is assumed to be the fixed saturation magnetization of the sample as explained on p. 351. Hence we can write

$$\mathbf{M} = \hat{\imath} M_x e^{i\omega t} + \hat{\jmath} M_y e^{i\omega t} + \hat{k} M_z$$

$$d\mathbf{M}/dt = \hat{\imath}\omega(i M_x + \hat{\jmath} M_y) e^{i\omega t}. \tag{23.3.5}$$

Substituting these expressions into eqn (23.2.2) and taking $x$ and $y$ components we find

$$i\omega M_x = -\gamma M_y B_z$$

$$i\omega M_y = \gamma(M_x B_z - M_z B_x).$$

After some manipulation the following ratio can be formed

$$\frac{M_x}{B_x} = \frac{\gamma^2 B_z M_z}{(\gamma B_z)^2 - \omega^2} = \frac{(\gamma B_z)^2}{(\gamma B_z)^2 - \omega^2} \cdot \frac{M_z}{B_z}.$$

Recalling from eqn (23.2.2) that $\gamma B_z = \omega_0$ and replacing $B$ by $\mu_0 H$ we obtain the a.c. susceptibility in the form

$$\frac{M_x}{H_x} = \chi = \frac{\chi_0}{1 - (\omega/\omega_0)^2} = \chi' \qquad (23.3.6)$$

where $\chi_0 = M_z/H_z$ is the d.c. or static susceptibility of the material.

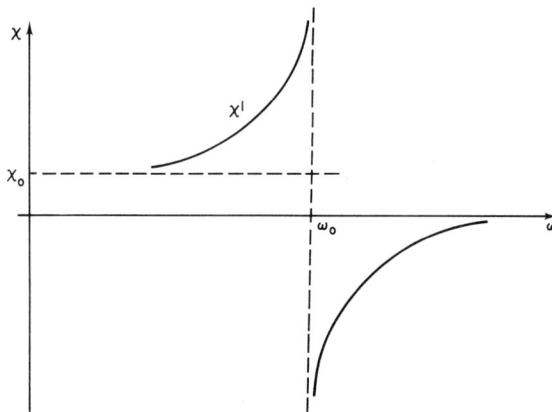

Fig. 23.2. Real part of the A.C. susceptibility plotted as a function of frequency.

From eqn (23.3.6), which is represented graphically in Fig. 23.2, it can be seen that the a.c. susceptibility approaches the static susceptibility $\chi_0$ when the frequency $\omega$ of the a.c. field tends to zero. On the other hand when $\omega \to \infty$, $\chi \to 0$ indicating that under these conditions the a.c. field hardly affects the magnetic state of the sample. However, when $\omega \to \omega_0$ the a.c. susceptibility becomes very large, indicating that a strong interaction is taking place.

The reader will have observed that in eqn (23.3.6) we have obtained a real quantity. The imaginary part of the a.c. susceptibility, $\chi''$, vanishes in this case. Thus eqn (23.3.6) is a description of the dispersive properties of ferromagnetic resonance but not of the energy dissipation that goes with it. The cause of this limited result lies in the original equation of motion, eqn (23.2.2). This is a second-order differential equation which does not include a term to account for any losses in the system under discussion, i.e. the spin system of the ferromagnetic substance. Readers interest in this aspect of ferromagnetic resonance are referred to specialist texts dealing with magnetism.

A reference back to the expressions preceding eqn (23.3.6) will disclose that the a.c. susceptibility is anisotropic. This is easily verified by forming the ratio $M_y/B_x$ which is found not to vanish. This means that the susceptibility and hence the permeability matrix (see Section 20.1) has off-diagonal elements, indicating that the material is anisotropic. It is emphasized that this property of ferrimagnetic substances is quite consistent with them being isotropic under static conditions. The a.c. anisotropy is related to the fact that the precessional motion of the spin system only interacts with r.f. fields circularly polarized in the correct sense. It is shown at length in texts dealing with electromagnetic waves that anisotropic media can deflect a wave from propagation along a straight line. This fact forms the basis of the use of ferrimagnetic substances in microwave circulators to be described in the next section.

## 23.4  Some Ferrite Microwave Devices

(i)  *Resonance isolator*

The resonant interaction between circularly polarized a.c. fields or waves and the precessing magnetization of a ferrite forms the basis of a microwave device which permits electromagnetic waves to travel freely in one direction while attenuating them in the opposite direction. Such a unidirectional transmission line is usually termed an *isolator*, because it is used to isolate signal sources from reflected waves. The functioning of the device can be explained with reference to Fig. 23.3, which shows schematically one particular type of waveguide isolator.

In Fig. 23.3(a) a cross-sectional view of the waveguide is presented. An elongated slab of ferrite is stuck to the wide wall of the rectangular waveguide, somewhat off the central position. A d.c. magnetic field, usually supplied by a permanent magnet, is applied as shown in a direction normal to the plane of the waveguide and the ferrite sample. The ferrite must be chosen in such a way that it can be magnetized to saturation by the field required to establish resonant conditions at the signal frequency $\omega$. According to eqn (23.2.2), $\omega = \omega_0 = \gamma B_{d.c.}$.

Figure 23.3(b) shows a plan view of the isolator with the magnetic field patterns of the dominant mode of propagation sketched in. Wave propagation in the waveguide is usually visualized in terms of the field patterns travelling along at the phase velocity. A moments reflection will convince the reader that such movement gives rise to an approximately circularly polarized a.c. field at any fixed point $A$ inside the ferrite slab. Since the a.c. field is normal to the d.c. field, correct conditions for resonant interaction with the precessing spin system exist, provided the sense of circular polarization is as shown in Fig. 23.1. This will be the case for the direction of propagation indicated in Fig. 23.3(b) and absorption of wave energy by the ferrite will take place. However, for the opposite direction of propagation, the sense of circular polarization will be incorrect for interaction and the wave energy will pass along the waveguide substantially unaffected. Thus, the device will act as a unidirectional transmission line, or a two-port microwave junction which allows energy to pass between the ports in one direction but not the other.

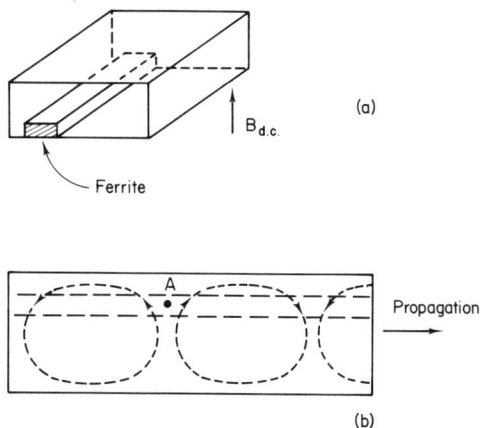

**Fig. 23.3.** Resonance isolator (schematic).

As pointed out in Section (23.2), the d.c. field to be applied for resonance varies with sample geometry. In particular, the flat slab shown in Fig. 23.3 will require a higher field than predicted by eqn (23.2.2) for a spherical sample. This is because for sample shapes other than spherical the required d.c. field is affected by the saturation magnetization $M_s$ of the material, the effect varying continuously with sample geometry.

The fact that $M_s$ affects the resonant d.c. field introduces an additional complication related to the temperature variation of $M_s$ (see Section 21.7). A ferrite device expected to operate over a range of temperatures would, in general, require an adjustable d.c. field. Since such provision is technologically not practicable, other means of avoiding the difficulty have been devised, including the design of sample shapes which are insensitive to changes of saturation magnetization. For details the reader is referred to specialist texts and papers.

(ii) *Circulator*

The *circulator* is a more complicated device than the isolator in the sense that it is a microwave junction with at least 3 ports and possibly more. Its function is to

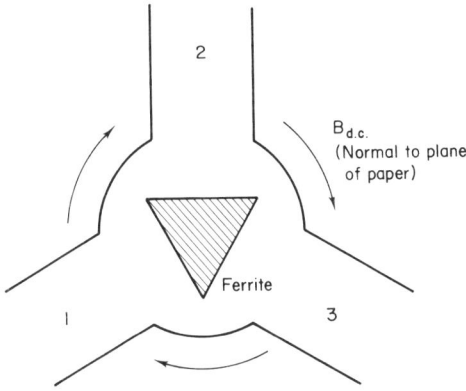

Fig. 23.4. Three-port ferrite circulator.

channel microwave energy towards one port rather than another in a manner which can be explained by reference to Fig. 23.4. Microwave energy, entering the junction along port 1, also enters the block of ferrite located in the centre. However, instead of propagating along a straight line, the waves are deflected towards port 2 by virtue of the anisotropic a.c. permeability of the ferrite medium (see p. 355). Hence, wave propagation is selectively *circulated* from port 1 to port 2 and, with careful design, it is possible to ensure that virtually no energy emerges at port 3. By the same argument, waves approaching the junction along port 2 will be deflected towards port 3, no energy entering port 1.

By virtue of this operation, the device can be used to separate incident and reflected waves. Thus, a signal source could be connected to port 1 and a (mismatched) load to port 2. The wave reflected by the load would then be circulated to port 3 where it could be monitored by a detector. An example of

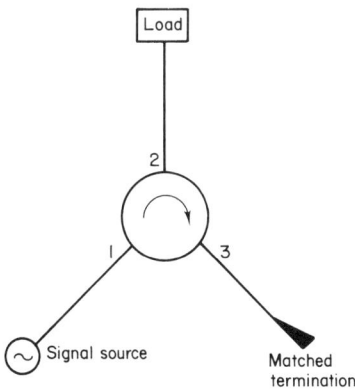

Fig. 23.5. Circulator used as isolator.

such use of the circulator is included in Fig. 20.9 where the load takes the form of a one-port resonant cavity.

The circulator can also be used as an isolator if connections are made as shown in Fig. 23.5. Any energy reflected by the load at port 2 is diverted into the matched termination connected to port 3 and dissipated there. Thus the signal source is protected from any reflected waves and effectively feeds energy into a matched line.

*Chapter 24*

# Diamagnetism

In the preceding chapters we have devoted a considerable amount of attention to magnetic materials which have a positive susceptibility. The value of the susceptibility varies widely between paramagnetic and ferromagnetic substances, but in every case the source of magnetic properties can be traced to permanent magnetic moments possessed by atoms or ions having incomplete electron shells. In the present chapter we consider briefly the third type of magnetic behaviour, first listed in Section 20.1, which is characterized by a negative susceptibility, and we shall find that this is due to induced circulating currents which may be macroscopic or microscopic in scale.

### 24.1 Electromagnetic Induction as Diamagnetism

It is well known that the application of a time-varying magnetic field to a closed, ring-shaped conductor, will induce a circulating current which in turn will set up a field of its own. Moreover, according to Lenz' law the direction of the induced field will be such as to oppose the applied field. The net result will be that the magnetic induction in the space enclosed by the ring will be less than the free-space value, or in other words the susceptibility of the space within the ring will be negative.

The situation is depicted schematically in Fig. 24.1. By Faraday's law the induced current is directly proportional to the time rate of change of the applied field if the resistance of the ring is much greater than its inductive reactance.

$$i(t) \sim \frac{dB}{dt}. \qquad (24.1.1)$$

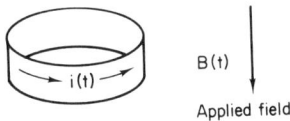

Fig. 24.1. Diamagnetic effect of induced currents.

Moreover, the induced field is proportional to the current. Hence, it will oppose the applied field only as long as the current continues to circulate. In a normal conductor, which has finite resistance, this is only possible if the applied field varies periodically with time. Otherwise the term $dB/dt$ must tend to zero sometime and the current must collapse according to the time constant $L/R$ of the ring. The applied field then penetrates to the space within the ring. Hence the diamagnetic effect of macroscopic currents in normal conductors is only a temporary phenomenon.

The situation is quite different in superconductors in which induced screening currents can be maintained indefinitely by virtue of the zero resistance of these materials. However, as superconductors will be dealt with at some length in the next chapter we shall not pursue the question any further now.

## 24.2 Diamagnetism

The application of a magnetic field will in general cause a rearrangement of the internal orbital currents within the atoms, which has the effect of opposing the applied field. Such rearrangement on an atomic scale is not subject to resistive dissipation like macroscopic conduction currents and gives rise to a small diamagnetic effect which persists as long as the applied field is maintained.

To obtain an estimate of this effect we go back to the basic equation of motion of electrons in a central field of force and extend it to include the effect of a magnetic field. The latter affects the motion of the electrons through the Lorentz force $\mathbf{F} = -e\,\mathbf{v} \times \mathbf{B}$ which is included in the Hamiltonian with the help of eqn (18.5.2), with $q = -e$, since $e$ in the present treatment is a positive physical constant denoting the magnitude of the electronic charge.

The basic Hamiltonian, eqn (13.1.3), can now be written in the form

$$
\begin{aligned}
H &= \frac{1}{2m_e}\,(\mathbf{p} + e\mathbf{A})^2 + V(r) \\
&= \frac{\mathbf{p}^2}{2m_e} + V(r) + \frac{e}{2m_e}\,(\mathbf{p} \cdot \mathbf{A} + \mathbf{A} \cdot \mathbf{p}) + \frac{e^2}{2m_e}\,\mathbf{A}^2
\end{aligned}
\qquad (24.2.1)
$$

where the first two terms are the original Hamiltonian and the remaining two terms include the effect of the magnetic field. To examine the latter more closely we assume that a d.c. field $B_z$ is applied along the $z$-axis. This simplifies the expressions to be considered without loss of generality. It is readily verified that a vector potential associated with this field has the form

$$
\mathbf{A} = -\hat{\imath}\tfrac{1}{2}yB_z + \hat{\jmath}\tfrac{1}{2}xB_z.
$$

Substituting this expression into the Hamiltonian (24.2.1) and writing $\mathbf{p} = -i\hbar\nabla$ we find

$$H = H_0 - \frac{ie\hbar B_z}{2m_e}\left(x\frac{\partial}{\partial y} - y\frac{\partial}{\partial x}\right) + \frac{e^2 B_z^2}{2m_e}(x^2 + y^2) \qquad (24.2.2)$$

where the original Hamiltonian is denoted by $H_0$. By reference to Section 14.1 the second term is found to contain the $z$-component of angular momentum $L_z$, and can be rewritten in the form

$$H = H_0 + \frac{\beta}{\hbar} L_z B_z + \frac{e^2 B_z^2}{8m_e}(x^2 + y^2) \qquad (24.2.3)$$

where $\beta$ is the Bohr magneton (see Section 15.1).

The second term can now be identified as the paramagnetic energy of the resultant magnetic moment of the atom, first introduced in Chapter 15 by an independent method. That method was less general than the present discussion because it assumed that the applied field can only interact with the magnetic moment. By contrast the third term above includes the effect of the field on the orbiting electronic charge and must therefore be the basis of any diamagnetic effect.

The contribution of the paramagnetic term to the energy levels of the atom was easily evaluated in Chapter 15 because it was found that it shared eigenstates with the original Hamiltonian $H_0$. However, this is not the case with the third term of eqn (24.2.3). As the reader can verify, this does not commute with $H_0$.

Despite this difficulty we can obtain an estimate of the diamagnetic effect on the assumption that it is small and does not significantly modify the eigenstates of the atom. We evaluate the average values or diagonal matrix elements of the diamagnetic term with respect to the eigenstates of the original Hamiltonian $H_0$ following the procedure of Chapter 12. In fact the detailed calculation can be bypassed with the help of a physical argument. The only factor to be averaged is $x^2 + y^2$ which represents the radius of the curved motion of the electron in the $xy$-plane. As a first approximation we assume that this will be of the order of the Bohr radius $a_0$ (Section 13.2). Hence the diamagnetic modification of the energy levels of an atom is approximately

$$W_{\text{dia}} = \frac{e^2 B_z^2}{8m_e} a_0^2. \qquad (24.2.4)$$

The equivalent magnetic moment giving rise to this is obtained by differentiation.

$$|\mathbf{m}|_{\text{dia}} = -\frac{\partial W_{\text{dia}}}{dB_z} = -\frac{e^2 a_0^2}{4m_e} B_z = -\beta \frac{ea_0^2}{2\hbar} B_z. \qquad (24.2.5)$$

This should be compared with the paramagnetic moment which is $\beta m$ for a state having magnetic quantum number $m$. Substitution of the requisite constants will disclose that $ea_0^2/2\hbar \cong 10^{-5}$. Hence even for the highest d.c. fields available in the laboratory ($\cong 10$ T) the diamagnetic moment is minute compared with the paramagnetic moment. The conclusion must be that the diamagnetic effect is swamped in atoms which have a resultant paramagnetic moment, be it orbital or spin in character. However, in substances made up of atoms or molecules having no resultant paramagnetic moment, weak diamagnetism can be detected experimentally.

*Chapter 25*

# Superconductivity

The subject of superconductivity is dealt with in the present chapter in a manner entirely dictated by expediency of presentation. The objective is to cover as many concepts and phenomena relevant to applications as possible, drawing on the minimum theoretical background. Thus the treatment opens with the London theory which should be relatively easy to assimilate for readers familiar with classical electromagnetism. The microscopic BCS theory is outlined only briefly, as it requires methods of quantum field theory to develop quantitatively. Throughout the chapter every effort is made to select material of interest to the engineer rather than the physicist, although the really basic concepts should be common ground to all.

### 25.1 Some Basic Concepts

The electrical resistivity of most metals and alloys decreases with temperature as a result of diminished scattering of electrons by lattice vibrations which die down at low temperatures. This trend continues until at a point scattering of electrons by lattice defects and impurities becomes the dominant resistive mechanism. As this is independent of temperature the resultant resistivity remains constant, as suggested in Fig. 25.1, and is usually referred to as the *residual* resistivity or resistance.

In many metals, alloys and intermetallic compounds the electrical resistance vanishes completely at a well-defined temperature $T_c$ called the *critical temperature*. Below this temperature the material is said to be *superconductive*. All materials which exhibit this transition from the *normal* conductive to the superconductive state are referred to as *superconductors*. The critical temperatures of known superconductors are very low. The values of $T_c$ for some typical elemental superconductors are listed in Table 25.1. Niobium has the highest critical temperature of $9 \cdot 3 \,^\circ K$ among the elements. A number of alloys are known having higher $T_c$ values, but even these barely exceed $20 \,^\circ K$ at the present state of knowledge.

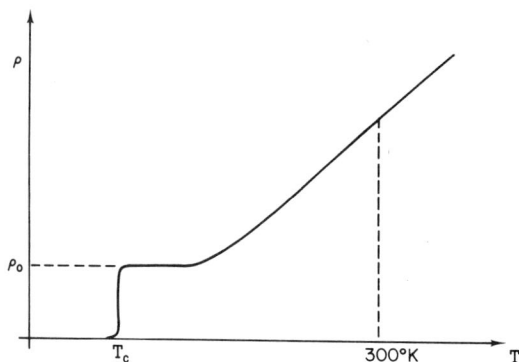

**Fig. 25.1.** Critical temperature of a superconductor.

The resistanceless state of a superconductor implies that the conduction electrons, or at least some of them, have undergone a change into a different phase which enables them to transport electric currents without being scattered by the lattice. The electrons which have entered the *superconductive phase* are usually called *superelectrons* by contrast with *normal electrons*. The density of superelectrons will be denoted by $n_s$, a number which is generally less than the density of all conduction electrons.

The resistanceless current which can be caused to flow in a superconductor is made up entirely of superelectrons. Its current density will be denoted by the symbol $\mathbf{J}_s$ to distinguish it from the normal current density $\mathbf{J}_n$. As a rule only the supercurrent will be present in a superconductor, but there are occasions

**Table 25.1**
Critical data of Elemental Superconductors

| Element | Critical Temperature $T_c[^\circ K]$ | $B_{co} = \mu_0 H_{co}$ [Gauss = $10^{-4}$ T] |
|---|---|---|
| Aluminium (Al) | 1·2 | 99 |
| Indium (In) | 3·4 | 276 |
| Lead (Pb) | 7·2 | 803 |
| Niobium (II) (Nb) | 9·3 | |
| Tantalum (Ta) | 4·5 | 830 |
| Tin (Sn) | 3·7 | 306 |
| Titanium (Ti) | 0·4 | |
| Tungsten (W) | 0·01 | |

when the normal current has to be taken into account. Thus in general the total current density in a superconductor takes the form

$$\mathbf{J} = \mathbf{J}_n + \mathbf{J}_s. \qquad (25.1.1)$$

The normal current $\mathbf{J}_n$ will be related to a local electric field $\mathbf{E}$ through the usual conduction relation $\mathbf{J}_n = \sigma\mathbf{E}$ where $\sigma$ is the normal state conductivity of the material.

The supercurrent $\mathbf{J}_s$ will be found to be the agency whereby the super-conductor is screened from the effect of magnetic flux, thus giving rise to diamagnetism. The equations governing the electromagnetic behaviour of supercurrents under certain conditions are introduced in the next section.

### 25.2  The London Theory

Long before the microscopic origins of superconductivity were understood the brothers F. and H. London proposed a pair of equations which were found to predict very well the electromagnetic properties of a group of materials now classified as type I superconductors. Although superconductors of practical importance are classified as type II and have properties which are not embraced by the London theory, it is nevertheless essential to start with an introduction of this subject since it provides a very convenient means of defining a number of important concepts and relations. In the present treatment we shall not attempt to justify or in any way "derive" London's equations. They will be stated as an *axiomatic supplement to Maxwell's equations* and then used in conjunction with the latter to obtain a description of the principal properties of type I superconductors. All the elemental superconductors listed in Table 25.1 are type I with the exception of niobium.

The London equations read as follows

$$\operatorname{curl} \mathbf{J}_s = -\frac{1}{\mu_0 \lambda^2}\, B \qquad (25.2.1)$$

$$\dot{\mathbf{J}}_s = \frac{1}{\mu_0 \lambda^2}\, E. \qquad (25.2.2)$$

They can be seen to relate the supercurrent density and its time rate of change to local magnetic and electric fields in the superconductor. $\mu_0$ is the permeability of free space and $\lambda$ is called the London *penetration depth*. Its significance will be dealt with presently as we apply the first equation to derive the diamagnetic property of superconductors.

As a first step in that direction we eliminate the current density from eqn (25.2.1) with the help of Ampére's law in the form $\operatorname{curl} \mathbf{B} = \mu_0 \operatorname{curl} \mathbf{H} = \mu_0 \mathbf{J}_s$. Taking the curl of both sides and using a well-known vector identity we find

$$\nabla \times (\nabla \times B) = \nabla(\nabla . \mathbf{B}) - (\nabla . \nabla)\mathbf{B} = -\nabla^2 \mathbf{B} = \mu_0 \operatorname{curl} \mathbf{J}_s \qquad (25.2.3)$$

where the relation $\nabla \cdot \mathbf{B} = 0$ has been applied. Comparing eqns (25.2.1) and (25.2.3) we can write

$$\nabla^2 \mathbf{B} = \frac{1}{\lambda^2} \mathbf{B}. \tag{25.2.4}$$

### 25.3 Perfect Diamagnetism and Superconductivity

Equation (25.2.4) will now be used to evaluate the magnetic field inside a superconducting slab of thickness $2a$ and infinite extent in the $yz$-plane, immersed in a tangential d.c. field $B_0$, as depicted in Fig. 25.2(a). From the

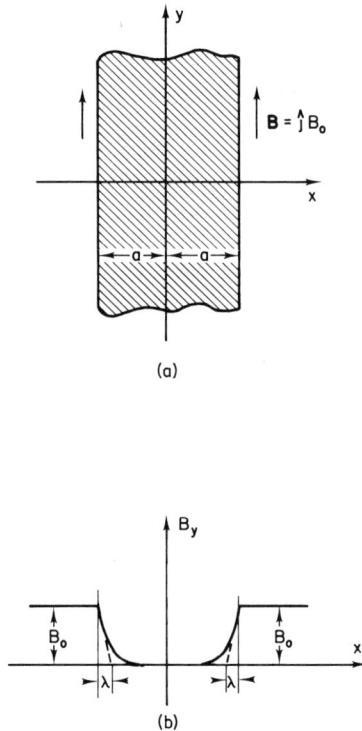

(a)

(b)

Fig. 25.2. Magnetic flux expulsion from a superconductor.

plane symmetry of the slab it follows that the magnetic field inside the superconductor can only vary along the $x$-axis. Hence the partial differential equation (25.2.4) reduces to the simple form

$$\frac{d^2 B_y}{dx^2} = \frac{1}{\lambda^2} B_y \tag{25.3.1}$$

having the solution

$$B_y(x) = A e^{-x/\lambda} + B e^{x/\lambda}. \qquad (25.3.2)$$

The constants of integration are determined from the boundary conditions $B_y(\pm a) = B_0$ and are found to be

$$A = B = \frac{1}{2} \frac{B_0}{\cosh (a/\lambda)}.$$

Hence the field inside the superconducting slab has the form

$$B_y(x) = \frac{B_0}{\cosh(a/\lambda)} \cosh(x/\lambda). \qquad (25.3.3)$$

If $a \gg \lambda$ a very close approximation to the above expression can be written down for values of $x$ near $\pm a$, that is near the surfaces of the slab. The following exponential formula is then obtained

$$B_y(x) = B_0 e^{-(a \mp x)/\lambda} \qquad (25.3.4)$$

where the $-$ve sign applies to positive values of $x$ and the $+$ve sign applies to negative values of $x$. This exponential decay of the magnetic field near the surface of the superconductor is shown in Fig. 25.2(b). The London *penetration depth* $\lambda$ is now identified as the characteristic constant of this decay. Experimental data and theoretical estimates of $\lambda$ both indicate that the penetration depth is of the order of $10^{-6}$ cm. Hence the exponential approximation holds for all bulk samples. However, in thin films whose thickness is comparable with $\lambda$, the exact expression must be used.

The exclusion of the magnetic field from the superconducting slab is of course due to screening supercurrents which are induced within the penetration depth. The current density can be evaluated with the help of Ampére's law. On taking $(1/\mu_0)$ curl $(\hat{j}B_y)$ we find

$$J_z = \frac{1}{\mu_0} \frac{\partial B_y(x)}{\partial x} = \frac{B_0}{\mu_0 \lambda \cosh (a/\lambda)} \sinh (x/\lambda). \qquad (25.3.5)$$

Forming again the approximations applicable when $a \gg \lambda$ the screening currents are found to fall off exponentially as the superconductor is penetrated.

$$J_z = \frac{B_0}{\mu_0 \lambda} e^{-(a-x)/\lambda} \quad \text{for } x \text{ near } +a$$

$$J_z = \frac{B_0}{\mu_0 \lambda} e^{-(a+x)/\lambda} \quad \text{for } x \text{ near } -a . \qquad (25.3.6)$$

As pointed out in the preceding chapter the magnetic state of the interior of the slab can be described in terms of a negative susceptibility. To do so we ignore the presence of the screening currents and postulate that the applied magnetizing force $H_0 = B_0/\mu_0$ is *also present inside the slab*. However, its effect is cancelled by a negative magnetization yielding zero induction $B_y$. Thus

$$B_y = 0 = \mu_0(H_0 + M)$$

$$M = -H_0 = -\frac{B_0}{\mu_0}. \qquad (25.3.7)$$

Since by definition (p. 299) $M = \chi H$ we find that

$$\chi = -1$$

and the relative permeability $\mu_r = 0$.

Within the penetration depth itself the susceptibility varies continuously from $-1$ to $0$ and the permeability from $0$ to $1$.

Although the foregoing arguments have been formulated on the example of a slab-shaped superconductor the same conclusions can be arrived at by treating a cylinder with a longitudinal applied field. Indeed Fig. 25.2 can be interpreted as applying to the cross-section of a cylinder of radius $a$ while the current densities of eqn (25.3.6) can represent a circumferential screening current. As in the case of the slab a distinction must be made between a macroscopic cylinder to which the concept of a small exponential penetration depth applies, and a microscopic cylinder whose radius is comparable with $\lambda$ and which may not be completely screened from the applied field.

From the foregoing it follows that the first of London's equations, (25.2.1), embodies the perfect diamagnetism of superconductors. We shall now find that the second equation (25.2.2) describes the perfect conductivity of superconductors and we shall use it to obtain an estimate of the London penetration depth $\lambda$.

As a first step the supercurrent density is expressed in the form

$$\mathbf{J}_s = -n_s e\mathbf{v} \qquad (25.3.8)$$

where $n_s$ is the density of superelectrons. Differentiation of the above relation with respect to time and substitution into eqn (25.2.2) : yields the result

$$-e\mathbf{E} = \mu_0 e^2 n_s \lambda^2 \dot{\mathbf{v}}. \qquad (25.3.9)$$

Now, under conditions of perfect conductivity the superelectrons experience no resistance to their motion and the effect of an electric field will be to accelerate them. Hence they will satisfy the following classical equation of motion

$$-e\mathbf{E} = m_e \dot{\mathbf{v}} \qquad (25.3.10)$$

where $m_e$ is the electron mass. Comparison of eqns (25.3.9) and (25.3.10) yields a relation between the penetration depth and known physical constants

$$\lambda^2 = \frac{m_e}{\mu_0 \, e^2 \, n_s}. \tag{25.3.11}$$

This expression gives a value of the order of $10^{-6}$ cm if $n_s$ is substituted on the assumption that there is one superelectron per atom. Measurements yield a value of the same order of magnitude thus confirming that the London theory provides a correct description of the electromagnetic properties of type I superconductors.

### 25.4 Flux Quantization and Electron Pairs

Perfect diamagnetism implies that a multiply connected superconductor is capable of trapping magnetic flux under suitable conditions. As an example refer to Fig. 25.3 which represents a cross-section through a hollow tube or ring. If a d.c. field is applied to the ring in the normal state, then the ring is cooled below the critical temperature $T_c$, and finally the applied field is removed, supercurrents will be induced in the ring which will set up a field. The trapped flux inside the ring or tube will correspond to the original external field. We shall now show that the trapped flux inside a multiply connected superconductor is quantized.

As a first step we establish a conservation law which applies to the flux contained within a superconducting ring together with the circulating current. We envisage a situation in which the magnetic conditions of the tube of Fig. 25.3 have been altered by the temporary application of a magnetic field. We apply Maxwell's equation curl $\mathbf{E} = -\partial \mathbf{B}/\partial t$ to the area $S$ within the dotted contour. Integration over the area yields the relation

$$\int_S \dot{\mathbf{B}} . d\mathbf{S} + \int_S \text{curl } \mathbf{E} . d\mathbf{S} = 0.$$

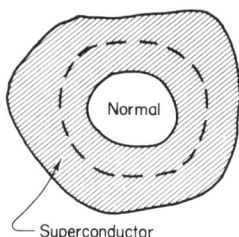

Fig. 25.3. Multiply connected superconductor.

The second term is transformed into a line integral by Stoke's theorem

$$\int_s \dot{\mathbf{B}} . d\mathbf{S} + \oint \mathbf{E} . d\mathbf{l} = 0.$$

Substituting for $\mathbf{E}$ from eqn (25.2.2) and integrating with respect to time we find

$$\int_s \mathbf{B} . d\mathbf{S} + \mu_0 \lambda^2 \oint \mathbf{J}_s . d\mathbf{l} = \text{const.} = \phi'. \tag{25.4.1}$$

The quantity $\phi'$ is called the *fluxoid*. The first (surface) integral represents the total flux enclosed within the dotted curve of Fig. 25.3. The second term is the line integral of the supercurrent around the dotted contour. In situations of practical interest the second term usually vanishes, as when Fig. 25.3 represents a very long tube with some flux trapped inside. If the dotted contour is placed outside the penetration depth $\lambda$, there will be no supercurrents to contribute to the second integral and it will vanish. Under these conditions eqn (25.4.1) represents just the flux trapped within the tube and in the penetration depth. However, in general the fluxoid includes the current circulation, and the symbol $\phi'$, is used to distinguish it from the magnetic flux proper. Equation (25.4.1) shows that, once established in a multiply connected superconductor, the fluxoid remains constant or is conserved.

As the next step we show that the fluxoid is *quantized* on the basis of Bohr's quantization condition. Using the magnetic vector potential $\mathbf{B} = \text{curl } \mathbf{A}$ the remaining surface integral of eqn (25.4.1) is transformed into a line integral with the help of Stoke's theorem. Substituting for the current density from eqn (25.3.8) we find

$$\oint \mathbf{A} . d\mathbf{l} - \frac{m_e}{e} \oint \mathbf{v} . d\mathbf{l} = \phi'.$$

This can be rewritten in the form

$$\frac{1}{e} \oint (m_e \mathbf{v} - e\mathbf{A}) . d\mathbf{l} = \frac{1}{e} \oint \mathbf{p} . d\mathbf{l} = \tag{25.4.2}$$

where the term $m_e \mathbf{v} - e\mathbf{A} = \mathbf{p}$ is recognized as the momentum of an electron (charge $-e$) moving with velocity $\mathbf{v}$ in a magnetic field represented by the vector potential $\mathbf{A}$.

The above integral is the *phase integral* which forms the basis of quantization of electron orbits in the Bohr model of the atom, from which we recall that

$$\frac{1}{h} \oint \mathbf{p} . d\mathbf{l} = n \quad (n = 1, 2, 3, \ldots). \tag{25.4.3}$$

Hence on the basis of the foregoing argument the fluxoid can only assume discrete values given by $\phi' = (h/e)n$.

Experimental measurements confirm the predicted flux quantization with one very important discrepancy. Flux measured in very small diameter ($10 \, \mu m$) superconducting tubes was found to be quantized in units of $h/2e$ rather than $h/e$. This is interpreted as meaning that superelectrons occur in pairs of charge $2e$. The superconducting state thus consists of electron pairs in bound states usually called *Cooper pairs*. The quantum of flux

$$\phi_0 = \frac{h}{2e} \qquad (25.4.4)$$

is sometimes referred to as a *fluxon* or more usually a *flux line*. It constitutes a new physical constant expressed in terms of Planck's constant and the electronic charge. Its value is readily found to be about $2 \times 10^{-15}$ Wb. The flux-line concept is crucial to the description of type II superconductors as we shall see presently.

## 25.5 The Superconducting Phase Transition and the Critical Field

The present section will deal with the effect of an applied magnetic field on the superconducting to normal phase transition. As the treatment is concerned with thermodynamic aspects of a phase change it will be based on the Gibb's free energy which remains continuous across such a change.

The formulation of the Gibb's function including the effect of magnetic fields requires a certain amount of care. Readers not familiar with this question are referred to texts covering the thermodynamics of magnetic phenomena, or dealing thoroughly with the thermodynamic background of superconductivity, some of which are listed at the end of this book. For our present purpose we shall be content to quote the necessary relationship and use it.

The application of a magnetic field to a magnetizable body causes a change in the Gibb's free energy of the body given by the expression

$$G(H_0) - G(0) = - \mu_0 \int_0^{H_0} M dH \qquad (25.5.1)$$

where $M$ is the magnetization of the material in the presence of an external magnetizing force $H$ and $G(H)$ is Gibb's function. The above expression assumes that the magnetic field is applied under conditions of *constant temperature and pressure* or that the material is incompressible. The vectorial notation has been dropped as it is assumed that the magnetic vectors are parallel.

From eqn (25.5.1) it follows that there is no change, certainly no increase, of the free energy of a *normal* non-ferrous metal on application of a field, because there is no magnetization: $M = 0$. Hence the integral vanishes and we can write

$$G_n(H_0) = G_n(0). \qquad (25.5.2)$$

However, a superconductor acquires negative magnetization, which will impart a finite value to the integral of eqn (25.5.1). The latter can be evaluated on substitution of $M = -H$ in accordance with eqn (25.3.7). Thus

$$\int_0^{H_0} M dH = -\int_0^{H_0} H dH = -\tfrac{1}{2}H_0^2. \qquad (25.5.3)$$

From this it follows that the application of a magnetic field increases the energy of the superconductor by the amount

$$G_s(H_0) - G_s(0) = \tfrac{1}{2}\mu_0 H_0^2 = \tfrac{1}{2}B_0^2. \qquad (25.5.4)$$

The implication of this result is that the application of an increasing magnetic field will destroy the superconducting state. For at some point the free energy of the sample in the superconducting state must exceed the normal state energy which remains unchanged in the presence of a field. At this point the material becomes normal, the screening supercurrents die down and the magnetic flux penetrates into the sample as suggested in Fig. 25.4. The value of the applied field at which the transition takes place is termed the *critical field*. It may be denoted by $B_c$ and given in T (Tesla) or by $H_c$ in units of A/m. Existing literature on superconductivity uses almost entirely $H_c$ because it is based on c.g.s. units.

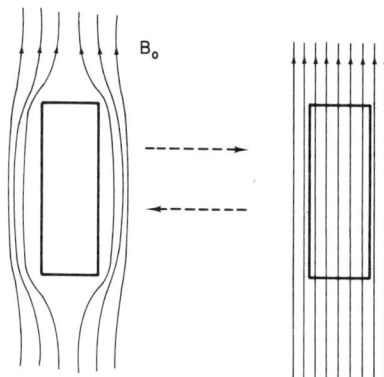

**Fig. 25.4.** Field induced superconducting to normal transition.

The destruction of superconductivity by a magnetic field is reversible. By this we mean that a sample will undergo a transition from the normal to the superconducting state when the applied field is reduced from a value above $H_c$. At this point the magnetic flux is expelled from the body of the sample, screening currents are restored and the susceptibility assumes the value $\chi = -1$. The reversibility of the transition and the phenomenon of flux expulsion were by no means obvious to the early explorers of superconductivity. It was not until experiments carried out by Meissner and Ochsenfeld that these facts were firmly established. For this reason the phenomenon of flux expulsion is frequently referred to as *Meissner effect*.

The transition is explained in terms of a *condensation energy* of conduction electrons. When normal electrons change into superelectrons or Cooper pairs (see p. 371) they condense into a lower energy state. The material is thus able to accept the magnetic energy represented by eqn (25.5.3). Since the negative magnetization which provides this energy represents in effect screening supercurrents, the extra energy is really the kinetic energy of moving superelectrons.

The concept of the critical field is paralleled by that of the critical current. As the field applied to a superconductor is increased, the screening currents increase until they reach the *critical current density* $\mathbf{J}_c$ at the instant the field reaches the critical value $H_c$. Under this condition the kinetic energy of the superelectrons is higher than the condensation energy and the Cooper pairs are broken up, thus rendering the conduction electrons normal.

The above considerations suggest that it is possible to relate the critical current and field values through Ampére's law, applied to the surface of the superconductor. This possibility was postulated as *Silsbee's hypothesis* and is verified for type I superconductors of simple geometrical shapes such as long cylinders. However, in the case of more complicated sample shapes, such as wires wound into coils, there is no simple relationship between critical currents and field. In the case of type II superconductors the position is further complicated by their specific properties. However, this question will be dealt with at a later stage.

The critical field increases with decreasing temperature, and it is found that experimental results follow very closely a quadratic law of the following form

$$H_c = H_{co}\left\{1 - \left(\frac{T}{T_c}\right)^2\right\} \qquad (25.5.5)$$

where $H_{co}$ is the critical field at the absolute zero of temperature. A critical field curve is depicted in Fig. 25.5. It forms a convenient reference whereby to define the state of a superconductor. A change of phase in the absence of a field is represented by movement along the temperature axis across the critical temperature $T_c$. In the presence of a fixed field phase transitions can still be effected by temperature variations represented by horizontal lines like $AB$. Finally at a fixed temperature a transition can be induced by increasing the applied field to its critical value defined by the crossing of the vertical line $BC$ with the critical field curve.

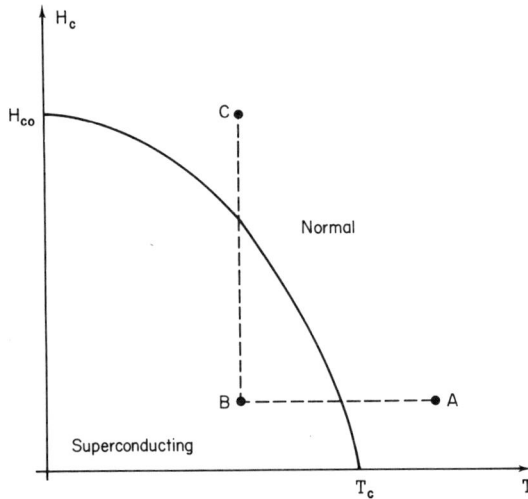

**Fig. 25.5.** Critical field curve of a type I superconductor.

The critical field curve of a type I superconductor is completely defined by eqn (25.5.5) provided the critical temperature $T_c$ and the critical field $H_{co}$ at $T = 0\,°K$ are given. These data are listed for a few typical elemental superconductors in Table 25.1. From this it will be seen that the values of the critical field are relatively low.

This observation has an important bearing on the possibility of constructing magnets which use superconducting coils. The passage of a current along a superconducting wire will set up a magnetic field which will increase in proportion to the transport current. When the field reaches the critical value superconductivity will be destroyed and the magnet windings will become resistive. As this will happen to type I superconductors at quite low fields, they are not very useful as magnet material. The difficulty has been overcome with the discovery and development of type II superconductors which can carry supercurrents in the presence of very high fields (10 T or more.)

We are now in a position to plot magnetization curves of type I superconductors. Figure 25.6(a) shows a $M/H$ graph. It should be noted that the

$M$-axis is negative in the upward direction, which is now the convential way of presenting magnetization data of superconductors. As the field $H$ is applied, the negative magnetization rises linearly in accordance with eqn (25.3.7). When the critical value $H_c$ is reached the magnetization drops suddenly to zero, as should

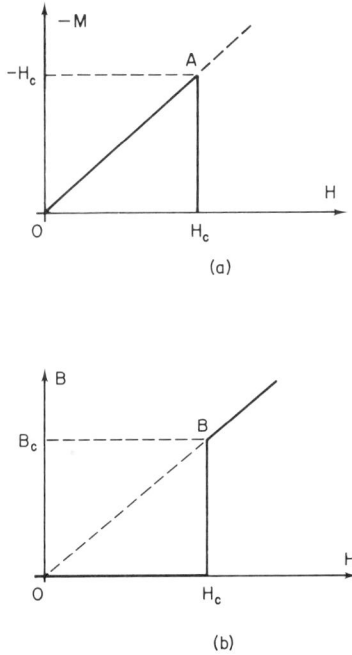

(a)

(b)

Fig. 25.6. Magnetization graphs of type I superconductors.

be the case in a normal metal, and remains at zero at higher fields. The line $OA$ represents the *perfect diamagnetic* line in $M/H$ coordinates.

A $B/H$ curve is readily constructed on the basis of eqn (25.3.7). At low applied fields $B = 0$ in the superconductor and the $B/H$ curve coincides with the $H$-axis. When the critical value $H_c$ is reached, flux penetrates the superconductor and assumes a value equal to the applied field as shown in Fig. 25.6(b). In $B/H$ coordinates a diamagnetic line is horizontal. The line $OB$ represents the magnetic state of free space, where $B = \mu_0 H$, which is a good approximation to a normal metal (non-ferrous).

The graphs of Fig. 25.6 are idealized in the sense that they apply to an infinitely long cylinder or flat slab with a tangential external field. For other sample geometries complications arise due to field fringing effects at the edges, see Fig. 25.4. The resulting distortion of field lines causes the value of the applied field to be locally higher at the edges than the average. A partial local transition to the normal state may be induced, thus modifying the magnetization curves. These effects will be further discussed in the next section.

## 25.6  Intermediate State of Type I Superconductors

As pointed out at the close of the preceding section the superconductive to normal transition takes on a modified form for sample geometries differing from

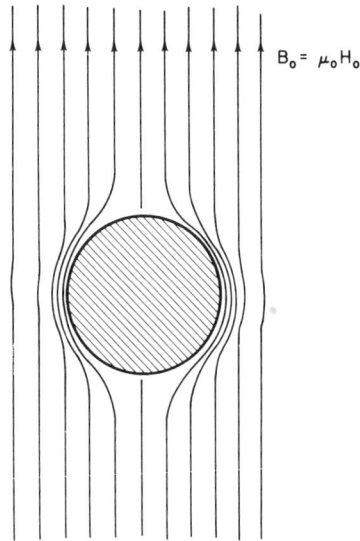

Fig. 25.7. Spherical superconductor in a magnetic field.

the long cylinder or slab. In bulbous shapes, e.g. ellipsoids or spheres, an *intermediate state* is observed consisting of interleaved superconducting and normal filaments. Since the magnetic and thermodynamic properties of the intermediate state provide a convenient introduction to type II superconductors we shall now consider them more closely.

We consider a spherically shaped superconductor as shown in Fig. 25.7, which also indicates the way a uniform applied field will be distorted by the screening supercurrents. We note that the field will be most intense along the equator. Indeed a magnetostatic calculation shows that the tangential field just outside the superconductor in the equatorial plane will be $H_m = \frac{3}{2}H_0 = \frac{3}{2}(B_0/\mu_0)$ where $H_0$ is the average uniform applied field, far away from the sphere or in the absence of it.

From the foregoing it follows that the equatorial field reaches the critical value long before the applied field $H_0$ itself. Hence the sphere can no longer remain completely superconducting. On the other hand the sphere cannot become completely normal. For if this were the case the applied field $H_0$ would appear inside the sphere and, since $H_0 < H_c$, the sphere would have to turn superconducting again.

In fact the superconductor breaks up into a *domain structure* of normal and superconducting *filaments* forming the *intermediate state*. The filaments or

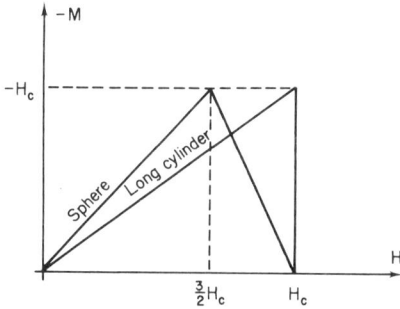

Fig. 25.8. The intermediate state of type I superconductors.

*laminar domains* are typically $10^{-2}$ cm in width, and are readily observed under a microscope by methods resembling those used to observe ferromagnetic domains. The fractions of the sample volume occupied by normal or superconductive domains adjust themselves to give an average magnetization consistent with an internal field equal to $B_c$ ($= \mu_0 H_c$) in the normal domains and zero in the superconducting domains.

The magnetic state of the sphere or indeed any ellipsoidal sample at any value of the applied field can be evaluated by methods of potential theory applied to magnetostatics. In the present context it will suffice to state the results for the typical spherical sample. These are summarized in the $M/H$ graph of Fig. 25.8. The diamagnetic magnetization of the sphere rises more steeply than that of the long cylinder, reaching the maximum value of $M = -H_c$ when the magnetizing force has the value $(\frac{2}{3})H_c$. After that the *average* magnetization of the filamentary structure of the intermediate state drops linearly, reaching zero at $H_c$.

## 25.7  Interphase Boundaries and their Surface Energy

The formation of domains means that *spatial transitions* between the normal and superconducting phases are present within one sample. *Interphase boundaries* are thus formed having surface energy. As it turns out that the interphase boundaries and their energy are of crucial importance to the understanding of type II superconductors we now consider them in some detail.

As one passes across an interphase boundary from a normal domain into a superconducting domain the magnetic field varies from a finite value to zero, while the density of superelectrons increases from zero to a fixed value, appropriate to the superconductor in question and the prevailing temperature.

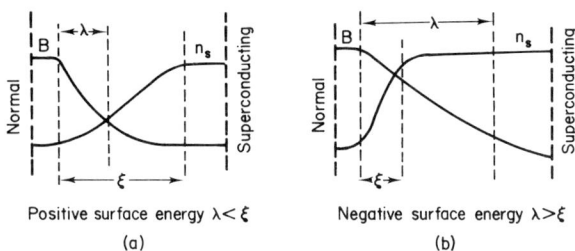

Positive surface energy $\lambda < \xi$          Negative surface energy $\lambda > \xi$
            (a)                                        (b)

Fig. 25.9. Two types of normal to superconducting interphase boundary.

The decrease of the field will be exponential, characterized by the London penetration depth $\lambda$, as suggested in Fig. 25.9. Now, it is important to realize that the density of superelectrons will *not increase according to the same law*. Their build up in the interphase boundary will depend on the size of individual Cooper pairs, which is not necessarily comparable with the penetration depth, but may be greater or smaller. The size of a Cooper pair is usually denoted by $\xi$ and called the *coherence length*. The name is due to the fact that electrons bound in Cooper pairs move in a correlated fashion and it is natural to refer to their size as a coherence length.

As depicted in Fig. 25.9 there are two types of interphase boundary distinguished by the relative magnitude of the penetration depth and coherence length. They occur in different materials and profoundly influence their superconductive properties.

### (i) *Positive surface energy* $\lambda < \xi$

Since Gibb's free energy remains continuous across a phase change, it will be the same in the normal and superconducting domains of a sample in the intermediate state. However, in a phase boundary for which $\lambda < \xi$ the free energy will be locally increased on he basis of the following considerations.

In a normal domain the magnetostatic energy is low while the energy of the electrons is high because they have not condensed into Cooper pairs (see p. 373). On passage through a phase boundary of the type shown in Fig. 25.9(a) the magnetostatic term in the free energy increases more rapidly than the reduction of the electron energy as a result of condensation. Hence the *surface free energy of the interphase boundary is higher than either that of the normal or superconducting phases*. Thus the formation of domains in a superconductor of this type is not favoured. The type I superconductors, discussed so far, are in this class. They form the intermediate state only in sample shapes, e.g. spheres, in which the magnetostatic energy is greater than in long cylinders or slabs. In such cases the magnetostatic energy of the superconducting sample is too great to be compensated by condensation energy of electrons, and the magnetic flux enters the material partially, forming the intermediate state despite the fact that some extra energy is required by the interphase boundaries.

(ii) *Negative surface energy* $\lambda > \xi$

An argument analogous to the above shows that in this case the *free energy of the interphase boundary is less than that of either the normal or superconducting phases*. Hence in superconductors in this class the formation of domains is favoured energetically, and the penetration of the field into the bulk of the sample can take place without destroying superconductivity. This is the principal characteristic of type II superconductors which enables them to remain superconductive in the presence of very high fields. This last property in turn renders them important in applications.

The structure of the normal and superconducting domains in type II materials differs profoundly from the irregular laminar domains found in the intermediate state of type I superconductors. The *mixed state* is the term used to describe the state of a type II superconductor containing a domain structure, to distinguish it from a type I material. The effect of the mixed state on the magnetic properties of type II superconductors will be dealt with in the next section.

## 25.8 Type II Superconductors

In this section a qualitative account of the most important properties of type II superconductors will be given with a view to providing a simple background to an understanding of their applications. It is emphasized that the properties to be discussed cannot be deduced from the London theory introduced in earlier sections and no quantitative relations can therefore be derived at this stage.

The salient properties of type II superconductors are best represented in their magnetization curves shown in Fig. 25.10. As a magnetic field is applied, the material is at first perfectly diamagnetic just like a type I superconductor. However, as the magnetizing force exceeds the *lower critical value* $H_{c1}$ a profound difference becomes apparent. The applied field penetrates gradually into the material which still exhibits a negative susceptibility, but the latter is now greater than $-1$ $(0 > \chi > -1)$. The penetration of the flux into the sample continues until at the *upper critical field* $H_{c2}$ the flux density in the material becomes the same as in the surrounding free space.

The state of the superconductor at applied fields in the range $0 < H < H_{c1}$ is referred to as the *Meissner state* to signify its perfect diamagnetism. The range $H_{c1} < H < H_{c2}$ corresponds to the *mixed state*. This is related to the

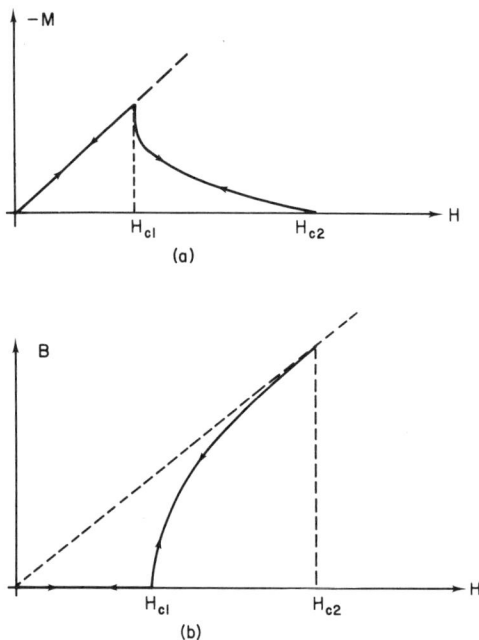

(a)

(b)

Fig. 25.10. Magnetization curves of a type II superconductor.

intermediate state of a type I superconductor in that it consists of a domain structure of normal and superconducting filaments. However, the detailed magnetic and energetic structure of the mixed state has some very characteristic properties.

In the first place the surface energy of the interphase boundaries is negative. Hence their formation is independent of sample shape, contrary to the intermediate state of a type I superconductor. Secondly the mixed state filaments are distinguished by complete regularity of both magnitude and arrangement. The magnetic flux enters the bulk of the material in the form of the quantized *flux lines* introduced in Section 25.4. Once in the sample the flux lines arrange themselves in a regular two-dimensional lattice having triangular symmetry as suggested in Fig. 25.11. The *flux line lattice* shares some of the properties of an ordinary crystal lattice in that it can display various defects, e.g.

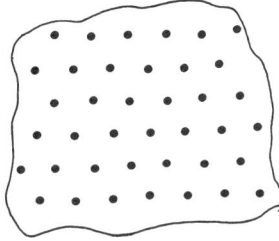

Fig. 25.11. Triangular flux line lattice. (Flux lines normal to plane of paper.)

dislocations. Such defects are usually caused by local variations of the superconducting parameters of the material, which affect the energy balance of the flux line in relation to its neighbours. The energy variations may assume the form of local minima which have the effect of *pinning* the flux line to the locality in question.

An individual flux line consists of a core of normal material, in which the magnetic flux is concentrated, surrounded by superconducting material. The latter supports persistent circulating supercurrents which confine the field to the normal core. The arrangement constitutes a *vortex,* a term frequently used as a synonym for flux line.

The detailed domain structure of flux lines is depicted in Fig. 25.12. At the top a representative line of vortices is shown. The first graph indicates the

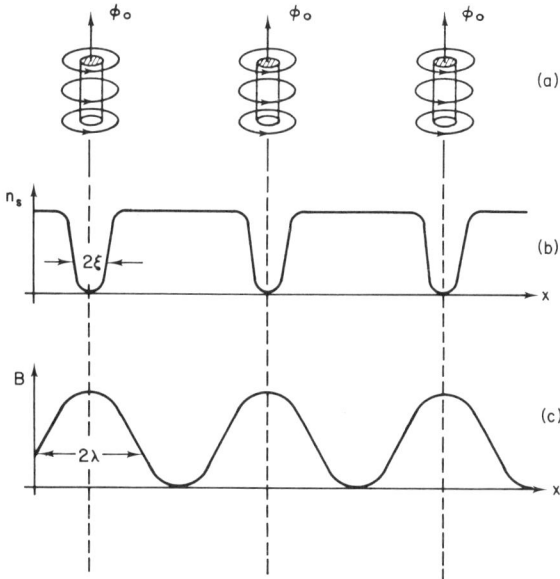

Fig. 25.12. Flux lines in the mixed state of a type II superconductor.

concentration of superelectrons $n_s$ along a line passing through the centres of the vortices. The radius of the normal core is approximately equal to one coherence length $\xi$. The second graph gives the variation of the magnetic field along the same line of vortices. The flux quanta which occupy the normal cores should be thought of as having "smeared" out boundaries, the flux density increasing from zero to its maximum value at the core centre over a distance of the order of one London penetration depth $\lambda$. As pointed out above the mixed state only exists by virtue of the negative surface energy of interphase boundaries associated with the fact that in a type II superconductor the penetration depth $\lambda$ is greater than the coherence length $\xi$.

The flux line lattice of Fig. 25.11 must be visualized as a dynamic arrangement in which the density of vortices changes with the value of the applied field. The spacing of the flux lines, or their lattice constant, will be large at applied field values only slightly in excess of the lower critical field $H_{c1}$ and will decrease with increasing field. When the upper critical field $H_{c2}$ is approached the lattice constant becomes comparable with the material para-meters $\lambda$ and $\xi$, the normal vortex cores tend to merge, thus squeezing out the superconductive matrix, until the whole sample becomes normal. The mixed state thus represents a wide range of magnetic conditions in the sample.

It is useful to note at this juncture the values of the critical fields of type II superconductors. The lower critical field is always low, up to about $10^3$ Gauss (0·1 T). The upper critical field on the other hand can assume very high values—in excess of 100 KGauss (10 T). It is the latter property which makes type II superconductors suitable for the construction of high-field magnets. Both the critical fields of a type II superconductor vary with temperature according to the same law as $H_c$ of type I materials (p. 373). The values quoted above apply at 0 °K.

The wide variation in the values of the critical fields suggests that there are type II superconductors having upper critical fields $H_{c2}$ only slightly in excess of the lower critical fields $H_{c1}$ and that there is a continuous transition from type I to type II properties. This is indeed the case and there is a special parameter whose values classify the two types of superconductivity. This is the $\kappa$-*value* of the superconductor. The dividing value is $\kappa = 1/\sqrt{2}$, type II superconductors having $\kappa$-parameters in excess of this value. It can be shown that $\kappa \sim H_{c2}/H_{c1}$. Thus the $\kappa$-value is a measure of the extent of the mixed state along the magnetic field axis.

The foregoing description assumes tacitly that the flux line lattice is uniform over the cross-section of the sample. Hence the *average* flux density is also uniform and, by Ampére's law, there are no net currents flowing in the superconductor. However, in the mixed state there is still a difference between the fields outside and inside the material and a screening current is necessary to account for that. This screening current flows within the London penetration depth, just like the currents in the Meissner state below $H_{c1}$. and diminishes in magnitude as the applied field approaches the upper critical value $H_{c2}$.

The uniform field distribution over the sample cross-section is a property of a few *ideal type II superconductors* which contain no obstacles to the free movement of flux lines. Most practical materials do not permit the flux lines to rearrange themselves freely in response to variations of the applied field.

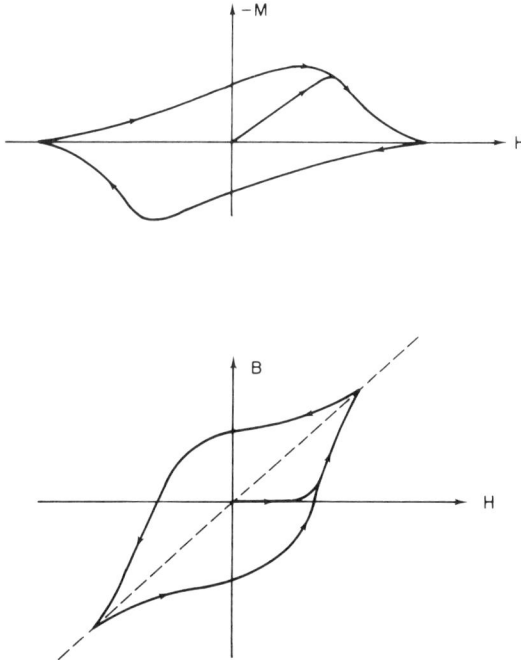

**Fig. 25.13.** Hysteresis loops of irreversible type II superconductor.

Magnetic field gradients are thus formed within the sample, allowing super-currents to flow in the bulk of the material. This is a highly desirable effect in the context of applications, because high supercurrents can be transported without necessitating excessive current densities in superconducting wires of sufficient cross-sectional area.

However, some concommitant effects of non-uniform magnetic field distributions are undesirable. A gradient in the flux line lattice can only exist if individual flux lines become *pinned* at *pinning centres*. The exact nature of pinning centres is not as yet fully understood, but it is clear that they are local energy minima to which a length of a flux line becomes "anchored". A definite force is then required to dislodge a pinned flux line, such force being usually provided by the magnetic pressure of a locally enhanced field gradient. Sometimes temperature variations may affect the local energy balance and thus cause the pinning or unpinning of flux lines.

Whatever the detailed causes of flux pinning, the latter, if present, introduces *irreversibility* into the magnetic properties of superconductors. Irreversibility appears in the form of hysteresis in the magnetization curves of superconductors and is thus analogous to magnetic hysteresis. A typical *hysteresis loop* of a type II superconductor is depicted in Fig. 25.13. This should be compared with Fig. 25.10 which gives the magnetization curves of an ideal, reversible specimen. The irreversible properties cause dissipation of energy in the material if the superconductor is subjected to a.c. fields or currents and constitute a major obstacle to the application of superconductors in a.c. power transmission.

## 25.9  Some Applications of Superconductors

The first application suggested by the discovery of superconductivity was in electromagnets which, it was hoped, would provide limitless magnetic fields without requiring any energy to be sustained. Such early hopes were soon dashed by the realization that elemental type I superconductors become normal in the presence of quite moderate fields (see Table 25.1). The possibility remained dormant for half a century until the discovery of type II superconductors.

The first useful magnets were made of niobium, one of the few elements having type II properties. It has an upper critical field $H_{c2}$ varying up to about 5 KGauss depending very much on the metallurgical treatment. This was enough to provide operating fields for microwave maser amplifiers. Several designs were evolved varying from simple solenoids to ingenious devices incorporating an iron magnetic circuit and a screening superconducting sheet used to confine the magnetic field to a predetermined space.

In this application a superconducting magnet is best operated in the persistent current mode. A circuit used to set up a persistent current is shown in Fig. 25.14. It should be noted that the superconducting coil forms a closed loop all immersed in liquid helium. Connections to an outside source of current, a battery, are via copper leads joined to the superconductor by pressure contacts. The thermal switch is used to raise the temperature locally above $T_c$ and thus to render the superconductor resistive. To set up a persistent current the battery circuit is closed while the thermal switch is activated. This forces the current to flow through the coil windings since the short-circuit link is resistive. The thermal element is next switched off, allowing the magnet loop to become all superconductive. As the battery is disconnected the coil current flows around the closed superconducting loop, thus maintaining a d.c. magnetic field, until it is extinguished by a repeated activation of the thermal switch. In this arrangement the battery must be capable of supplying the full current required by the magnet. As in some applications this may be very high, recourse may be had to a "flux pump" which enables a high current and field to be set up with the help of a low current source. Interested readers are referred to specialist texts dealing with superconductors for further details.

A large number of superconducting magnets, representing a great variety of designs, are used to provide high magnetic fields required in nuclear experiments

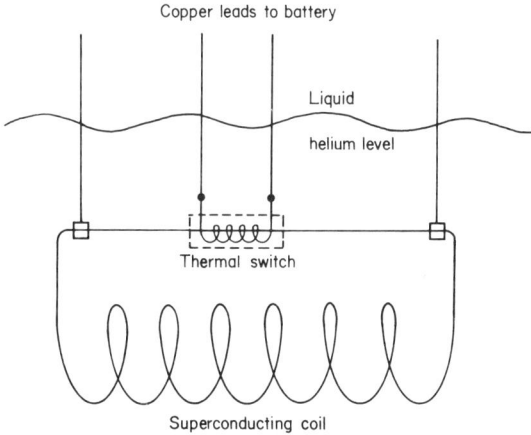

Fig. 25.14. Superconducting magnet circuit.

and in electrical machines. Practically all high field magnets use superconducting alloys such as niobium-zirconium and niobium-titanium. These alloys have upper critical fields approaching 100 KGauss. They are ductile and can be drawn into wires and fabricated into composite cables which incorporate other metals, mainly copper. A further advantage of the alloys is that their composition can be used to adapt their properties to individual applications. There are some super-conducting intermetallic compounds which have even higher critical fields, notably niobium-tin, $Nb_3Sn_2$. However, this material is brittle and very difficult to fabricate in the form of a coil and has therefore found little application.

A successful application of superconducting magnets has been made in the design of homopolar electrical machines based on the Faraday's disc principle. The high magnetic field available makes it possible to construct d.c. motors or generators which are smaller and cheaper to manufacture than conventional machines of similar power. They are particularly suitable where high d.c. currents are required, e.g. in aluminium smelting, or high torques at low speeds, as used to drive large pumps in oil refineries.

The application of superconductors to a.c. machines is complicated by a.c. losses due to the irreversible properties of most practical type II superconductors, explained at the close of the preceding section.

## 25.10 Outline of the Microscopic Theory of Superconductivity

The description of superconductors given in preceding sections dealt largely with their macroscopic properties such as bulk magnetization and large-scale screening currents. These properties were deduced from the London theory which was specifically formulated to coordinate large-scale superconducting phenomena. For this reason the London theory can neither illuminate the microscopic origins of superconductivity nor can it predict properties based directly on them, e.g. the existence of an energy gap or quantum interference phenomena.

The microscopic theory of Bardeen, Cooper and Schrieffer (BCS for short) is based on the possibility of an attractive interaction between pairs of electrons, which enables them to form bound states, the Cooper pairs mentioned at the end of Section 25.4. Before the formulation of the microscopic theory of superconductivity the formation of bound states could only be conceived between particles having opposite charges, e.g., a proton and electron or ions of opposite polarity. The attractive interaction between pairs of electrons is made possible through the medium of the lattice, which consists of positively charged ions situated at lattice sites. These ions are free to move, within limits, around their mean positions. A passing conduction electron displaces neighbouring ions from their rest positions by virtue of electrostatic attraction. The result is a temporary local concentration of positive charge which exerts an attractice force on another electron. The net outcome is an attractice interaction between two electrons, which are able to form a bound state.

The foregoing picture is, of course, greatly oversimplified and cannot be made the basis of any quantitative predictions regarding the dynamics of the electron pairs and their binding energy. To obtain mathematical results recourse must be had to elaborate methods of quantum field theory as applied to solids. Readers interested in pursuing the subject are referred to specialist texts, some of which are listed at the end of this book. We now proceed to quote some of the principal results and explain their practical significance.

The BCS theory predicts that movements of the individual electrons of a bound pair are characterized by equal and opposite momenta

$$\mathbf{p} = \pm \hbar \mathbf{k}$$

where $\mathbf{k}$ is the individual electron wave vector. This constitutes coherent or correlated motion about the centre of mass of the superelectron pair. Thermal motion of a Cooper pair appears as random motion of its centre of gravity. A *superconductive transport current* appears as a drift movement of the centre of mass of a bound pair.

In the process of forming a bound pair each electron gives up a quantity of energy $\Delta$, or $2\Delta$ per Cooper pair. This is the *energy of condensation* between the normal and superconductive phases mentioned earlier.

To break up a bound Cooper pair the energy $2\Delta$ must be imparted to it. This activation energy is equivalent to the existence of an *energy gap around the Fermi level* of the metal, similar to that in a semiconductor. The states below the gap are occupied by condensed superelectrons, while the states above the gap are occupied by normal electrons.

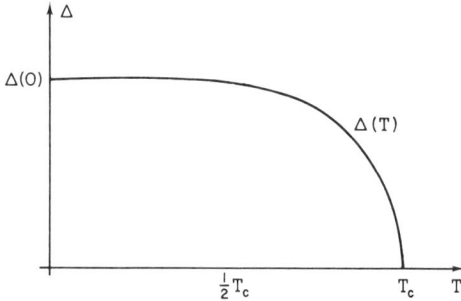

Fig. 25.15. Temperature variation of the energy gap in a superconductor.

The magnitude of the energy gap decreases monotonically with temperature as suggested in Fig. 25.15. The BCS theory predicts that near the critical temperature the energy gap varies according to the approximate relation

$$2\Delta(T) \cong 2\Delta(0)\left(1 - \frac{T}{T_c}\right)^{1/2} \tag{25.10.1}$$

where $2\Delta(0) \cong 3\cdot52\ kT_c$ is the gap at $0\,°K$. Thus the energy gap is directly related to the critical temperature of the superconductor. Moreover it hardly deviates from the maximum value of $2\Delta(0)$ until temperatures of about $0\cdot5\ T_c$ are reached.

The microscopic theory also predicts variation with temperature of the London penetration depth $\lambda$. This is found to have the approximate form

$$\lambda(T) \cong \lambda(0)\left\{1 - \left(\frac{T}{T_c}\right)^4\right\}^{-1/2} \tag{25.10.2}$$

where $\lambda(0)$ is the penetration depth at $0\,°K$. Thus the penetration depth increases rapidly as the critical temperature is approached.

Finally it can be shown from the BCS theory that the density of superelectrons $n_s$ varies with temperature in the same way as the energy gap in Fig. 25.15. At $0\,°K$ all conduction electrons have condensed into Cooper pairs and occupy states below the gap. As the temperature approaches $T_c$, an increasing fraction of superelectron pairs are broken up by the thermal agitation of the lattice and excited as normal electrons to states above the gap.

An experimental verification of the energy gap and a measure of its magnitude can be obtained by the methods applicable to semiconductors and described in Section 3.10. Electromagnetic radiation of frequency exceeding the value $f = (2\Delta)/h$ is absorbed by the superconductor. It is found that values of the threshold frequency lie in the upper microwave range corresponding to gaps of the order of $10^{-4}$ eV. This is about three or four orders or magnitude lower than typical semiconductor gaps.

Perhaps the most striking verification of the energy gap concept is provided by tunnelling experiments. Electron tunnelling can be observed between a superconductor and a normal metal or between two different superconductors. The potential barrier of Section 18.3 is provided by a very thin layer of insulating material ($\cong 50$ Å). The sandwich usually consists of thin films of the desired materials evaporated under high vacuum conditions.

Figure 25.16(a) depicts the arrangement of a superconductor and normal metal in equilibrium (no applied voltage). The Fermi level is constant throughout and occupies an approximately central position in the energy gap of the superconductor and the insulator. The energy gap of the latter is not marked explicitly because it is several orders of magnitude greater than the super-conductor gap. As there are no available electron states in the insulating layer over the extent of the drawing, the layer acts like the potential barrier of Section 18.3. Also suggested in the figure are the densities of states, and it should be noted that these differ fundamentally from semiconductors. The states are crowded together near gap edges in the superconductor.

For simplicity let us assume that the temperature is 0 °K and hence all electrons are confined to states below the Fermi level $E_F$ which means that in the superconductor there are only superelectrons in he lower band. Any of these electrons which may tunnel across the barrier find all states at their energy level fully occupied by the normal conduction electrons, and are unable to establish themselves in the normal metal. Hence no current can flow across the junction.

Let us now consider the consequences of applying a voltage to the junction. Because of the high conductivity of the metals, the entire voltage drop appears across the insulating layer, causing a lowering of the Fermi level in the normal metal relative to the superconductor (assuming the correct polarity of the applied voltage). As soon as the voltage reaches the value $V = \Delta/e$, the Fermi level in the normal metal coincides with the lower edge of the energy gap in the superconductor. Tunnelling electrons can now occupy states just above the Fermi level and a current begins to flow, rising rapidly as the applied voltage is increased. The resulting $I/V$ characteristic is shown in Fig. 25.16(b). At temperatures in excess of absolute zero the sharp onset of the tunnelling current is "smeared" out by the effect of the normal electrons occupying states above the gap. Finally above the critical temperature of the superconductor the tunnelling current is only limited by the thickness of the barrier and obeys an ohmic law.

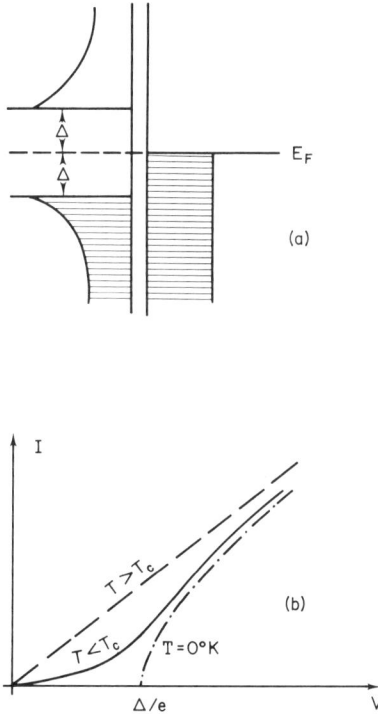

Fig. 25.16. Electron tunnelling between a superconductor and a normal metal.

Tunnelling between two superconductors with differing energy gaps is of practical importance because the $I/V$ characteristic includes a negative resistance region which can be made the basis of amplifiers and oscillators. A tunnel junction of this type is shown in Fig. 25.17 together with its $I/V$ graph. It is not profitable to discuss what happens at $0\,°K$ because the interesting features are not apparent. We therefore consider the application of a voltage to the junction at a temperature above absolute zero but below the lower critical temperature of the two materials. Assuming that the applied voltage is such as to lower the Fermi level on the right side relative to the left side, a gradually increasing current will be observed due to electrons tunnelling across the barrier to occupy some of the few vacant states below the gap $2\Delta_2$. This current will reach its maximum when the lower edges of the energy gaps of the two superconductors are at the same level on the energy axis. The corresponding applied voltage is $(\Delta_1 - \Delta_2)/e$ Volts. Increasing the voltage above this value causes the energy gap $2\Delta_2$ to be opposite the lower band of the left-hand superconductor. The tunnel current now decreases since the number of available states in material 2 is

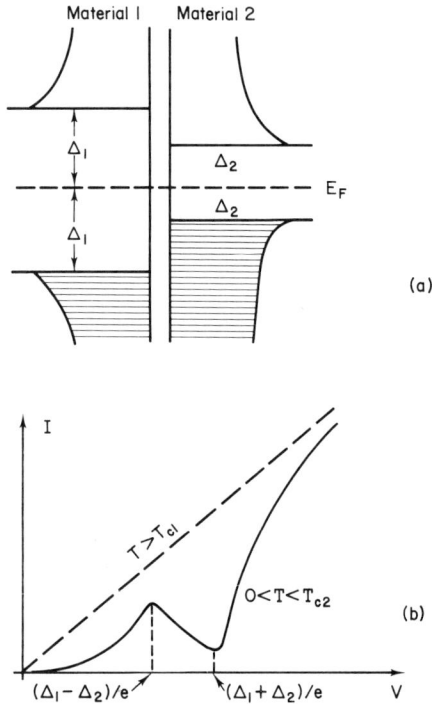

Fig. 25.17. Electron tunnelling between two different superconductors.

effectively decreased. This condition persists until the upper band of material 2 approaches the lower band of material 1 on the energy axis. Numerous empty electron states become available and the current rises steeply. When the applied voltage reaches the value $(\Delta_1 + \Delta_2)/e$ the upper gap edge of material 2 coincides with the lower gap edge of material 1. As can be seen from Fig. 25.17(b) the $I/V$ characteristic has a negative resistance section between the two turn-around voltages.

Although the microscopic theory of superconductivity takes as the starting point individual electrons and their interaction through the medium of lattice vibrations or phonons, it leads to the conclusion, supported by experimental evidence, that the population of superelectrons in a superconducting specimen can be described by a wave function entirely analogous to the wave function of the electron beam of Chapter 18. Thus we can write, by extension of eqn 18.1.2,

$$\psi(\mathbf{r}) = A(\mathbf{r})\, e^{\,i\mathbf{k}\cdot\mathbf{r}} \qquad (25.10.3)$$

where $|\psi(\mathbf{r})|^2 = |A(\mathbf{r})|^2$ is proportional to the local density of superelectrons, and can be normalized to be equal to it if necessary. The above equation is more general than the wave function of the electron beam in two ways. First $A$ is no longer constant but varies with position. This reflects the possibility that the density of superelectrons may vary from point to point within a specimen, as in the mixed state of a type II superconductor. Secondly the wave vector, $\mathbf{k}$ is assumed to vary with position, thus allowing the possibility that the super-electron momentum and velocity may change from point to point, as in a bent superconducting wire carrying a supercurrent.

More important than the differences is the similarity between eqns (18.1.2) and (25.10.3). Both of them describe a travelling wave with a unique phase, as defined by the exponential phase factor. Such a wave is said to be *coherent*. In the case of the electron beam the coherence is due to the uniform momentum (eqn (18.1.4)) imparted to the individual electrons by the accelerating voltage. In the case of the superelectrons the coherence has its origin in the bound state of Cooper pairs into which the normal conduction electrons have condensed. Hence the wave function of eqn (25.10.3) is frequently referred to as the *condensate* wave function. It is emphasized that the wave coherence extends over the whole sample, which may be miles of superconducting wire, just as it extends over the length and breadth of an electron beam in a cathode ray tube.

The coherent nature of the wave function suggests that it may be possible to arrange for superelectrons to display interference properties entirely analogous to electron beams, or any kind of coherent wave motion, for that matter. Such possibilities have been treated theoretically and experimentally and their existence has been verified. Some of the results take the form of interference effects which are measured on a macroscopic scale. Other results include the generation of microwave signals in the centimetre and millimetre range. Collectively they represent macroscopic *quantum interference phenomena*.

*Chapter 26*

# Lasers

The principle of amplification by stimulated emission of radiation was introduced in Chapter 20 on the example of the paramagnetic maser which operates in the microwave range of frequencies. That device embodies the principle in a particularly simple or pure form and is thus a suitable vehicle of exposition. However, the principle of stimulated emission assumes greater practical importance when applied at much higher frequencies—in the visible or near-visible range. The devices which operate in this range are called LASERS for "light amplification by stimulated emission of radiation". The principle of stimulated emission will be elaborated in the present chapter as required and additional material will be included with a view to understanding some of the applications of lasers.

### 26.1 Population Inversion at Light Frequencies

The principle of the three-level maser explained in Section 20.6 is applicable at any frequency and there is no theoretical reason why it should not be used to amplify and generate coherent signals at visible wavelengths. Although this general expectation is valid there are practical reasons why useful lasers embody significant modifications of the basic principle. The object of this section is to explain some of these modifications.

The basis of net stimulated emission is population inversion which can be secured in a three-level system according to the schemes shown in Fig. 20.10. At microwave frequencies the three levels are the result of splitting of the ground level and even at low temperatures ($\cong 4 \cdot 2$ °K) the population *differences* are only very small fractions of the *total* populations. Hence it is not difficult to equalize or saturate the two levels of the pump transition, whenever the transition probability is adequate, at reasonable microwave power levels ($\cong 10$ mW).

At optical frequencies the spacing of the three levels on the energy axis is three or four orders of magnitude greater, and the Maxwell-Boltzmann distribution predicts very small populations in the upper two levels as the reader can easily check. Indeed it is perfectly adequate to assume as a first approximation that only the lowest level is populated under equilibrium conditions. If at the same time the relaxation rates between all three levels are of the same order of magnitude it is impossible to achieve population inversion according to the scheme of Fig. 20.10(b), since even if one succeeds to saturate the $E_1 \rightarrow E_3$ transition the population of level $E_2$ will remain negligibly small.

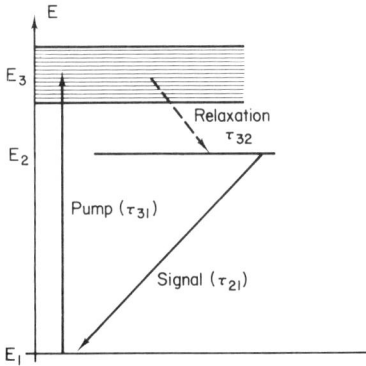

Fig. 26.1. Three-level laser system.

However, population inversion can be achieved according to the scheme of Fig. 20.10(a) because only a fractional transfer of the population from $E_1$ will create an excess population in $E_3$ over $E_2$.

In both cases a monochromatic source of visible radiation of considerable power is required. Moreover, the frequency of the pump source must have exactly the value $f_{13} = (E_3 - E_1)/h$. These requirements make a laser of this type impracticable.

The requirement of a monochromatic source of pump power is avoided by the use of a system of levels in which the uppermost energy state consists of a band of many closely packed levels as suggested in Fig. 26.1. This shows a three-level arrangement which gives rise to laser action in a few rare cases, e.g. ruby. Population inversion between levels $E_2$ and $E_1$ can be obtained if the relaxation rate between states $E_3$ and $E_2$ is very much faster than between $E_2$ and $E_1$ or $E_3$ and $E_1$. In terms of lifetimes this means that $\tau_{32} \ll \tau_{31}, \tau_{21}$. As this implies that individual atoms which arrive at level $E_2$ stay there for an exceptionally long time, this state is said to be *metastable* (see Section 9.5). The application of pump power, possibly in the form of intense white light, excites atoms or ions from the ground level to any one of the states in the band $E_3$. The excited atoms relax quickly to level $E_2$ where they stay for a relatively long period of time. As a result an excess population builds up and laser amplification or oscillation can be obtained at the frequency $f_{12} = (E_2 - E_1)/h$ if the active material is contained in a suitable resonator.

The requirement of a metastable state, suitably positioned within the energy level scheme, together with an intense source of optical pump power makes the three-level laser a relatively rare possibility. The difficulties are overcome by the use of a *four-level system* which is illustrated in Fig. 26.2. As

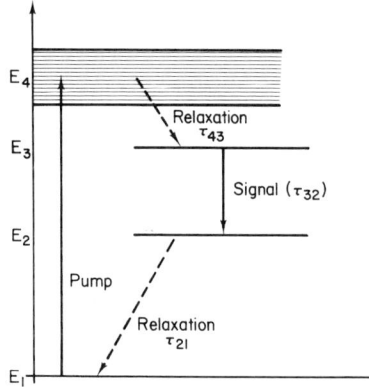

Fig. 26.2. Four-level laser system.

before, the pump power is applied between the ground level and a band of numerous states at the level $E_4$ while a high lying level $E_3$ is allowed to be populated by relaxation from $E_4$. However, by contrast with the three-level system the signal transition is not to the ground state but to the level $E_2$ which is sufficiently high above the ground state to be effectively empty. This implies that $E_2 - E_1 \gg kT$. Under this scheme it is unnecessary to depopulate the ground state, a small amount of pump power being sufficient to excite a fraction of the system population to $E_4$. Inversion of the $E_3$ and $E_2$ populations is then readily obtained without the need for the state $E_3$ to be metastable. The amount of signal power available from the four-level system depends ultimately on the rate at which the lower laser level is emptied to the ground state. If this relaxation rate is too low increasing pump power will have the effect of increasing the total populations of both levels $E_3$ and $E_2$ without affecting their relative magnitudes. In this sense the relaxation time $\tau_{12}$ constitutes a "bottleneck" in the laser output. In solid laser materials this relaxation takes place via lattice interactions which provide a fast rate. In gases the process is usually much slower, unless the relaxation takes the favourable form of recombination of excited ions, as in mercury or argon. Otherwise the final transition to the ground state demands collisions with the walls of the gas container (tube) whose probability is limited by the geometry of the container.

The absorption of infrared and optical radiation in semiconductors was first discussed in Section 3.10. The possibility of reversing the process to obtain radiation was touched on in Section 8.4 in connection with electroluminescence, a process whereby incoherent radiation can be obtained. As a preliminary to laser action or stimulated emission of coherent radiation we now consider more closely the question of population inversion in a semiconductor.

In Chapter 3 the populations of electrons and holes in semiconductors under equilibrium conditions were found to follow the Fermi-Dirac distribution, the Fermi level being positioned somewhere in he energy gap. By *population inversion in a semiconductor* we understand a departure from these equilibrium conditions in the direction of an excess concentration of free conduction electrons and holes. Methods of securing inversion in the context of laser action will be discussed in more detail in a subsequent section, but for the present we note that an excess population of free charge carriers is always present in the vicinity of a *p-n* junction as a result of injection of minority carriers. This being so it would be natural to expect some radiation to be emitted at *p-n* junctions as a result of excess electrons dropping from the conduction band to the valence band. However, as explained in Section 8.4, conditions are not always favourable to radiative processes by virtue of momentum conservation laws which must be observed. Hence in many cases electrons are able to give up their energy to the lattice in the form of heat without emitting any photons. It is beyond our scope to analyse the conditions in which electrons can conserve their momentum in transitions and are thus enabled to emit radiation. For our purpose it will be sufficient to state that these conditions are related to the detailed form of the energy bands, which in turn depends on the crystal structure. Moreover, of the semiconductors known currently, the III to V compounds have the most favourable band structure, gallium arsenide being the best developed.

To secure effective population inversion by injection of minority carriers a heavily doped *p-n* junction is used as shown in Fig. 26.3. The doping is sufficient to cause the Fermi level to be situated within the conduction and valence bands of the *n*-type and *p*-type material respectively. The junction is thus similar to the tunnel diode of Section 8.3 but the impurity concentration is not high enough to introduce a significant negative resistance section into the $I/V$ characteristic. Figure 26.3(a) represents the equilibrium condition with the Fermi level constant across the junction. The *n*-type material is more heavily doped as indicated by the position of $E_F$, deeper inside the conduction band than inside the valence band in the *p*-type material.

The application of a forward bias $V$ to this junction is represented in Fig. 26.3(b). The concentration of injected electrons in the *p*-type material is sufficient for more than 50% of available states above $E_c$ to be occupied. Thus we are faced with a situation in which there are two Fermi levels in the vicinity of the junction: one, denoted by $E_{FC}$, indicates 50% occupation of states by electrons in the conduction band, while the second, denoted by $E_{FV}$, indicates 50% occupation of states by holes in the valence band. This state of affairs is characteristic of population inversion in a semiconductor and prevails in a flat region of material about 1 $\mu$m thick. The distribution of excess electrons over

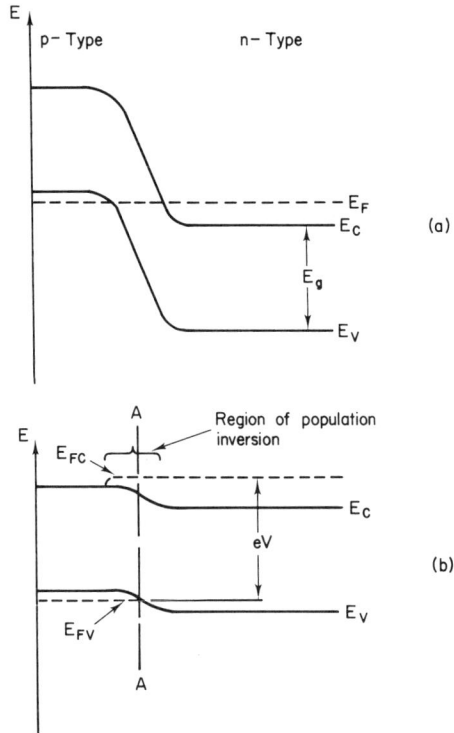

Fig. 26.3. Energy level diagram of a semiconductor laser.

states of the conduction band follows a Fermi-Dirac curve centred on the *quasi Fermi level $E_{FC}$*, also referred to as *imref*. Holes in the valance band are distributed similarly around the *quasi Fermi level $E_{FV}$*, also called an *imref*.

The distribution of electrons and holes along a typical cross-section of the junction, say $AA$, is depicted in Fig. 26.4, which is simplified by the omission of peculiarities of the density-of-states function due to heavy doping (see Section 8.3). The effect of incident radiation on this system depends on frequency. If the frequency is such that $hf > E_{FC} - E_{FV}$ only absorption will be observed since there are few electrons above $E_{FC}$ which can be stimulated to make transitions downwards to empty states below $E_{FV}$, which are also sparse. On the other hand if the radiation frequency is in the range

$$E_g < hf < E_{FC} - E_{FV} \qquad (26.1.1)$$

there is a high density of states in the conduction band occupied by electrons which can be stimulated to make transitions to numerous empty states in the valence band, and net stimulated emission can be obtained. According to eqn

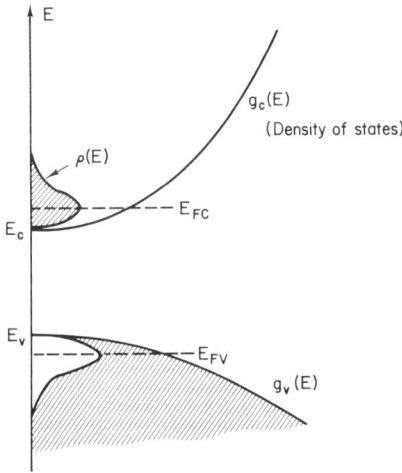

Fig. 26.4. Schematic representation of population inversion in a semiconductor (shaded areas represent filled states).

26.1.1 stimulated emission will be obtained from a semiconductor at a frequency corresponding approximately to the energy gap. In this context the energy gap $E_g$ can be expected to have a somewhat modified value due to heavy doping (see Section 8.3). On the other hand stimulated electron transitions may take place to an impurity level situated just above the valence band in which case the emitted frequency will be less than $E_g/h$.

To obtain coherent stimulated emission it is not enough to secure population inversion. A quantity of active material must be enclosed within an optical resonator or must itself form one. Before taking up the question of optical resonators in the next section it must be emphasized that transitions between atomic energy levels always cover a finite band of frequencies referred to as *linewidth*. Although the energy levels, as evaluated from the time-independent Schroedinger equation, may be discrete, the transition itself is a time-dependent effect which requires a finite interval of time. Hence, the energy emitted or absorbed is subject to the uncertainty of eqn (18.6.2) with a consequent uncertainty in the frequency. In the case of radiative transitions in semi-conductors there is an additional effect due to the finite bands of levels between which transitions can take place. Similarly transitions in a gas are affected by the thermal motions of gas atoms and molecules which further widen the linewidth.

## 26.2  Optical Resonators

One of the primary considerations in the design of resonators for masers and lasers is the number of resonant modes that a cavity can support. It is shown at length in texts dealing with electromagnetic theory that if a cavity is to support only one or two of the lowest-order modes its dimensions must be comparable with the wavelength of the signal. At microwave frequencies this principle leads naturally to the design of cavities which operate in he lowest-order mode at the signal frequency. Such cavities are large enough to contain a volume of the active material which can yield a useful amount of power. Moreover the frequency separation of the resonant modes is much greater than the linewidth of the transition used for maser action. Hence the device can only operate in one, easily predictable, well-defined mode. This desirable condition is impossible to realize at optical frequencies because a single mode cavity would be of microscopic size and could not contain a sufficient quantity of the active material. To obtain a useful amount of power an optical resonator must be of macroscopic size. Hence it will be capable of sustaining a large number of modes in the neighbourhood of the signal frequency. At the same time the spacing of the modes in the frequency spectrum is very much less than the typical linewidth of an optical transition, and a laser tends to operate in many resonant modes simultaneously. However, for various practical reasons it is necessary to confine laser action to as few modes as possible and various practical measures are taken to reject unwanted modes. As a first step towards a quantitative treatment of this question we recall some facts regarding the density of resonant modes and the linewidth of transitions between atomic energy levels.

It is shown at length in texts dealing with electromagnetic radiation that the *mode density*, the number of resonant modes per unit frequency range and unit volume of a closed cavity, is given by the expression

$$p(f) = \frac{8\pi}{c^3} f^2 \qquad (26.2.1)$$

where $f$ is the frequency and $c$ is velocity of light. The above function is entirely analogous to the density of states discussed in Section 17.4 and can be derived by a similar method, modified to apply to electromagnetic waves in a cavity instead of electron waves in a potential well. It should be remembered that eqn (26.2.1) applies only to resonators which are large compared with the wavelength of the radiation and assumes that the dielectric constant of the medium filling the cavity is equal to unity.

With the help of the mode density function it is possible to obtain an estimate of the number of modes which cover the linewidth of an atomic transition. A typical linewidth is about $\delta f = 3 \times 10^{10}$ Hz in the optical range of the spectrum, say at $f = 3 \times 10^{14}$ Hz. Hence we find $p(f)\, \delta f = 2 \times 10^9$ modes per

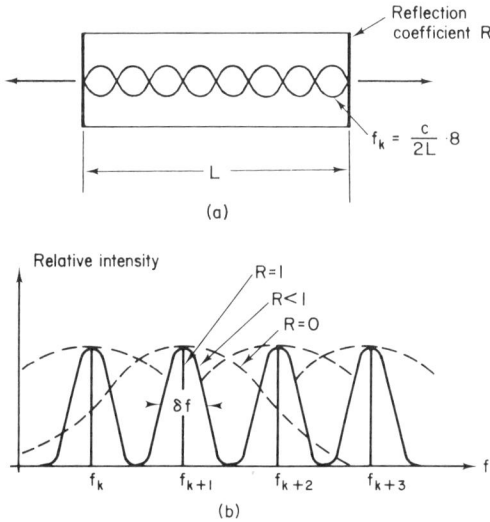

(a)

(b)

Fig. 26.5. (a) Fabry-Perot cavity and one of the "temporal" modes. (b) Resonant mode spectrum.

$cm^3$. This is a very large number of modes, most of which would participate in the atomic transition, thus causing undesirable interference effects. It is therefore imperative to devise a resonator which will suppress most of them and yet retain some modes having high $Q$-factors which can sustain laser action. It is found that these requirements are satisfied by partially open resonators, consisting of two mirrors facing each other. The path between the mirrors provides a resonant space for some modes while others are damped out by radiation or absorption at the non-reflecting walls. The simplest arrangement consists of two plane parallel mirrors which form the *Fabry-Perot cavity*. As the next step we consider this arrangement in greater detail.

As suggested in Fig. 26.5 a resonant condition exists when a standing wave is set up between the reflectors. This requires the following condition to be satisfied

$$k\,\frac{\lambda}{2} = L; \quad k = 1, 2, 3, \ldots \tag{26.2.2}$$

where $L$ is the distance between reflectors and $\lambda$ is the radiation wavelength. The frequencies of the resonator modes are then given by the expression

$$f_k = \frac{c}{\lambda} = \frac{c}{2L} k \qquad (26.2.3)$$

where $c$ is the velocity of light in the medium filling the resonator. In the case of gas lasers this is very nearly the velocity of light in free space. From the above relation it is seen that adjacent modes are separated in frequency by the interval

$$\Delta f = \frac{c}{2L}. \qquad (26.2.4)$$

The mode separation in a laser which is 1 m long is 150 MHz. Thus there are only 200 modes within a transition linewidth of $3 \times 10^{10}$ Hz. Additional effects, to be discussed later, make it possible to restrict laser action to a single resonator mode, if required by a particular application. However, in many cases, e.g. applications to welding, multimode operation of a laser is quite acceptable.

The above discussion of the Fabry-Perot cavity is restricted by the tacit assumption that the transverse distribution of the electromagnetic field patterns of the various modes is identical. In fact the conditions resemble those in a microwave waveguide, where the modes of propagation have different field patterns and different group and phase velocities. Consequently eqn (26.2.4) gives the frequency separation of modes having the same transverse field configuration and hence the seme velocity. The distinction can be emphasized by referring to the transverse field configurations as *spatial modes*, and to the longitudinal standing waves defined by eqn (26.2.2) as *temporal modes*. To obtain a well-defined beam it is necessary to operate in just one spatial mode. This is achieved by making the transverse dimension of the laser cavity much smaller than its length. It can be shown that under these conditions the lowest-order spatial mode suffers the least loss due to diffraction at the laser aperture, and tends to remain as the mode of operation while higher-order modes are damped out by this less mechanism.

Another assumption implicit in the above discussion concerns the value of the reflection coefficient of the mirrors in Fig. 26.5. Only if the reflection coefficient equals unity, $(R = 1)$, is it possible to assign the precise values of eqn (26.2.3) to the mode frequencies, because under these conditions a photon is constrained to traverse the length of the cavity for an indefinitely long interval of time. Its energy and hence frequency is then precisely defined on the basis of the uncertainty principle eqns (18.6.2) and (18.6.1). This idealized case implies that no radiated power can leave the cavity and there is in effect no useful laser output. In practice the reflection coefficient of the mirrors is less than unity, say 0·95, to allow some of the radiation to escape. As a result a photon remains in the cavity for a finite interval of time and therefore its energy and frequency are only defined within the range laid down by the uncertainty principle. This means that the resonant modes of eqn (26.2.3) have frequencies spread over a finite bandwidth which is related to the reflection coefficient.

To obtain an estimate of this bandwidth we note that the average number of transits between mirrors having reflection coefficient $R$ is given by the expression $1/(1 - R)$. Thus the lifetime of a photon in the resonator is

$$\delta t = \frac{L}{c} \cdot \frac{1}{1 - R}. \tag{26.2.5}$$

Substitution of this value into eqn (18.6.1) now yields the bandwidth of a mode

$$\delta f \cong \frac{1}{\delta t} = \frac{c}{L}(1 - R) = 2\Delta f(1 - R). \tag{26.2.6}$$

The last expression follows from eqn (26.2.4). It should be carefully noted that in the present context we denote the uncertainty (or bandwidth) of Chapter 18 by the symbol $\delta f$ while reserving the symbol $\Delta f$ for the mode separation. The mode spectrum of the Fabry-Perot resonator is depicted in Fig. 26.5(b). As expected eqn (26.2.6) predicts a zero bandwidth in the extreme case when $R = 1$. At the opposite extreme of $R = 0$ the bandwidth is comparable with the mode separation. Realistic, operational values of the reflection coefficient lie somewhere between these extremes, but anything less than 0·9 implies some degree of overlap between the resonant modes and a truly monochromatic output cannot be expected from the laser.

The reflection coefficient of the mirrors is closely related to the $Q$-factor of the cavity. A high reflection coefficient means that there is a high ratio of stored to lost energy which by definition means a high $Q$-factor. In this context lost energy means largely the radiation output emerging through the mirrors. Associated with a high $Q$-factor is a narrow resonance bandwidth $\delta f$. The position is thus completely analogous to the properties of microwave cavities whose resonant bandwidths are given by $\delta f = f_0/Q$.

Since the operation of the laser depends very much on the reflecting properties of the mirrors it is natural to use them to select any desired atomic transition for operation in cases where there is more than one possibility. Thus the helium-neon laser can be made to operate in the visible or infrared bands by designing the mirrors to have the necessary reflection coefficient at the desired frequency.

The foregoing discussion of the Fabry-Perot cavity is intended as a typical and simple example of optical resonators of which there are several varieties currently known. In most cases curved rather than flat mirrors are used, a fact which makes their analysis more complicated but facilitates their adjustment. Readers interested in the subject are referred to specialist references, some of which are listed at the end of this book.

### 26.3  Laser Oscillators and Amplifiers

The passage of an electromagnetic wave through any medium which affects its
intensity or power can be described by the expression

$$I(x) = I_0 \, e^{\alpha x} \tag{26.3.1}$$

where $I_0$ is proportional to the square of the wave amplitude at some reference
point, and $I(x)$ is the same quantity at a distance $x$ from the reference point. The
index $\alpha$ is thus a measure of the loss or gain sustained by the wave in unit
distance. If gain is present $\alpha$ is a positive quantity and the wave grows
exponentially with distance $x$. In a resonator of the Fabry-Perot type the
distance travelled by the wave includes return journeys between the mirrors. If
the medium contains atoms or ions having a pair of levels separated by an energy
interval equal to the photon energy, the wave will stimulate transitions as
explained in Section 20.4, and absorption or emission will take place depending
on relative level populations. Accordingly the amplification factor $\alpha$ can be
expected to be proportional to the population difference $n_2 - n_1$, where $n_2$ is
the population of the upper level. At the same time emission or absorption will
be proportional to the transition probability $P$. Hence we can write

$$\alpha = c(n_2 - n_1)P \tag{26.3.2}$$

where $c$ is a coefficient of proportionality which is constant under specified
conditions of operation.

The above equation emphasizes the relation between population inversion
and gain. When $n_2 > n_1$ $\alpha$ is positive and radiation grows exponentially with
distance in the active medium. Otherwise $\alpha$ is negative and radiation decays.

The foregoing relations are well exemplified by the travelling wave maser of
Fig. 20.13. A low power signal entering the input terminal will emerge amplified
according to the law of eqn (26.3.1) provided population inversion is secured by
the input of pump power to the structure. In the absence of pump power the
signal decays between the input and output terminals. In the microwave case
pump power takes the form of a coherent wave obtainable from a tunable
electronic source, e.g. a klystron. As mentioned in Section 26.1 such a facility is
not available at optical frequencies. Alternative methods of pumping will be
discussed in the next section and in the meantime we assume that population
inversion is present.

Although most lasers currently available are in the form of oscillators,
amplifiers analogous to the travelling wave maser can be constructed. Solid laser
materials such as ruby are an example. The crystal is prepared in the form of a
rod and a beam of light to be amplified is passed through it in a longitudinal
direction. Assuming that population inversion has been secured, light of the

correct frequency will be amplified. However, the amplification obtainable from a single traverse along a rod of reasonable length, say 10 cm, is very small and may in fact be negative if a fraction of the beam energy is lost as a result of partial reflection at the end faces. Measures must, therefore, be taken to eliminate reflections, or "match" the two media. One possibility is to use non-reflective coatings. Another, very effective method is applicable when the radiation to be amplified is in the form of a plane polarized wave. In this case the end faces are not normal to the axis of the rod but are inclined at an angle to ensure incidence of the beam at the Brewster's angle as suggested in Fig. 26.6.

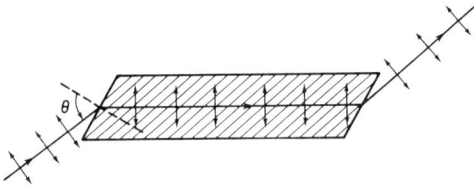

**Fig. 26.6.** Amplification of light which is polarized in the plane of the diagram. Brewster's angle $\theta$ ensures reflection-free incidence at end faces.

*Brewster's angle* is defined by the relation $\tan \theta = k_2/k_1$ where $k_2$ and $k_1$ are indices of refraction of the laser material and the space outside, respectively. It is shown in texts dealing with electromagnetic waves that under these conditions the incidence is reflection free if the wave is polarized in a plane normal to the end face of the laser rod.

An amplifier of the above type can be made into an oscillator if a fraction of the output power is fed back to the input with the help of an arrangement of mirrors. However it is much more convenient and efficient to provide internal feedback by the use of a resonator of the type described in the preceding section. The radiation is bounced back and forth between the mirrors and is all the time subject to amplification according to eqn (26.3.1). The oscillations are started by fluctuations inside the material and build up to a steady state by the same processes which apply to any feedback oscillator. Power is allowed to escape through one or both end mirrors which are made partially transparent for this purpose. Self oscillations can only be sustained if a balance is secured between the internal gain and the power extracted through the mirrors. This means that the reflection coefficient must not be too low, otherwise the aggregate average distance covered by a photon within the active medium will be too short to ensure sufficient amplification. Typically between 20 and 100 journeys between the mirrors are necessary to sustain oscillations which means that reflection coefficients in the range 0·95 to 0·99 are required.

The practical method of assembling a resonator of the Fabry-Perot type depends on the laser material. In he case of a solid crystal such as ruby a rod-shaped specimen is prepared and the end faces are polished and coated to become mirrors having the requisite reflection coefficient at the operating frequency. The procedure can be applied to semiconductor lasers but in this case the active material is in the shape of a flat slab rather than a rod.

Gaseous laser materials must be contained within a tube. The mirrors cannot in this case be incorporated in the ends of the tube because of alignment difficulties. The mirrors must be parallel to a high degree of precision which cannot be achieved within the flexibility of glass tubes. For this reason the reflectors of gas laser resonators are mounted separately on an optical bench as depicted in Fig. 26.7. In this arrangement the glass end plates of the tube can be

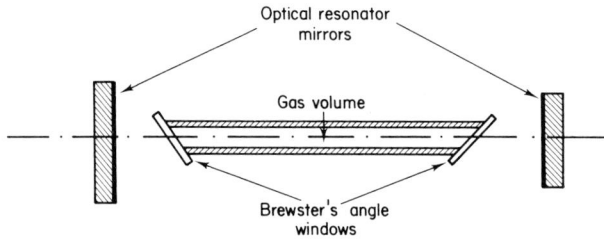

**Fig. 26.7.** Gas laser tube with Brewster's windows and separately mounted resonator mirrors.

the cause of unwanted reflections, which are eliminated by orienting the plates at the Brewster's angle to the laser beam. In addition to passing light polarized in a plane normal to the plane of the end plate without reflection, Brewster's windows reflect completely light polarized in a plane at $90°$. Such light leaves the laser tube in a transverse direction to the tube and is thus not subject to amplification. This has the additional desirable effect of restricting laser operation to resonator modes polarized in one plane only.

### 26.4  Pumping Methods

It was emphasized in Section 26.1 that the pumping process presents considerable problems because there are no tunable sources of coherent radiation at optical frequencies. Indeed it is an important objective of current research to develop tunable lasers which could fulfil this function. In the meantime recourse must be had to methods which do not require coherent radiation or no radiation at all.

*Optical pumping* of a three-level system represents the most direct extension of the microwave maser principle. Two practical limitations must be borne in mind. Firstly the light sources available, e.g. incandescent lamps, high pressure mercury lamps, xenon arc lamps, provide light over a broad band and are therefore only effective with systems having a band of excited levels as illustrated in Figs 26.1 and 26.2. Secondly the output of these sources is concentrated in the infrared band and the lower part of the visible spectrum, relatively little radiation being available in the violet and ultraviolet parts of the spectrum. It follows from this that laser substances which have a pumping band in the upper part of the spectrum are unlikely to be pumped effectively.

Optical pumping must be used in conjunction with solid insulating laser materials since none of the other methods, to be discussed below, is applicable. Such materials take the form of dielectric crystals containing metal ions as impurities whose energy levels are used for laser action. The best known of these is ruby $(Al_2O_3 - Cr^{3+})$ which will be described in greater detail in the next section. It should be noted that the energy level scheme differs from crystal to crystal and from that of the free ion.

Another group of solid laser materials is provided by rare earth ions embedded in insulating crystals. The energy level configurations of the ions differ significantly depending on whether they are divalent or trivalent. The divalent ions, such as dysprosium in calcium fluoride $(CaF_2 - Dy^{2+})$, have pump bands extending from the infrared to the middle of the visible spectrum and are thus very susceptible to optical pumping. The trivalent ions, such as neodymium in calcium tangstate $(CaWO_4 - Nd^{3+})$, have pump bands in the ultraviolet and cannot therefore be very effectively pumped by available sources.

An interesting situation may arise when two species of ions are embedded in the same lattice. Pumping may then be effected indirectly by an *energy exchange* between the ions. An example of this type is provided by yttrium-aluminium-garnet doped simultaneously with chromium and neodymium. The energy-level scheme of $Cr^{3+}$ in YAG is similar to that in ruby, see Fig. 26.11. Population inversion of a transition in the red part of the spectrum can be secured by optically pumping a band of levels lying somewhat higher in the visible range. Spontaneous emission by the relaxing chromium ions then pumps the $Nd^{3+}$ ions, resulting in population inversion of a transition lying in the infrared.

One of the principal advantages of gas lasers is that population inversion in a gas can be secured without the need for a separate source of radiation. Indeed only a d.c. or a.c. voltage source is required to set up a glow discharge according to the principles discussed in Chapter 9. Population inversion between a pair of levels is then obtained through one or more of the ionization and excitation processes present in a gas discharge. A detailed description of a typical example, the helium-neon laser, will be given in the next section. In the meantime we list some of the processes leading to population inversion.

Perhaps the simplest process is *excitation of inert gas atoms by electron collisions*. As can be seen by reference to Table 16.1 the ground states of inert gases are $p$ states. In view of the selection rule $\Delta l = \pm 1$ excitation can only take place to *s* or *d* states as suggested in the diagram of Fig. 26.8, which indicates that each set of values of the quantum numbers *n* and *l* corresponds to a group of states which are not degenerate in a many-electron atom. Population inversion is easily achieved and maintained because all the excited states are metastable. Consequently a large number of laser transitions has been observed in this system.

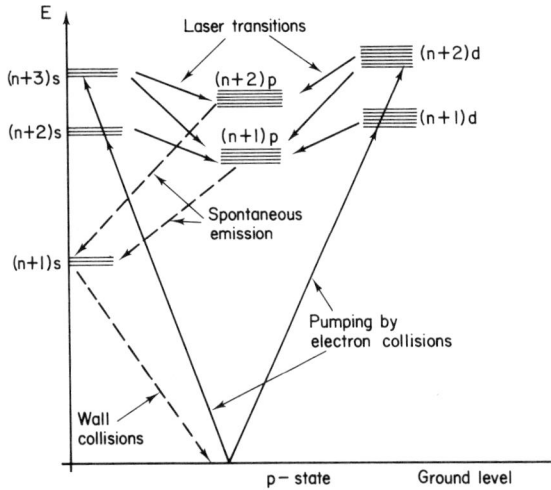

**Fig. 26.8.** Schematic energy level scheme of inert gas lasers.

A somewhat more elaborate pumping scheme is obtained as a result of *dissociation of polyatomic molecules* in collisions with excited atoms. Usually one of the dissociated atoms appears in its own ground state while the other appears in an excited state, whose population becomes inverted. The bombarding atoms are usually inert gases which have themselves been excited by electron collisions. A considerable number of laser materials can be pumped by this process. Figure 26.9 illustrates the method on the example of atomic oxygen obtained by dissociation of oxygen molecules.

A rare but effective pumping possibility occurs as a result of a *direct transfer of excitation energy in an inelastic collision* between an excited atom of one species and a ground level atom of another species. A transition of this type is only probable when the energy levels of the colliding atoms are very close to each other, within a range of 1 kT. The most extensively studied example of this type is the helium-neon laser which is described in greater detail in the next section where a simplified energy level diagram is given in Fig. 26.12.

Population inversion by injection of charge carriers across a semiconductor *p-n* junction was described in the opening section of this chapter. At this juncture we note that this method requires only a low voltage d.c. source of pumping energy. It is thus even more advantageous technologically than a gas discharge which is sustained by a potential of hundreds of volts. A more detailed description of the gallium arsenide laser is given in the next section.

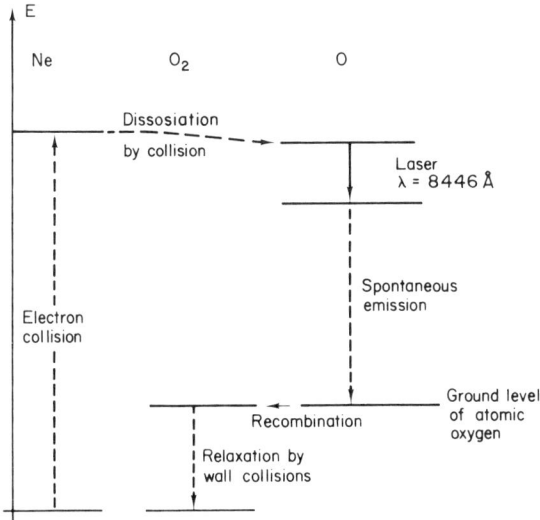

Fig. 26.9. Simplified energy level scheme of the oxygen laser.

## 26.5 Some Typical Lasers

Various aspects of laser operation have been dealt with in preceding sections to the point where we are sufficiently equipped to assemble the elements and consider some working devices. Although the objective of laser operation has an impressive unity and simplicity, the generation and amplification of monochromatic light, the achievement of this objective is based on a great diversity of physical phenomena and technological means. In the present section we attempt to exemplify this diversity through the description of three typical lasers: (i) ruby laser, (ii) helium-neon laser and (iii) gallium arsenide laser.

### (i) Ruby laser

Ruby provides an example of a solid state dielectric laser which must be pumped optically, as explained in the preceding section. The resonator is prepared in the form of a cylindrical rod, typically 10 cm long and 0·5 cm in diameter, cut from a single crystal, artificially grown boule. The cylindrical envelope is given a ground finish which absorbs and scatters incident radiation as explained in Section 26.2. The end faces are polished and oriented accurately to form an optical resonator, and coated with gold to give partly transparent mirrors at wavelengths in the red part of the spectrum.

The ruby rod is mounted in the focus of an eliptic reflecting enclosure as suggested in Fig. 26.10. In the other focus a tubular source of optical pump

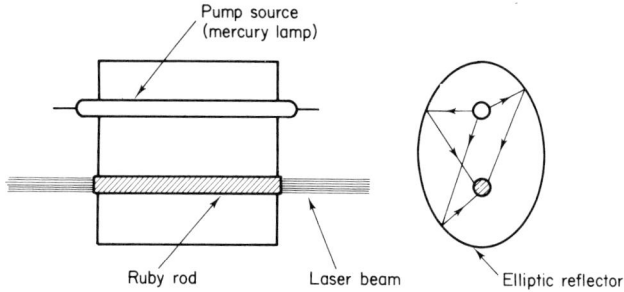

Fig. 26.10. Schematic layout of a ruby laser.

radiation is situated, its output being directed towards the ruby by reflection from the eliptic wall. For continuous operation mercury high pressure lamps are used, while for pulsed operation xenon flash tubes are employed. The power supplies required to drive the pump sources are adapted correspondingly.

The energy level scheme of $Cr^{3+}$ in ruby is shown in Fig. 26.11. There are two absorption bands within the optical band which allow effective pumping by gaseous sources of light. The excited chromium ions relax quickly (lifetime $\tau_{32} < 10^{-7}$ sec) to two discrete levels whose lifetime to the ground state is about 3 msec. On the other hand the relaxation between the two levels is very rapid thus ensuring that they remain in thermal equilibrium. The relationship of relaxation rates is so favourable in this case that population inversion with respect to the ground state is readily achieved and laser action between the lower of the two discrete states and the ground state takes place yielding red light of wavelength $\lambda = 6943$ Å. Laser oscillation does not usually take place from the higher level because of its lower population. It will be recalled from Section 20.5 that the ground level has a zero field splitting of about 12 GHz. At low temperatures (77 °K) this gives rise to two distinct laser lines, but at room temperature the linewidth of the laser line is about 300 GHz, which prevents resolution, and only one, wide line is observed.

Optimum doping of ruby crystals for laser use is found to be about 0·035% $Cr^{3+}$. Lower doping yields less power while higher doping causes interactions between adjacent chromium ions which modify the energy level scheme and the relaxation rates.

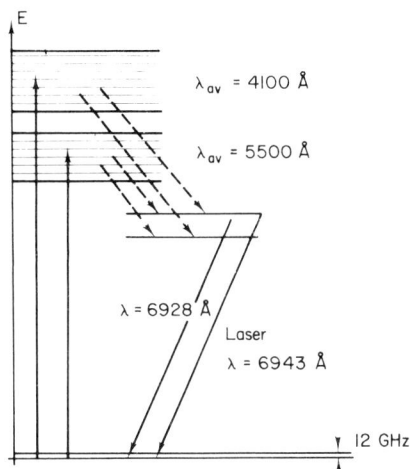

Fig. 26.11. Energy level scheme of $Cr^{3+}$ in ruby.

## (ii) Helium-neon laser

A simplified energy level scheme of the helium-neon laser is shown in Fig. 26.12. Of the multiplicity of excited helium levels the two shown in the diagram are metastable and tend to be preferentially populated by electron collisions in a glow discharge, which act over a broad band of energies. The excited helium

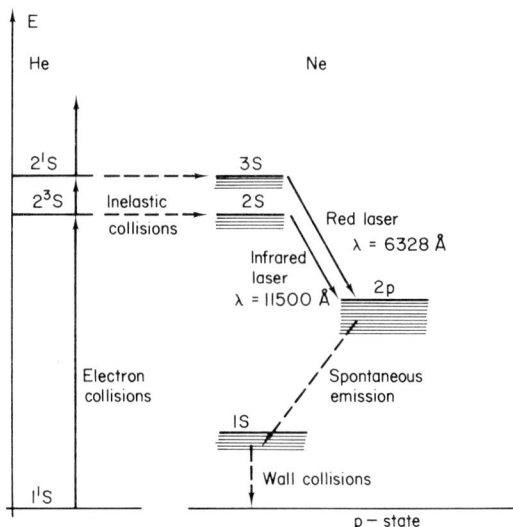

**Fig. 26.12.** Simplified energy level diagram of the helium-neon laser. Numerous other laser transitions are not shown.

atoms then give up their energy via inelastic collisions to neon atoms occupying the ground state. As explained in the preceding section this is possible by virtue of the close coincidence of the 2S and 3S states of neon with the metastable states of helium. The lifetimes of the 2S and 3S states are much longer (0·1 μsec) than the lifetime of the 2p states (0·01 μsec) the conditions for population inversion are therefore present. The resulting laser action can take place at the red and infrared wavelengths marked in the diagram. Selection of one or the other transition is effected through properly designed mirror coatings. Emptying of the lower laser level is somewhat complicated by the fact that spontaneous emission leads only to the 1S state, transitions to the ground level being forbidden. Now, the 1S state happens to be metastable and the only mechanism whereby neon atoms can relax to the ground state is wall collisions. This is confirmed by the observation that the gain of He-Ne lasers is inversely proportional to the tube diameter. We find here a typical example of the pumping "bottleneck" mentioned in Section 26.1.

Because of this and for the reasons connected with mode suppression (see p. 400) the He-Ne laser takes the form of a long tube, up to 1 m long and about 5 mm diameter. The tube is invariably provided with Brewster's windows and mounted in line with separate mirrors as shown in Fig. 26.7. Typical partial pressures of the gas mixture are 1 mm Hg of He and 0·1 mm Hg of Ne. The gas discharge can be maintained with the help of either a d.c. voltage source or a r.f. generator of about 30 MHz. The latter is more convenient in that only electrodes mounted externally on the tube are required, as against sealed in electrodes necessary with the d.c. discharge. On the other hand the r.f. source causes interference with other electronic equipment.

The mirrors are provided with screw adjustments designed to facilitate alignment which is effected through special optical procedures before the discharge is switched on. If mirrors for operation in the red part of the spectrum are used, a beam of red light appears as soon as the discharge is established throughout the length of the laser tube. In the case of infrared mirrors an infrared detector or an image converter tube must be used to ascertain the presence of a beam.

### (iii) *Gallium-arsenide laser*

Although semiconductor lasers can be pumped optically, their principal advantage lies in the fact that only a low voltage d.c. source is required to apply a forward bias to a *p-n* junction and to produce population inversion of the type outlined in Section 26.1. Figure 26.13 depicts the basic structure of a junction laser in the form of a Fabry-Perot cavity. The thickness of the dice is a small fraction of a millimetre while the lateral dimensions are about 1 mm. The thickness of the active region is about 1 μm, the requirement of a long and thin resonator is thus met (see Section 26.2). The resonator is effectively open at sides other than the reflecting ones by having the exposed faces roughened, while the boundaries between the active region and the inactive material are themselves absorbing.

A GaAs laser junction is usually prepared by diffusion of acceptor atoms, say Zn, into $n$-type material having impurity concentration in the range $10^{17}$ to $10^{19}$ cm$^{-3}$. The junction must be mounted on a heat sink, preferably diamond, to prevent the temperature from rising too high as a result of ohmic losses caused by the injection current.

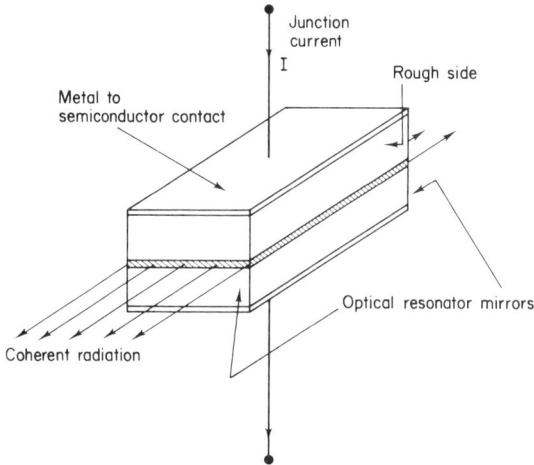

**Fig. 26.13.** Simplified structure of a semiconductor injection laser.

Laser operation only sets in after the junction current has been increased above a threshold value, required to produce a sufficient concentration of injected minority carriers which results in population inversion as indicated in Fig. 26.3. The value of the threshold current is much lower at low temperatures ($< 100\,^\circ$K) since the form of the Fermi-Dirac distribution and low rate of recombination favour inversion. As the junction approaches normal temperatures the threshold current must be increased steeply. Since the ohmic losses are also increased, a stage is reached when it becomes impossible to operate the laser cotinuously and only pulsed operation is permissible.

The wavelength of the emitted radiation is closely related to the bandgap as modified by heavy doping. In the case of GaAs this is about 8600 Å at 77 $^\circ$K, increasing with temperature in accordance with the narrowing of the energy gap.

# Short List of Collateral and Further Reading

## Part I

Anderson, J. C. "Dielectrics". Chapman and Hall, London (1964).

Howatson, A. M. "An Introduction to Gas Discharges", Pergamon Press, Oxford (1965).

Kittel, C. "Introduction to Solid State Physics". (Second edition.) Wiley, New York (1963).

Shive, J. N. "Semiconductor Devices". Van Nostrand, New York (1959).

Smith, R. A. "Semiconductors". Cambridge University Press (1964).

Sze, S. M. "Physics of Semiconductor Devices". Wiley, New York (1969).

## Part II

Fast, J. D. "Entropy". Philips Technical Library (1962).

Kuhn, H. G. "Atomic Spectra". Longmans, Green, London (1964).

Landau, L. D. and Lifshitz, E. M. "Quantum Mechanics". Pergamon Press, London (1959).

Mandl, F. "Quantum Mechanics". Butterworths, London (1957).

Messiah, A. "Quantum Mechanics". (2 vols.) North Holland Publishing Company, Amsterdam (1965).

Schiff, L. I. "Quantum Mechanics" (Second edition.) McGraw-Hill, New York (1955).

Smith, R. A. "Wave Mechanics of Crystalline Solids". Chapman and Hall, London (1961).

## Part III

Bozorth, R. M. "Ferromagnetism". Van Nostrand, New York (1956).
413

414 SHORT LIST OF COLLATERAL AND FURTHER READING

Lax, B. and Button, K. J. "Microwave Ferrites and Ferrimagnetics". McGraw-Hill, New York (1962).

Morrish, A. H. "The Physical Principles of Magnetism". Wiley, New York (1965).

Newhouse, V. L. "Applied Superconductivity". Wiley, New York (1964).

Rose-Innes A. C. and Rhoderick, E. H. "Introduction to Superconductivity". Pergamon Press, Oxford (1969).

Röss, D. "Lasers". Academic Press, London (1969).

Siegman, A. E. "Microwave Solid State Masers". McGraw-Hill, New York (1964).

Yariv, A. "Quantum Electronics". Wiley, New York (1967).

## General

Bleaney, B. I. and Bleaney, B. "Electricity and Magnetism". (Second edition.) Oxford University Press, Oxford (1965).

Coulson, C. A. "Waves". (Fifth edition.) Oliver and Boyd, Edinburgh (1949).

Hlawiczka, P. "Matrix Algebra for Electronic Engineers". Iliffe Books, London (1965).

Kittel, C. "Thermal Physics". Wiley, Chichester (1970).

Pippard, A. B. "Elements of Classical Thermodynamics". Cambridge University Press, Cambridge (1964).

# Some Physical Constants

Electronic charge, $e = 1 \cdot 6021 \times 10^{-19}$ coulomb

Electron rest mass, $m_e = 9 \cdot 1085 \times 10^{-31}$ kg

Proton rest mass, $M = 1 \cdot 673 \times 10^{-27}$ kg

Planck's constant, $h = 6 \cdot 6252 \times 10^{-34}$ joule-sec

Boltzmann's constant, $k = 1 \cdot 3804 \times 10^{-23}$ joule $(^\circ K)^{-1}$

Velocity of light, $c = 2 \cdot 9979 \times 10^8$ m . $(sec)^{-1}$

$\mu_0 = 4\pi \times 10^{-7} = 1 \cdot 2566 \times 10^{-6}$ H . m$^{-1}$

$\epsilon_0 = \dfrac{1}{36 \times 10^9} = 8 \cdot 8542 \times 10^{-12}$ F . m$^{-1}$

Avogadro's constant, $N_0 = 6 \cdot 0247 \times 10^{23}$ (g . mole)$^{-1}$

One mole = molecular or atomic weight (g)

Gas volume per mole = $2 \cdot 2421 \times 10^{-2}$ m$^3$

Gas constant $R = 8 \cdot 317$ Joules $(^\circ C)^{-1}$ (mole)$^{-1}$

Radius of first hydrogen orbit, $a_0 = 0 \cdot 529$ Å

Rhydberg energy, $R$ . $hc/e = 13 \cdot 605$ eV

Bohr magneton, $\beta = 9 \cdot 273 \times 10^{-24}$ Am$^2$

Quantum of magnetic flux, $\phi_0 = 2 \cdot 07 \times 10^{-15}$ Wb

$J = 4 \cdot 1855$ joules (calorie)$^{-1}$

One electron volt = $1 \cdot 6021 \times 10^{-19}$ joule

At $4 \cdot 2 \,^\circ K$, $kT = 3 \cdot 61 \times 10^{-4}$ eV

At $27 \,^\circ C$ (= $300 \,^\circ K$), $kT = 0 \cdot 0258$ eV

One micron ($\mu$m) = $10^{-6}$ m

One Angstrom (Å) = $10^{-10}$ m = $10^{-4}$ $\mu$m

# Examples

Some of the examples collected below may require reference to the tables in the text for additional data. Others yield results which should be compared with tabulated information.

Most of the examples constitute direct applications of the text, but in a few cases an element of extension is included.

### Examples on Part I

1. Evaluate the spacing between the two lowest energy levels of an electron in a one-dimensional potential well. Also find the number of levels within 1 eV of the lowest level.

Calculate the number of allowed energy states within the first Brillouin zone of a one-dimensional lattice of length 1 cm having a lattice spacing of 1 Å.

$$(1.24 \times 10^{-14} \text{ eV}, 1.6 \times 10^7, 10^8)$$

2. Germanium, silicon and cadmium sulphide are transparent to light of wavelengths less than 18 700 Å, 11 400 Å and 5200 Å respectively.

Calculate the photon energies at these wavelengths in terms of electron-volts and compare with Table 1.1.

3. Plot to scale on a sheet of graph paper the Fermi-Dirac distribution function for the temperatures $T = 0\,°\text{K}, 300\,°\text{K}, 1000\,°\text{K}$. Calibrate the energy axis in eV relative to the Fermi level.

On each curve mark the values of energy at which 95% and 5% of the available states are actually occupied by electrons. Also mark on the graph the energy gap of germanium assuming that it is centred on the Fermi level.

4. At 300 °K the intrinsic resistivity of germanium is 58 ohm-cm, and the Fermi level is 0.009 eV below the centre of the band-gap. Calculate the "density-of-

states" effective masses of electrons and holes in terms of the free-electron mass
($M_c = 4, M_v = 1$).

$$(0.22, 0.39).$$

5. Calculate the number of free electrons per unit volume of intrinsic silicon at room temperature. Compare the result with the number of atoms per unit volume of silicon. Carry out the above calculation using the full Fermi-Dirac function and its exponential approximation.

$$(7.8 \times 10^9 \ cm^{-3}).$$

6. A specimen of p-type silicon contains $3 \times 10^{15}$ acceptors per $cm^3$. Assuming these are fully ionized, determine the position of the Fermi level with respect to the valence band at 300 °K. Show that the assumption of complete ionization is reasonable. Calculate the specimen resistivity. At what temperature would half the acceptors be ionized?

$$(0.212 \ eV, 4.16 \ ohm\text{-}cm, 65 \ °K).$$

7. A sample of n-type germanium at 300 °K contains $N_d$ donor impurities per $cm^3$. If 10% of the donors are unionized, what is the value of $N_d$? Calculate the resistivity. If acceptors are added, will the resistivity increase or decrease?

$$(7 \times 10^{16} \ cm^{-3}, 0.024 \ ohm\text{-}cm).$$

8. Assuming that all impurities are ionized at 300 °K, determine the concentrations of mobile electrons and holes in silicon if $N_d = 10^{16} \ cm^{-3}$ and $N_a = 2 \times 10^{16} \ cm^{-3}$. Check the assumption of complete ionization.

$$(1.9 \times 10^4, 10^{16}, \text{about 7% of acceptors are unionized}).$$

9. Evaluate the impurity concentration in a p-type sample of germanium at room temperature given that the measured Hall voltage is 2.5 mV in the presence of a transport current of 10 mA and an applied magnetic field of 0.4 T. The thickness of the slab is 1 mm as measured in a direction parallel to the applied field. ($10^{16} \ cm^{-3}$).

10. Use the mobilities given in Table 3.1. to determine the velocities of holes and electrons at 300 °K for an electric field of 100 V/cm in (a) germanium and (b) silicon.

11. The ionization energy of a donor impurity atom can be estimated by applying the Bohr model of the hydrogen atom. The ionized donor is treated as a fixed positive charge and a single electron is considered as moving within its field of force. The host crystal is assumed to provide a medium of dielectric constant different from free space. The ionization energy is identified with the first excited level of this configuration.

Evaluate this energy for an impurity atom embedded in germanium which has a dielectric constant of 16. Also find the radius of the ground-state orbit and compare the ionization energy with $kT$ at room temperature.

$$(0.052 \text{ eV}, 8.5 \text{ Å.}).$$

12. Compute the number of thermal electrons in the C.B. of an ideal intrinsic semiconductor having an energy gap of 1 eV. Do the same for an ideal insulator having a gap of 6 eV. Assume the materials to be at room temperature.

$$(1.14 \times 10^{11} \text{ cm}^{-3}).$$

13. An unsymmetrically doped germanium $p$-$n$ junction has an acceptor concentration of $10^{16}$ per $cm^3$ on the $p$-type side, and a donor concentration of $10^{18}$ per $cm^3$ on the $n$-type side. All impurities can be considered to be fully ionized.

Calculate the contact potential of the junction under conditions of equilibrium.

$$(0.43 \text{ V}).$$

14. Plot to scale the current-voltage curve for a germanium $p$-$n$ junction having an impurity concentration of $10^{16}$ per $cm^3$. Cover only the range of bias voltages between $-0.5$ Volt and $+0.5$ Volt.

15. A silicon $p$-$n$ junction has a reverse saturation current of 200 $\mu A/cm^2$. Given that the cross-sectional area of the junction is 1 $mm^2$ find the forward bias voltage that must be applied to obtain a forward current of 5 mA. $(0.22 \text{ Volts})$.

16. For the junction of problem 13, evaluate and plot the injected minority-carrier concentration in the $p$-type region as a function of distance from the barrier. Also find the diffusion current of minority carriers crossing the barrier from the $n$-type into the $p$-type region. Assume lifetime of electrons to be 50 $\mu sec$, and take values of diffusion constants as follows $D_n = 90$, $D_p = 45$ $cm^2$ $sec^{-1}$.

State the saturation reverse current for this junction under conditions of reverse bias. $(44 \ \mu A/cm^2)$.

17. A silicon *p-n* junction of 1 mm cross-section carries a forward current of 5 mA. Assuming that the doping of the junction is symmetrical (impurity concentration = $10^{16}$ per $cm^3$), plot roughly the following quantities as a function of distance from the barrier:

    (a) concentration of majority carriers on both sides of the barrier;
    (b) concentration of diffusing minority carriers on both sides of the barrier (concentration of thermal minority carriers may be neglected);
    (c) diffusion current of minority carriers on both sides of the barrier,
    (d) drift current of majority carriers on both sides of the junction.

Mark on the diagram the following numerical values: concentration of holes close to the barrier on the *n*-type side; concentration of electrons on the *p*-type side; diffusion lengths on both sides and diffusion currents at the junction. Take the lifetime of minority carriers to be 50 $\mu$s, and the diffusion constants $D_n = 30, D_p = 13$ $cm^2$ $sec^{-1}$.

(0·226 x $10^{16}$, 0·226 x $10^{16}$, 0·025 cm, 0·045 cm, 1·9 mA, 3·1 mA)

18. Sketch the same curves and calculate the same quantities as in problem 17 for a silicon *p-n* junction having unsymmetrical doping of $10^{16}$ acceptors per $cm^3$ and $10^{15}$ donors per $cm^3$. The junction cross-section is 1 $mm^2$ and it carries a forward current of 5 mA.

19. A laser beam of red light is made to shine on several photocathodes in succession. Given that the wavelength of the laser radiation is 6350 Å, determine which of the following cathode materials will emit electrons; lithium (2·6 eV), (1·36 eV), strontium (2·0 eV). The figures in parentheses give the work function of the materials.

20. Two lasers, operating at different frequencies, are used to illuminate in turn a composite photocathode. The cathode is known to have a quantum yield of 6% at the wavelength of the lower frequency laser (11 500 Å) and 4·5% at the wavelength of the higher frequency laser (6350 Å).

If photocurrents of 5·6 $\mu$A and 2·3 $\mu$A are measured in the two cases, evaluate the power in the laser beams, assuming that all the power is intercepted by the cathode. Find the spectral sensitivity of the cathode at the two frequencies.

(0·1 mW, 56 mA/W, 23 mA/W).

21. Evaluate the frequencies (and wavelengths) of electromagnetic radiation which makes intrinsic germanium and cadmium sulphide photoconductive.

A sample of intrinsic germanium absorbs e.m. radiation, of frequency above the photoconductive threshold, at the rate of 100 mW. Assuming that all this light excites electrons into the conduction band, evaluate the ratio of the conductivities of the sample when irradiated and when in equilibrium at room temperature. The lifetime of electrons and holes can be taken to be 400 $\mu$sec.

$$(1\cdot6 \times 10^{14} \text{ Hz, } 18\ 700 \text{ Å, } 5\cdot8 \times 10^{14} \text{ Hz, } 5200 \text{ Å, } 20).$$

## Examples on Part II

1. Compute the energy of a linear oscillator which is in an eigenstate of energy (quantum number $n = 1$) and which oscillates at a frequency of $10^{13}$ Hz.

Find the spring constant of the oscillator given that its mass is that of a proton.

$$(9\cdot9 \times 10^{-21} \text{ joules, } 66 \text{ g/cm}).$$

2. Evaluate the energy of a classical oscillator of mass 1 g, elastic constant 1 g/cm, oscillating with an amplitude of 1 cm. Find its frequency and the quantum number of a quantum mechanical oscillator having the same energy.

$$(4\cdot9 \times 10^{-5} \text{ joules, } 31\cdot3 \text{ rads/sec, } 1\cdot49 \times 10^{28}).$$

3. Verify by direct integration for the state $n = 1$ that the average potential energy of the linear oscillator is given by the expression:

$$\langle V \rangle_n = \tfrac{1}{2}(n + \tfrac{1}{2})\hbar\omega_c.$$

4. Formulate the position probability distribution function for the classical linear harmonic oscillator and plot it graphically in normalized form.

$$\left(\frac{1}{\pi\sqrt{(a^2 - x^2)}}\right).$$

5. Plot to scale the probability distribution for the state $n = 2$ of the linear oscillator. Only the maxima and points corresponding to 10% of the exponential factor need be located numerically.

Superimpose on this graph a plot of the position probability density of a classical oscillator having the same energy.

6. Prove that the effect of the operator $xp_x - p_x x$ on an arbitrary function $f(x)$ is equivalent to multiplying that function by the constant factor $i\hbar$. This is expressed by the equation

$$xp_x - p_x x = i\hbar = [x, p_x]$$

which illustrates the general fact that *linear operators do not commute*.

The foregoing relation is satisfied by every generalized co-ordinate and the conjugate momentum. The relation, usually denoted by a square bracket, is called a *commutator*.

7. The following operators for the oscillator are given by definition

$$A = \sqrt{\frac{1}{2\hbar m \omega_c}} \, p_x - i \sqrt{\frac{K}{2\hbar \omega_c}} x$$

$$A^* = \sqrt{\frac{1}{2\hbar m \omega_c}} \, p_x + i \sqrt{\frac{K}{2\hbar \omega_c}} x$$

Using the result of Example 6, establish the following commutation relation for the new operators

$$AA^* - A^* A = 1$$

Also prove that the Hamiltonian of the oscillator can be expressed in terms of the new operators as follows:

$$H = \frac{\hbar \omega_c}{2} \, (AA^* + A^* A)$$

Using the above commutation relation, *verify* that the effect of the new operators on the eigenstates is given by the equations:

$$A^*|n\rangle = \sqrt{n+1}\,|n+1\rangle$$

$$A|n\rangle = \sqrt{n}\,|n-1\rangle$$

The operators raise or lower the quantum number of the oscillator by unity. Hence they are called *creation* and *annihilation* operators respectively.

8. Include the time-dependent factor in the energy eigenfunctions of the linear oscillator. Assuming that the squared modulus of the resulting function is still to be interpreted as a normalized position probability density, determine the value of the constant of integration. Verify that the time dependent wave functions remain eigenstates of energy of the oscillators for all time (hence the name *stationary states.*)

9. Plot a calibrated graph of the radial probability distribution for the state of hydrogen characterized by the quantum numbers $n = 2, l = 0$. Note the orders of magnitude of the radius and probability in the neighbourhood of the maximum.

Mark the energy levels of hydrogen on an axis calibrated in frequencies and reciprocal wavelengths. Read off the axis some frequencies and wavelengths (in Å) of the Balmer series.

10. Prove the following commutation relations for co-ordinates and momenta of a particle in three-dimensional Cartesian co-ordinates:

$$xy - yx = 0, \quad p_x p_y - p_y p_x = 0;$$
$$xp_z - p_z x = 0, \quad yp_y - p_y y = i\hbar;$$

The foregoing results are examples of the general rule that *dynamical variables associated with separate degrees of freedom of a system always* commute.

$$q_s p_r - p_r q_s = \delta_{rs} i\hbar$$

11. Prove the following commutation relation for the $z$-component of angular momentum and the co-ordinate $\varphi$ in spherical co-ordinates.

$$\varphi L_z - L_z \varphi = i\hbar$$

$\varphi$ and $L_z$ are an example of a generalized co-ordinate and the conjugate momentum whose commutators always equal $i\hbar$ (see problem 10).

12. Compare the magnitude of the angular momentum of an electron orbit having quantum number $l = 1$ with the angular momentum of a governor ball on a steam engine. Take the mass of the ball as $0.5$ kg, its speed of rotation as 100 rev/min, and assume that it moves at a distance of 10 cm from the axis.

$$(1.48 \times 10^{-34} \text{ kg m}^2 \text{ sec}^{-1}, 5.2 \times 10^{-2} \text{ kg m}^2 \text{ sec}^{-1}).$$

13. Prove that the components of angular momentum satisfy the following commutation rules:

$$L_x L_y - L_y L_x = i\hbar L_z$$
$$L_y L_z - L_z L_y = i\hbar L_x$$
$$L_z L_x - L_x L_z = i\hbar L_y$$

Since no two components commute, they cannot share a set of simultaneous eigenstates. Hence it is impossible to arrange one measurement which would yield a predictable value for both components.

14. Write out in full the matrix of the $z$-component of angular momentum. Fix the quantum number $l$ at the value 1 and allow the quantum number $m$ to vary subject to this limitation. For the same values of the quantum numbers write down the matrix of $L^2$. Why do these matrices commute?

Give a graphical representation of these results.

15. Given the matrices

$$S_x = \tfrac{1}{2}\hbar \begin{bmatrix} 0 & 1 \\ 1 & 0 \end{bmatrix}, \quad S_y = \tfrac{1}{2}\hbar \begin{bmatrix} 0 & -i \\ i & 0 \end{bmatrix}, \quad S_z = \tfrac{1}{2}\hbar \begin{bmatrix} 1 & 0 \\ 0 & -1 \end{bmatrix}$$

verify that they satisfy the commutation rules applicable to components of angular momentum (see problem 13).

Form the matrix

$$S^2 = S_x{}^2 + S_y{}^2 + S_z{}^2$$

and check that it commutes with the matrices $S_x$, $S_y$ and $S_z$.

The above matrices are frequently referred to as the Pauli spin matrices. They provide an example of the *isomorphism between the algebra of linear operators and matrices*.

16. Given the operators

$$L^+ = L_x + iL_y$$
$$L^- = L_x - iL_y$$

prove that their commutator $[L^+, L^-]$ equals $2\hbar L_z$.

Use this result to *verify* that the effect of the operators $L^+$ and $L^-$ on the states $|lm\rangle$ is given by the equations

$$L^+|lm\rangle = \hbar \sqrt{(l-m)(l+m+1)}\,|lm+1\rangle$$

$$L^-|lm\rangle = \hbar \sqrt{(l+m)(l-m+1)}\,|lm-1\rangle$$

The operators $L^+$ and $L^-$ are called *step up and step down operators* respectively.

17. Using the results of Problem 16, write down the matrices of the operators $L^+$ and $L^-$ with respect to the states $|lm\rangle$ for $l = 1$ and $|m| \leqslant l$. Hence deduce the matrices of the operators $L_x$ and $L_y$ in the same representation, utilizing the isomorphism between operators and matrices (see problem 15).

18. Prove the following commutator relations for angular momentum:

$$L^+ L_z - L_z L^+ = -\hbar L^+$$
$$L^- L_z - Lz\, L^- = \hbar L^-$$

Next operate with $L_z$ on the states $L^+|m\rangle$ and $L^-|m\rangle$ and use the foregoing relations to show that the effect of $L^+$ and $L^-$ on the state $|m\rangle$ is to raise and lower its quantum number by unity respectively.

19. Write out in full the complete set of quantum numbers for the ground state of sodium. Identify shells of electrons and label them, using spectroscopic symbols. Evaluate the ground-level energy of the system using the approximate theory dealt with in the text. Compare with the ground-level energy of hydrogen.

20. Practice the writing down of complete sets of quantum numbers for atoms of perfect gases, paying due attention to the exclusion principle and to the fact that only ground states are considered. Use spectroscopic symbols to label shells of electrons.

21. Combine an orbital angular momentum of quantum number $l = 2$ with a spin angular momentum, using the vector addition method. Write out in full all the possible quantum numbers $j$ of the resultant momentum and all the possible quantum numbers $m$ of the $z$-component.

22. Add the angular momenta of 3 electron orbits belonging to the quantum number $l = 1$. When selecting the $z$-components of these angular momenta, observe the exclusion principle and prove that the resultant is zero. Note that the inclusion of spin angular momenta still leaves the resultant angular momentum at zero, provided the exclusion principle is observed.

23. Given that the force on a magnetic moment in an inhomogeneous field is expressed by the formula (**m** . grad) **B**, show that the deflection of an atomic beam in a Stern-Gerlach experiment is

$$\frac{l^2}{2mv^2} m_z \frac{\partial B}{\partial z}$$

where $l$ = path of atoms in the field

$m$ = mass of atoms

$v$ = velocity of beam.

The magnetic field $B$ is assumed to have no components except one in the $z$-direction, which is perpendicular to the beam.

    Evaluate the deflection of a beam of neutral sodium atoms energing from an oven at 900 °K for a path length of 20 cm and a field gradient of 10 T/cm. (5·2 mm).

24. State what values can be given to the quantum number $J$ of the resultant angular momentum of a $d$-electron. Evaluate the Landé $g$-factor in each case.

25. Evaluate the commutator of the Hamiltonian and momentum for the linear oscillator and the electron wave in a potential well. What conclusions would you draw regarding eigenstates of these operators? ($i\hbar Kx$; 0).

26. Solve Schroedinger's equation in a one-dimensional potential well in terms of sine and cosine functions. The wave functions should satisfy the following boundary conditions: (i) they must vanish at the boundaries of the well: (ii) the position probability density $|\Psi|^2$ must be a symmetric function with respect to the centre of the well.

Evaluate the momentum and energy eigenvalues of these states and check that they are the same as those obtained from exponential wave functions.

## Examples on Part III

1. Evaluate the populations of the three levels of a simple paramagnetic substance consisting of (spin) magnetic moments of quantum number $j = 1$ at a temperature of 4·2 °K, and at applied fields of 0·1, 1·0 and 5·0 Tesla. Assume that the number density of paramagnetic ions is $10^{20}$ cm$^{-3}$. Is it possible to show these populations on a properly scaled graph?

$$(3\cdot33 \times 10^{19}, 3\cdot24 \times 10^{19} \text{ cm}^{-3} \text{ at } 0\cdot1 \text{ Tesla}).$$

2. Plot an approximate room-temperature magnetization curve for gadolinium sulphate octahydrate, $Gd_2(SO_4)_3 8(H_2O)$, by using the low field and saturation approximations. The density of the paramagnetic $Gd^{3+}$ ions should be estimated from the chemical formula.

Evelute the susceptibility in the linear region at 4·2 °K, 77 °K, 300 °K and 900 °K and plot it on a calibrated graph.

3. Evaluate the magnetization, the susceptibility and relative permeability of a paramagnetic ruby crystal at 4·2 °K. The applied magnetic field of 0·5 Tesla is parallel to the c-axis of the hexagonal crystal lattice. The four energy levels of ruby are due to the resultant spin of quantum number $\frac{3}{2}$ possessed by the chromium impurity ions. In this orientation the energy levels are identical with spin states according to the scheme: $E_1 \rightarrow -\frac{3}{2}$, $E_2 \rightarrow -\frac{1}{2}$, $E_3 \rightarrow \frac{1}{2}$, $E_4 \rightarrow \frac{3}{2}$. The spacing of the levels relative to $E_1$ is 25·5 GHz, 39·0 GHz, 41·7 GHz. The volume of a hexagonal unit cell is approximately 800 Å$^3$ and a chromium ion is contained in one out of 100 unit cells. Also evaluate the saturation magnetization of ruby at a high field and low temperature.

$$(59 \text{ A/m}, 1\cdot2 \times 10^{-4}, 350 \text{ A/m}).$$

4. A 3-level maser system has a pump transition of 12 GHz and a signal transition of 3 GHz. The latter is between the upper two levels. Assuming that the pump transition is effectively saturated and that the relaxation rates are such as to preserve the equilibrium population of level 2, evaluate the population difference between levels 2 and 3 per $cm^3$ of maser material. The maser material may be assumed to have $10^{19}$ active ions per $cm^3$ and is operated at $4\cdot2\,°K$.

$$(1\cdot1 \times 10^{17}\ cm^{-3}).$$

5. For the maser of problem 4, the transition probability between levels 2 and 3 is $10^{-5}$ at an input signal of $10^{-13}$ W. Evaluate the output power of the amplifier and its gain in db per $cm^3$ of active material.

$$(2\cdot3 \times 10^{-13}\ W,\ 18\cdot3\ db).$$

6. Show that the power required to saturate two paramagnetic levels $E_1$ and $E_2$ is given by the expression

$$W = (E_2 - E_1)(n_{10} - n_{20})\frac{1}{\tau}$$

where $n_{10}$, $n_{20}$ are equilibrium populations of the levels and $\tau$ is the relaxation time.

Evaluate the saturating power for a transition separated by 20 $GH_z$, given a relaxation time of $10^{-3}$ sec at $4\cdot2\,°K$, and a concentration of the active ions of $10^{19}\ cm^{-3}$.

# Subject Index

## A

A.C. discharge, 154
Allowed bands, 6
Alloys
  critical temperature, 363
  normal state, 363
  residual resistivity, 363
  superconductivity, 363
Amplifier, 85
Angular momentum, 207, 211
  addition, 246, 251
  angular probability distributions, 215
  average values, 214
  commutation rules, 423
  conjugate momentum, 423
  eigenstates, 214
  eigenvalue equation, 214
  eigenvalue relation, 213
  eigenvalues, 214
  of electron orbits, 216
  generalized coordinate, 423
  Hund's rule, 249
  interacting momenta, 246
  magnitude, 250
  Pauli spin matrices, 424
  quantized in direction, 216
  quantum number, 213
  resultant, 246
  vector resultant of momenta, 247
Anode, 82
Antiferromagnetism
  dipolar lattices, 342
  exchange energy, 342
  ferromagnetism among the elements, 342

  interlocking sublattices, 341
  Néel temperature, 342
  saturation magnetization, 345
Arc discharge, 152
  cold-cathode, 153
  by high field emission, 152
  by thermionic emission, 153
Atoms
  electronic polarizability, 165

## B

Bohr model of atom, 3
Bose-Einstein distribution, 295
Bragg diffraction, 276
Bragg reflection, 13
Brillouin function, 305
Brillouin zone, 13

## C

Cathode, 82
  cold, 96
  oxide-coated, 83
Charge carrier
  collisions, 63
  concentration, 34, 53
  density, 31, 43
  density gradient, 52
  diffusion, 52
  emission, 97
  excess local concentration, 52
  excess minority carrier, 53, 54, 56, 99
  excess minority carrier lifetime, 54, 56

429

Charge carrier—*cont.*
   minority carrier
      excess concentration, 53
      excess generation, 54
      recombination, 53, 99
Charge carrier diffusion
   concentration gradient, 52
   diffusion constant, 53
   diffusion current, 52, 60, 63
Charge transport in gases, 139
   diffusion current, 140
Chromium ion, 250
   paramagnetism, 301
   spectroscopy, 312
   zero field splitting, 313
Circulator, 356
   anisotropic a.c. permeability, 357
   as isolator, 358
   matched termination, 358
Conduction, 3
Conduction band, 6
Conductor, 7
Contact potential
   in thermocouple, 86
   in thermoelectric power generation,
      86
   for two solids, 85
Copper
   as conductor, 7
   energy bands, 7
Crystal defects, 19
   dislocations, 20
   Frenkel defects, 20
   interstitial impurities, 20
   Schottky defects, 20
   substitutional impurities, 19
   vacancies, 20
Crystalline solid, 4, 8
   energy band concept, 13
Curie law, 306
Curie-Weiss law, 334
Current flow
   electric current density, 279


                    **D**

de Broglie equation, 3, 10, 259, 272
Density of levels, 12
Density-of-levels function, 31, 90
Density of states, 12, 36, 46, 90, 266

Density-of-states function, 30, 31, 33,
      267
   free electron model, 267, 268
Diamagnetism
   diamagnetic moment, 362
   diamagnetic term contribution, 361
   electromagnetic induction, 359
   Lorentz force, 360
   paramagnetic moment, 362
   paramagnetic term contribution,
      361
   perfect, 368, 369
   in superconductor, 365, 368
   weak, 362
Diamond
   energy bands, 6, 7
   insulating properties, 7
Dielectrics
   absorption coefficient, 160
   anisotropic, 160
   breakdown, 171
   breakdown strength, 158
   dielectric constant, 159, 162, 167
   dielectric loss, 158
   discharge breakdown, 174
   dispersion, 167
   electric displacement, 159
   electric field, 159
   electric flux density, 159
   electric susceptibility, 159
   field emission breakdown, 174
   forced resonance, 162
   high temperature breakdown, 173
   internal field, 168
   internal vibrations, 168
   isotropic, 160
   loss angle, 160
   low temperature breakdown, 173
   non-polar, 161
   polar, 161
   polarizability, 159
   polarization, 159, 161
   refractive index, 160
   relative permittivity, 159
   relaxation, 169
   relaxation time, 169
   at resonance, 167
   thermal breakdown, 173
Dirac notation, 195, 209, 232
   associative law, 196

commutative law, 196
distributive laws, 196
eigenvalue relation, 210
for linear oscillator, 195
off-diagonal matrix elements, 210
orthonormality conditions, 195, 210
pointed brackets, 195
superposition, 231

### E

Eigenfunction, 183, 184
of linear oscillator, 184
normalized, 184
orthogonality, 184
wave character, 186
Einstein relations, 62, 65
Elastic collisions, 134
energy transfer, 134
Electrodes in gas discharges, 138
field emission, 139
photoelectric emission, 139
secondary emission, 139
thermionic emission, 138
Electroluminescence, 124
injection of minority carriers, 124
intrinsic excitation, 124
Electroluminescent devices, 121
Electromagnetic waves, 275
exponential decay, 275
evanescent mode, 275
in waveguide, 275
waveguide cut-off, 275
Electron beam, 271
accelerating voltage, 272
as communication signal, 281
boundary conditions
charge density, 273, 276
continuous sets of eigenvalues, 273
diffraction, 273
energy eigenvalues, 272
incident wave, 274
interference, 273
interference with crystal lattice, 276
Lorentz force, 280
magnetic vector potential, 280
plane wave equation, 273

position probability, 275
position probability density, 273
potential barrier, 275
potential step, 274, 275
reflection, 273, 274
reflection coefficient, 274
standing-wave pattern, 276
thermal energy, 275
transmission coefficient, 275
uncertainty principle, 281
wave function, 274
wave function normalization, 273
wave number, 272, 275
wave numbers continuum, 272
Electron distribution, 286
macrostate, 289
microstate, 289
Electron energy levels
quantization, 11
Electron motion
collisions, 16
in solids, 3, 14
standing waves, 10
Electron reflection, 273
Electron wave function
antisymmetrical, 284
exclusion principle, 284
symmetrical, 284
Electron waves, 3, 10, 13
coherent, 391
condensate wave function, 391
constant of separation, 258
cyclic boundary conditions, 257, 258
de Broglie relation, 259
electron wave function, 278
energy eigenvalues, 259
exclusion principle, 260
$k$-space, 259
"match", 278
matrix elements of momentum, 258
momentum eigenvalues, 259
momentum space, 259
orthogonal set of states, 258
periodic boundary conditions, 257, 258
phase constant, 257
position probability density, 258
spatial state, 260

Electron waves—*cont.*
  transmission coefficient, 279
  wave number, 257
Electrons
  conduction, 9, 16, 20, 45
  donor, 20
  effective mass, 14, 31, 36, 267
  exchange interaction, 338
  free, 9
  "heavy", 15
  intrinsic, 35
  kinetic energy, 11
  "light", 15
  normal, 364
  as particle, 3
  in potential well, 258
  reflection at potential step, 273
  relation of wavelength to momentum, 3
  scattering, 363
  spin, 221
  state, 284
  in superconductive phase, 364
  superelectrons, 364, 373
  thermal, 23
  thermal energy of excitation, 20
  in three-dimensional potential well, 261
  tunnelling, 277
  valence, 9, 16
  as wave, 3
Electrons, conduction
  collisions, 16
  drift velocity, 16
  electric current in external field, 16
  mobility, 16
  thermal velocity, 16
Electrons in three-dimensional potential well, 261
  constant energy spheres, 265
  constant energy surface, 265
  electron states, 269
  energy eigenvalues, 262, 265
  energy levels, 266
  Fermi sphere radius, 265
  Fermi surface, 265
  Gaussian distribution, 270
  k-space, 263, 264
  momentum eigenvalues, 264, 266
  momentum space, 263, 264, 268

  separation of variables, 261
  spatial motion, 267
  spatial state, 264
  thermionic current, 268
  vacuum level, 268
  wave vector, 262
Energy bands, 3, 13
  allowed, 6
  conduction band, 6
  energy gap, 7
  forbidden gap, 6
  occupation, 6
  valence band, 6
Energy gap, 7
Energy level diagram, 16
Energy level distribution
  gaps, 12
Energy levels
  discrete, 3, 4
  splitting, 5
Energy spectrum
  discontinuity, 13
Exchange interaction
  for dipolar sublattices, 344
  indirect exchange, 342, 343
  superexchange, 342, 343
Exclusion principle, 12
Extrinsic saturation, 37


**F**

Fabry-Perot resonator, 399, 402
  gain, 402
  population difference, 402
  population inversion, 402
  transition probability, 402
Fermi-Dirac distribution, 24, 26, 28, 30, 31, 37, 37, 269, 289, 292
  available states, 24, 30
  as step function, 33
  thermal equilibrium, 24, 26
Fermi-Dirac function, 27, 34
  in *n*-type material, 28
Fermi energy, 63
Fermi level, 25, 26, 28, 33, 35, 36, 37, 41, 43, 63, 64, 119, 260, 265, 387
  variation in compensated case, 42
  variation with impurity density, 42

variation with temperature, 41
Fermi sphere, 265
   radius, 265
Fermi surface, 265
Ferrimagnetic materials
   a.c. anisotropy, 355
   anisotropy, 355
Ferrimagnetic resonance, 352
Ferrites, 345
Ferromagnetic materials
   anisotropy energy, 323
   antiferromagnetism, 322
   Bohr magnetons per atom, 338
   crystalline anisotropy, 323
   exchange interaction, 322
   polycrystalline, 326
   single-crystal, 327
   temperature variation of magnetiz-
      ation, 333
Ferromagnetic resonance, 350, 354
   a.c. susceptibility, 352
   effect of a.c. field, 352
   demagnetization effects, 351
   dispersion relation, 353
   magnetization, 351
   precession of magnetization, 352
   saturation magnetization, 351
   susceptibility, 353
Ferromagnetism
   Bloch walls, 323
   dipolar interaction, 321
   domain walls, 322, 323
   domains, 322
   exchange constant, 340
   exchange energy, 340
   exchange field, 331
   exchange integral, 340
   exchange interaction, 321, 338
   hysteresis, 329
   internal field, 333
   magnetization energy, 329
   magnetostatic energy, 322
   molecular field, 333
   molecular field constant, 333
   Néel wall, 324
   reversible domain wall movement,
      326
   saturation magnetization, 331, 335,
      337
   spin magnetic moment, 321

spontaneous local magnetization,
   322
FET see Field-effect transistor
Field-effect transistor, 111
   drain, 111
   gate electrode, 111
   metal-insulator-semiconductor, 112
   metal-oxide-silicon, 112
   source, 111
Field emission, 96, 139
   electron tunnelling, 96
   from cold cathode, 96
   gettering, 139
   in Schottky effect, 96
   sputtering, 139
Forbidden gap, 6
Fourier transform, 282
Free-electron model, 8
Free electrons, 8, 9
   effect of electric field, 9
   effect of magnetic field, 9
   movement in crystals, 8, 9
   thermal excitation, 9


G

Gadolinium ion, 308
Gallium arsenide
   band structure, 395
   electroluminescence, 125
   lamps, 125
Garnets, 345, 346
   compensation point, 346
   Curie temperatures, 347
   rare earth iron, 346
   saturation magnetization, 346
   sintering, 346
   yttrium iron, 346
Gas discharges
   a.c. discharge, 154
   arc discharge, 152
   breakdown, 146
   breakdown voltage, 146
   corona discharge, 154
   glow discharge, 149
   ionization coefficients, 146
   primary current, 146
   primary electron, 146
   secondary electron, 146

Gas discharges—*cont.*
    self-sustaining discharge, 146
Gas laser
    electron collision excitation, 405
    excitation energy direct transfer, 406
    glow discharge, 405
    helium-neon, 409
    polyatomic molecule dissociation, 406
Gases
    collision cross-section, 133
    collision, elastic, 131
    collision, inelastic, 131
    collision probability, 132
    collisions, 130
    as dielectrics, 162
    effect of electromagnetic radiation, 311
    energy absorption, 311
    as insulator, 155
    ionization processes, 132
    mean free path, 130
    polar, 163
    relaxation processes, 311
    relaxation rate, 312
    stimulated emission, 312
    uncharged particles, 140
    weakly ionized, 140
Gas-filled rectifier, 156
Gaussian distribution, 270
Germanium
    energy-level diagram, 16, 17
Gibb's free energy, 371
Glow discharge, 149
    abnormal glow, 150
    cathode dark space, 151
    cathode fall, 151
    in fluorescent lighting, 155
    in voltage stabilization, 155
    normal glow, 150
    positive column, 151
    sputtered electrode, 155
    striations, 151
    virtual anode, 152
    voltage reference, 156

**H**

Hall coefficient, 48
Hall effect, 47

Hall effect devices, 116
Hall field, 48
Hermite polynomial, 183
High field emission
    in arc discharge, 152
Holes, 16
    current density, 16
    density, 17
    diffusion length, 60
    injection, 58
Hund's rule, 249
Hydrogen atom
    angular probability distributions, 207
    average radius, ground state, 207
    Balmer series, 423
    degeneracy, 209
    eigenfunctions orthogonal sets, 204
    eigenvalue equation, 209
    energy eigenstates, 204
    energy level splitting, 225
    lifting of degeneracy, 226
    magnetic energy eigenstates, 225
    median radius, 207
    orthogonality relation, 205
    principal quantum number, 207
    probability distribution, spherical symmetry, 207
    radial probability density, 205
    radial wave function, 202, 203
    simultaneous eigenstates, 218, 222
    spectral line splitting, 225
    spin magnetic moment, 223
    spin operator, 223
    Zeeman effect, 226
    Zeeman energy, 226
Hysteresis
    anisotropy forces, 330
    coercive force, 329
    domain wall pinning, 330
    hysteresis loop, 329
    hysteresis loss, 331
    initial magnetization curve, 329
    irreversible energy loss, 331
    remanent flux, 329

**I**

Inelastic collisions, 136
    attachment, 137

charge transfer, 137
    of first kind, 136
    ground state, 136
    ionizing collisions, 136
    Penning effect, 137
    principle of detailed balancing, 136
    recombination, 137
    of second kind, 136
Insulators, 158
Ionic crystals
    as dielectrics, 169
    interlocking sublattices, 168
Ionizing collisions
    autoionization, 137
    cumulative ionization, 137
    thermal ionization, 136


J

Junction laser
    threshold current, 411
Junction transistor, 100
    alloyed, 110
    base, 100
    base layer, 103
    base region, 100
    collector multiplication, 105, 108
    collector saturation current, 107
    constant current source, 101
    current gain, 107
    emitter input impedance, 106
    emitter junction, 101
    emitter-to-collector transport efficiency, 105
    incremental impedance, 107
    minority carrier emitting efficiency, 104
    n-p-n, 100, 103, 108
    p-n-p, 103, 108
    quasi-static conditions, 106
    transit time, 106
Junction triode, 100
    collector, 100


L

Laguerre polynomial, 202
Larmor precession, 348

angle of precession, 350
    forced resonance, 350
    Larmar frequency, 349
    precession frequency, 349
    proton resonance, 350
Laser, 315, 392
    amplifier, 402
    bottleneck, 394, 410
    four-level, 394
    gallium arsenide, 395, 410
    gas, 404, 405
    helium-neon, 409
    ion energy exchange, 405
    junction, 410
    metastable state, 393
    mode separation, 400
    monochromatic output, 401
    multimode operation, 400
    optical pumping, 404
    oscillator, 402
    population inversion, 393, 395, 405, 406
    pump power, 393
    ruby, 407
    three-level, 394
Laser amplifier
    Brewster's angle, 403
Laser oscillator
    feedback, 403
Lattice
    conduction electrons, 9, 16
    core electrons, 9
    defects, 363
    free electrons, 9
    impurities, 363
    periodic potential, 13
    valence electrons, 9
Lattice forces, 8
Lattice potential
    periodic, 10
    potential well, 10, 256
    of solids, 256
Lattice spacing, 13
Lattice vibrations, 45, 363
Legendre differential equation, 201
Legendre function, 201, 203, 204
    orthogonality conditions, 204
Linear oscillator, 181
    annihilation operator, 422
    average position, 187

Linear oscillator—*cont.*
  commutator, 282, 422
  creation operator, 422
  dynamical state, 184
  dynamical variables, 188, 189
  eigenfunction, 184
  eigenstate, 184
  eigenvalue, 190, 194
  energy levels, 191
  energy quanta, 191
  generalized coordinate, 422
  Hamiltonian operator, 190
  matrix statistical average, 194
  momentum, 188
  momentum average value, 189
  operator, 189
  operator diagonal matrix, 194
  operator matrix elements, 192
  phonon, 191
  position probability density, 184, 186
  stationary state, 422
  uncertainty range, 281
  zero point energy, 191
Liquids
  as dielectrics, 168
  ionic polarizability, 165
  supercooled relaxation time, 171
London theory, 365
  equations, 365
  penetration depth, 365, 367, 368, 378, 387
Lorentz field, 47
Lorentz force, 47, 48, 280

**M**

Magnetic materials
  angular momentum quantum number, 335
  anisotropic, 300
  Curie temperature, 336, 337
  Curie-Weiss law, 334
  diamagnetic, 300
  external applied field, 334
  ferromagnetic, 300
  induction, 299
  isotropic, 300
  magnetic field, 299
  magnetic flux density, 299
  magnetic susceptibility, 299
  magnetization, 299
  magnetizing force, 299
  paramagnetic, 300
  permeability of free space, 299
  permeability tensor, 300
  relative permeability, 299
  saturation magnetization, 335
  susceptibility tensor, 300
  total applied field, 333
  Weiss constant, 334, 337
Many-electron atom
  angular momenta addition, 246
  angular momenta, different species, 251
  angular momenta magnitude, 250
  chromium ion, 250
  eigenstate, 237
  eigenvalues, 237
  electron configuration, 239
  electron shell, 240
  electron sub-shell, 240
  excited, 246
  excited state, 239
  ground state, 239
  Hamiltonian, 234
  Hund's rule, 249
  incomplete shell, 240
  interacting angular momenta, 246
  ionized, configuration, 246
  Landé $g$-factor, 253
  magnetic moment, 251
  magnetic moment magnitude, 251
  Pauli exclusion principle, 239
  resultant angular momentum, 246
  spectroscopic notation, 239
  spectroscopic splitting factor, 251
  unpaired electron, 240, 249
Maser, 315
  cavity, 318
  coherent emission, 319
  incoherent emission, 319
  slow wave structure, 318
  spontaneous emission, 319
  travelling-wave, 318, 402
  use of paramagnetic crystals, 320
  with superconducting magnets, 320
Matter wave, 271
Maxwell-Boltzman distribution, 29, 293

for classical gas, 29
constant A, 303
constant A for one particle, 294
partition function, 294, 303
state sum, 294, 303
Metal-to-semiconductor contact, 66, 76, 90, 110
alloying process, 90
breakdown conditions, 93
depletion layer, 91
image charge, 93
for $n$-type semiconductor, 94
Ohmic contact, 94
for $p$-type semiconductor, 95
rectifying contact, 91
Schottky barrier, 92
surface states, 90, 93
under forward bias, 92
under reverse bias, 92
Metals
critical temperature, 363
normal state, 363
residual resistivity, 363
superconductivity, 363
Microcircuit, 85
Microwave spectrometer, 314
reflection coefficient, 315
spectroscopic absorption, 315
MIS see Field-effect transistor metal-insulator-semiconductor
Momentum
discreet values, 11
relation to wavelength, 3
MOS see Field-effect transistor metal-oxide-silicon

N

$N_c$ state, 35
Niobium
critical temperature, 363
Non-polar atoms
as dielectrics, 161
Non-polar molecules
electronic polarizability, 165
local field, 162
molecular polarizability, 162
$N_v$ state, 36

O

Ohmic conduction, 140
Opto-electronic devices, 120
solar cell, 124
Operators
isomorphism, 424
step down, 424
step up, 424
Optical resonator
atomic transition linewidth, 399
cavity $Q$-factor, 401
Fabry-Perot cavity, 399
mode bandwidth, 401
mode density, 398
photon lifetime, 401
reflection coefficient, 400, 401
spatial mode, 400
temporal mode, 400
uncertainty principle, 401
Orbital magnetic moment, 216
Bohr magneton, 218
gyromagnetic ratio, 217
magnetic moment $z$-component, 218
magnetic moment operators, 216

P

Paramagnetic energy level, 301
Paramagnetic materials
average magnetic moment, 304
Brillouin function, 305
Curie law, 306
effect of a.c. field, 309
effect of crystal lattice, 302, 309
effect of d.c. field, 302
effect of electromagnetic radiation, 309
effect of electrostatic lattice forces, 308
energy-level transitions, 310
forbidden transitions, 311
ions, 301, 309
magnetic moment, 307
magnetic moment, magnitude, 307
magnetization, calculation, 303
net magnetization, 304
orbital quenching, 309
perturbation theory, 310

Paramagnetic materials—*cont.*
  saturation magnetization, 307
  spectroscopic splitting factor, 308
  transition probability, 310
Paramagnetic solids, spectroscopy
  chromium, 312
  paramagnetic levels, 312
  rare earths, 312
  selection rule, 312
  spectral line splitting, 312
Paramagnetic susceptibility, 302
Particles
  boson, 286
  composite, 285
  exclusion principle, 285
  fermion, 286
  indistinguishability, 283
  non-localized, 283
Particles, state occupation, 283
Particles, wave function
  antisymmetrical, 284
  exclusion principle, 284, 285
  symmetrical, 284
Paschen's law, 146, 148, 152
Pauli exclusion principle, 239
Penning effect, 137
  ionization, 137
  metastable state, 137
Periodic table, 240
Photocathode, 86
  photoemission, 86
Photocell, 88
Photoconductor, 122
Photodetector, 123
  depletion layer photodiode, 123
  photoconductive gain, 123
  response time, 123
Photodiode
  avalanche, 124
  depletion layer, 123
  *p-i-n*, 123
Photoelectric emission, 86, 139
  between metal and semiconductor, 89
  between two metals, 87
  photocurrent, 87
  radiation quanta, 86
  retarding potential, 88
  threshold frequency, 87
Photoexcitation, 137

Photoionization, 137
Photon, 87
  absorption, 120
  emission, 120
Photovoltaic effect, 77
*P-n* junction, 62, 66, 77, 86, 119
  avalanche breakdown, 78
  barrier capacitance, 75
  barrier layer, 67
  breakdown, 78
  breakdown voltage, 78
  contact potential, 68, 86, 98
  density of injected electrons, 71
  depletion layer, 67
  depletion layer, unsymmetrical, 98
  diffusion current, 69
  electron sink, 67
  as electron source, 70
  electron tunnelling, 119
  as emitter of charge carriers, 97
  emitter junction, 98, 101
  energy-level diagram, 68
  in equilibrium, 68
  excess minority charge carriers, 99
  graded, 123
  in integrated microcircuit, 77
  immobile charge, 67
  isolated, 107
  majority holes, 99
  minority carrier emitting efficiency, 100, 104
  *n*-type side, 68
  in parametric amplifier, 77
  as photodetector, 123
  in photodiode, 78
  *p*-type side, 68
  rectifier equation, 74
  reverse bias, 72
  reverse saturation current, 101, 107
  saturated extrinsic condition, 66
  saturation current of electrons, 69
  saturation current of holes, 71
  in solar cell, 78
  as source of holes, 72
  space charge, 67
  symmetrical, 76
  total current density, 74
  tunnelling, 79
  under forward voltage, 74
  unequally doped, 97

unsymmetrical, 76
Zener breakdown, 79
*P-n* junction depletion layer
as minority electron sink, 69
*P-n* junction, forward biased
forward current, 74
*P-n* junction, reverse biased
reverse saturation current, 73
*P-n-p-n* devices, 125
anode, 125
cathode, 125
doping profile, 125
"forward blocking", 127
forward "on" condition, 128
gate, 128
"holding current", 128
"reverse blocking" condition, 128
"reverse breakdown", 128
as semiconductor-controlled rectifiers, 125
"switching" voltage, 127
thyristor, 129
Polar molecules
as dielectrics, 161
dipolar polarizability, 165
orientational polarizability, 165
Potential well, 256
three-dimensional, 261

Q

Quantum theory
antisymmetric wave function, 186
Dirac notation, 195
eigenfunction, 183
eigenvalue, 183
eigenvalue equation, 183
even parity function, 186
odd parity function, 186
operator, 189
quantum number, 183
spin-orbit coupling, 227
symmetric wave function, 186
wave function, 186

R

Radiative transitions, 121
Auger process, 122

edge emission, 121
extrinsic, 121
interband, 121
intraband, 121
intrinsic emission, 121
quantum efficiency, 122
trapping levels, 121
traps, 121
Resonance
bandwidth, 166
line, 166
linewidth, 167
Resonance isolator, 355
circular polarization, 356
saturation magnetization, 356

S

Schottky defect, 20
Schottky effect, 84, 96
Schroedinger equation, 177
application to linear oscillator, 181
central field approximation, 235
commutators, 219
commuting operators, 218
cyclic boundary conditions, 200
eigenstate degeneracy, 209
eigenstate mixtures, 231
eigenvalue equation, 233
Hamiltonian, many-electron atom,
234
for hydrogen atom, 197
incompatible dynamical variables,
220
Laplacian operator, 198
linear superposition of eigenstates,
228
normalization, 232
orthogonality, 232
position probability distribution,
178
principle of uncertainty, 220
probability amplitudes, 231
probability density, 200
separation constant, 180, 182, 199
separation of variables, 179
spherical coordinates, 197
Semiconductor, intrinsic
absorption, 46

Semiconductor—*cont.*
  absorption edge, 46
  electromagnetic activation, 45
  energy gap, 46
  opaque, 45
  photoconductivity, 45, 46
  photon activation, 45
  thermal activation, 45
  threshold frequency, 45
  transmission, 46
  transparent, 45
Semiconductor, *p*-type
  electron diffusion current, 62
  electron diffusion length, 62
  electron sink, 61
  electrostatic potential, 63
Semiconductors, 7
  charge-carrier concentration, 34, 53
  charge-carrier density, 31, 43
  charge-carrier diffusion, 52
  charge-carrier injection, 406
  charge continuity equation, 54
  charge sink, 54
  charge source, 54
  conduction electron density, 16
  conductivity, 16, 45
  covalent-bonding scheme, 20
  current density, 16
  degenerate, 117
  doped, 21
  doping, *n*-type, 21
  doping, *p*-type, 22
  electron affinity, 91
  electron current density, 16
  energy-band scheme, 20
  energy band tilting, 64
  energy bands, 64
  energy gap, 38
  equilibrium conditions, 52, 63
  excess-charge concentration, 65
  excess holes recombination, 56
  extrinsic, 38, 49, 50
  free electron density, 16
  heavy doping, 43
  hole current density, 16
  hole density, 17
  hole diffusion length, 60
  hole-electron pair lifetime, 51
  hole injection, 58
  impurity, 37

  imref, 396
  injection current density, 60
  intrinsic, 36, 42, 45, 50
  intrinsic condition in pure, 15, 17
  intrinsic sample, 38
  linewidth, 397
  local electric field, 64
  mobilities ratio, 49
  non-degenerate, 117
  non-equilibrium conditions, 53
  non-uniform doping, 63
  *n-p* product, 37
  *n*-type, 20, 37, 41
  optical phenomena, 45
  *p-n* junction, 52
  population inversion, 393, 395, 406
  potential gradient, 16
  *p*-type, 21, 37
  quasi Fermi level, 396
  resistivity, 45
  thermal excitation probability, 38
  thermodynamic potential, 63
  work function, 82
Semiconductors, doped
  acceptor density, 38
  band-edge tailing, 118
  complete compensation, 39
  donor density, 38
  extrinsic carriers, 23
  extrinsic conduction, 23
  impurity concentration, 37
  intrinsic carriers, 23
  intrinsic characteristics, 39
  intrinsic conduction, 23
  ionized impurities, 40
  majority carriers, 23
  minority carriers, 23
  *n*-type, 20, 23, 40
  *p*-type, 21, 23, 40
  saturated extrinsic case, 39
  thermal activation, 23
  thermal electrons, 23
  for tunnel diode, 43
Secondary emission, 95, 139
  coefficient, 95
  electron impact, 139
  neutral atom impact, 139
  positive-ion impact, 139
Self-sustaining discharge
  breakdown, 146

breakdown voltage, 146
primary current, 146
primary electron, 146
secondary electron, 146
Semiconductor-controlled rectifier, 125
Semiconductor impurities, acceptor, 21
acceptor atom, 22
acceptor impurities, 20
immobile negative charge, 22
positive-charge carriers, 22
$p$-type doping, 22
Semiconductor impurities, donor, 20
donor electron, 20
donor impurities, 20
donor level, 20
immobile positive charge, 20
$n$-type doping, 21
Silsbee hypothesis, 373
Solids
crystalline field energy, 256
as dielectrics, 168
ionic polarizability, 165
lattice potential, 256
nuclear closed electron shells, 255
nuclear core, 255
potential well, 256
valence electrons, 255
Spontaneous emission, 138
metastable state, 138
Standing waves, 10
Stern-Gerlach beam-splitting, 253
Stimulated emission, 138, 315
angular momentum eigenstates, 317
energy eigenstates, 317
population inversion, 315
pump transition, 316
pumping principle, 315
pumping signal, 315
rate equation, 316
relaxation, 316
signal transition, 316
superposition principle, 317
transition probability, 315
Superconductivity, 386
BCS theory, 386
condensation energy, 387
electron attractive interaction, 386

electron bound state, 386
energy gap, 387
microscopic theory, 386
superconductive transport current, 386
Superconductor, type I
diamagnetic magnetization, 377
domain structure, 377
intermediate state, 376
laminar domains, 377
superconducting filaments, 377
Superconductor, type II
in a.c. machines, 386
in a.c. power transmission, 384
alloys, 385
in electromagnets, 384
field gradient, 384
flux line lattice, 380, 382
flux lines, 380
flux pump, 385
for high-field magnet, 382
in homopolar electrical machine, 385
hysteresis, 384
hysteresis loop, 384
ideal, 383
irreversibility, 384
$\kappa$-value, 382
lattice constant, 382
lower critical field, 379
Meissner state, 380
mixed state, 379, 380
niobium-tin alloy, 385
niobium-titanium alloy, 385
niobium-zirconium alloy, 385
persistent current mode, 384
pinned flux line, 383
pinning centres, 383
upper critical field, 380
vortex, 381
Superconductors, 363
$B_c$, 372
coherence length, 378
condensation energy, 372
Cooper pairs, 371, 373, 378
critical current, 373
critical current density, 373
critical field, 372
critical field, at absolute zero, 373
critical field curve, 374

Superconductors–*cont.*
electron pairs, 371
electron tunnelling, 388
field-fringing effect, 375
flux line, 371
fluxoid, 370
fluxon, 371
$H_c$, 372
interphase boundaries, 377
laminar domain, 377
London equations, 365
London penetration depth, 365, 367, 368, 378, 387
Meissner effect, 373
multiply connected, 370
negative magnetization, 368, 372
negative surface energy, 379
negative susceptibility, 367
normal current, 365
normal current density, 364
normal electrons, 364, 387
perfect conductivity, 368
perfect diamagnetic line, 375
perfect diamagnetism, 368
phase integral, 370
positive surface energy, 378
quantized fluxoid, 370
quantum of flux, 371
resistanceless state, 364
screening supercurrent, 367
Silsbee hypothesis, 373
spatial transitions, 377
supercurrent, 364
superelectron density, 364, 387, 391
superelectrons, 364, 373, 387
superelectrons, kinetic energy, 373
type I, 365, 369, 373
type I magnetization curve, 374
type II, 371, 379
wave coherence, 391
Superelectrons
quantum interference phenomena, 391

**T**

Thermionic emission, 82, 138, 275
in arc discharge, 152
thermionic current, 82

Thermionic emission equation, 268
Thermionic valve, 85
Thermistor, 114
Thyristor, 129
controlled rectification, 129
Tight-binding approach, 4, 6
Townsend discharge, 141
avalanche condition, 142, 143
coefficient of secondary emission, 144
first ionization coefficient, 142
primary electrons, 143
secondary effects, 144
secondary electrons, 145
secondary emission, 145
second ionization coefficient, 144
Transistor, 52, 85
Triode, 84
control grid, 84
Tunnel diode, 43, 117
negative resistance, 120
valley current, 120

**U**

Uncertainty principle, 281
conjugate coordinates, 281

**V**

Vacant state, 16
Vacuum diode, 82
saturation current, 84
Schottky effect, 84, 96
space-charge limited current, 84
temperature limited current, 84
Valence band, 6

**W**

Weakly ionized gas, 140
Ohmic conduction, 140
saturation current density, 141
Work function, 81

**X**

X quantization, 369
X-rays, 95

**Z**

Zeeman effect, 226
  anomalous, 227

magnetic field perturbing effect,
  226
perturbation theory, 226